Jean-Pierre Lefort

Basement Correlation Across the North Atlantic

English by M. S. N. Carpenter

With 77 figures

Springer-Verlag
Berlin Heidelberg New York London Paris Tokyo

Dr. Jean-Pierre Lefort

Institut de Géologie
Université de Rennes I
Campus de Beaulieu
F–35042 Rennes-Cedex

ISBN-13:978-3-642-73352-9   e-ISBN-13:978-3-642-73350-5
DOI: 10.1007/978-3-642-73350-5

Library of Congress Cataloging-in-Publication Data

Lefort, Jean-Pierre. Basement Correlation Across the North Atlantic/Jean-Pierre Lefort; English by
M. S. N. Carpenter. p.cm. – Bibliography: p. – Includes index.
ISBN-13:978-3-642-73352-9 –1.Stratigraphic correlation–North Atlantic Ocean Region.2.Geology–North
Atlantic Ocean Region. I. Title. – QE652.55.N67L44 1988 – 551.7'009182'1 – dc 19 – 88-29491 CIP

© Springer-Verlag Berlin Heidelberg 1989
Softcover reprint of the hardcover 1st edition   1989

Typesetting: Overseas Typographers

2132/3020-543210 – Printed on acid-free-paper

"It is as if we were trying to repair the fragments of a newspaper by looking at the details of tearing and then confirming the accuracy of our fit with the printed text. If the lines are re-constituted, we may assume that these fragments were originally part of the same sheet; a single line would suffice to prove the likelihood of our reconstruction".

A. Wegener

Foreword

Only a relatively small number of research workers have been interested in the collection and interpretation of geological or geophysical data relevant to the understanding of concealed basement and the recognition of the broad structural units bordering the North Atlantic. The fruit of this considerable effort, undertaken by no more than 30 scientists from Germany, the U.S.A., Canada, Ireland, Great Britain and France, is still poorly known despite the temporary collaboration of various working parties brought together by the International Geological Correlation Programme N° 27. Rather few, however, were really interested in all the possibilities of trans-Atlantic correlation. The British, Canadians and Irish are particularly interested in correlating the Northern Appalachians with the Caledonides whereas the Canadians and French have been mainly involved in an attempt to fit together the Newfoundland Banks with the Iberian peninsular and the rest of Western Europe. The very few American and French workers in this field have tried to compare orogenic belts on the West Coast of Africa with the Southern Appalachians. Even though small-scale correlations remain incomplete, it is nevertheless possible to go beyond regional studies in order to begin answering some important questions on why Caledonian effects are detectable within the Hercynian domain and why Hercynian deformation is locally overprinted onto Taconic or Acadian structures. In fact, certain workers interested in trans-Atlantic markers have considered that comparisons and correlations between deep structures do not constitute an end in itself, but represent rather a means by which ancient plate sutures may be revealed. Actually, four or five sutures are thought to exist beneath the present-day margins of the North Atlantic Ocean and various segments of these sutures are now recognized off West Africa, Western Europe and along the eastern seabord of N. America. Taking account of the fragments of plate suture already known on land, the trend of concealed sutures can be used to better understand the imbrication of Cadomian, Caledonian, Hercynian and Appalachian structures. Of course, such results run against the long-established concepts of Hercynian effects being restricted to Europe and W. Africa, a Caledonian orogeny occurring only in the British Isles and Scandinavia and "Appalachian" deformation found exclusively in the U.S.A. and Canada.

In the present work, a compilation of available information, concerning the concealed basement is brought together; this information is probably incomplete because it is widely dispersed throughout the literature and also because the rare data extracted from oceanographic cruise reports and drill-hole descriptions contain few explicit results on the nature of the concealed basement.

Certain other deliberate omissions have been necessary due to the confidential nature of geological and geophysical data belonging to Oil Companies. The author has occasionally made use of this valuable source in his general conclusions but takes care not to divulge detailed information on the regional scale.

However, it would seem that the overall effect of absent data – whether due to incomplete compilation or a real gap in information – is unlikely to result in a false general interpretation.

The final outcome of this study is the result of numerous overlapping observations which, furthermore, appear consistent with established palaeo-magnetic reconstructions of the Atlantic-bordering continents during the Palaeozoic.

One of the most difficult aspects of the present work is the need to include discussion of all scientific disciplines likely to provide information on the nature of the concealed basement. This widely-based approach, even though sometimes lying outside the normal expertise of the author, will nevertheless prove worthwhile if it provokes future workers to undertake more detailed investigations.

Finally, an attempt has been made to overcome the considerable language barrier which is apparent in many recent geological publications. In this way, due consideration has been given to all articles, whether appearing in English, French, Spanish, Portuguese or German. The author has also contributed the results of his own research undertaken over a period of fifteen years in many areas from Labrador to Florida and Rockall to Mauritania.

Rennes, November 1988   J.-P. Lefort

Contents

The Search for Concealed Continental Basement Beneath the Margins and Coastal Basins of the North Atlantic: Methods and Motives

The continental basement marginal to the Atlantic Ocean basin is never directly observed in outcrop. This concealed border zone is sometimes only about 50 km wide, as on the W. African or Iberian margins; elsewhere, off the Canadian Coast, the continental margin is nearly 700 km wide. Apart from these extreme cases, it appears there is a gap of about 500 km, between known basement areas in both sides of the North Atlantic, when account is made for ocean floor spreading. The missing information must be found if correlations (other than pure speculation) are to be established between basement terrains in Europe, Africa and America. The term "basement" in this study, is, above all, taken to mean rocks of Pre-Cambrian and Palaeozoic age — the oldest basement terrains found in the studied area are Lewisian or equivalent in age (max: 2900 Ma). Triassic events affecting the basement are well marked during the early stages of opening of the North Atlantic; this aspect of basement history is not treated in the present study because such effects often obscure the structural relationships of pre-Mesozoic basement terrains.

1.1 The Methods Available

The search for offshore basement, or even basement concealed beneath the Mesozoic and Tertiary cover rocks of passive margins and coastal basins, has only recently become of interest to the scientific community. When a review is made of reported offshore occurrences of pre-Mesozoic basement, it becomes apparent that the older data were often acquired by chance during mining and oil exploration. Many drill-holes were originally intended only to penetrate cover formations and dredge-hauls were initially designed to recover recent sediments during marine geological surveys.

At the end of the 19th Century, the sparse body of data available was of little use since the paucity of fossils in offshore samples tended to discourage the few geologists who were interested. As far as igneous and metamorphic rock-types were concerned, the classic approach was to make petrographic comparisons on the regional scale. However, this approach was hampered by the idea that some samples did not come from in situ outcrops and had probably been transported from elsewhere. In order to overcome such doubts, certain workers used to undertake systematic pebble counts in dredge-hauls; a "probable" basement map could then be drawn up in areas where fragments of cover rock had not been recovered. In general, the problems of direct sampling at sea were so great that basement recognition was often based on bottom topography; if the relief was well-accentuated, then submerged formations would be assigned to "pre-Mesozoic basement", "Hercynian basement", "Caledonian basement", "Pre-Cambrian basement" or even "basement" with no qualifier.

Of course, even nowadays it is sometimes difficult to assign an unfossiliferous and isolated sample to a known formation. However, a systematic search for microfossils and detailed petrographic observations are generally sufficient to characterize the studied samples. In most cases, it is possible to assign samples to formations that have already been described and this may suggest correlations on the large scale. This kind of operation is, however, only meaningful if the distances involved are not too great.

Geochronological and geochemical studies are still rather limited in number due to the relatively small size and altered character of available samples.

Even though sampling at sea has undergone considerable progress in terms of handling and performance, such methods are nevertheless rather primitive in principle. The main disadvantage is that samples are generally torn off without any record of orientation with respect to bedrock. This means that small-scale tectonic conclusions cannot normally be drawn from undersea samples. It is only through direct diver observation, or the use of submersibles with viewing ports, that structural measurements can be made. Side-scan sonar methods and underwater television/photography can also provide data which yield results comparable to on-land techniques.

Commercial drill-holes into basement, whether offshore or not, provide only stratigraphic and petrographic information, since the process of core-cutting produces chips without any orientation. Even if the occasional core-section has been recovered with its original orientation, these samples are usually too infrequent to enable a comparison between tectonic style in the borehole and the surrounding area.

Thus, it should be borne in mind that the petrographic, stratigraphic and geochemical data quoted in this study are always of higher quality compared with the available structural information for the same zone; this is particularly the case if preliminary geophysical surveys have not preceded the selection of the sampling area.

Without any doubt, geophysical methods have brought the greatest contribution to our knowledge of the concealed basement. Geophysical surveys, in fact, have enabled the extrapolation of surface information to greater depth as well as the correlation of well sections which are separated by several tens of km. Here again, information on deep structure is often a by-product of exploration undertaken to improve knowledge of the cover rocks. A great amount of data available today has been obtained during attempts to define the depth to formations of economic interest rather than from the detailed mapping of structure in the basement itself. Although seismic techniques are amongst the most widely used in the investigation of recent cover formations, sparker and airgun surveys yield little useful information on the basement. This is because the areas we are interested in are all located in orogenic belts, where intrusions and highly deformed zones produce complex diffraction patterns and render any seismic reflection records impossible to interpret. High energy seismic reflection studies, however, have been developed recently which are capable of revealing deep thrust structures within the crust. The presence of deep thrusting has challenged certain concepts of orogenesis. Since these deep seismic reflection profiles are rather rare, other techniques, usually considered as ancilliary, have come to the fore. The most useful techniques in this category are seismic refraction, magnetism and gravity. The magnetic and gravimetric anomalies recorded on the continental margins are more closely related to structures within the basement than structures in the cover.

This is generally true, other than in basins, grabens and horsts, due to the small lateral variation of recent sedimentary facies and the relative flatness of the roof of the basement. In addition, the gradual changes in sedimentary thicknesses recorded on continental margins and coastal basins tend not to disturb the magnetic and gravimetric anomalies arising from basement structures.

The use of geophysical data for the interpretation of deep structures leads to an approach which is the reverse of that normally used in the field; one proceeds from the regional scale before more detailed information from drill-holes and dredges are integrated into the picture. The local material is often used to date or characterize the nature of objects already mapped by geophysics. The correlation between large-scale and small-scale types of information is certainly one of the major difficulties in studying the concealed basement because mesoscale (10-100's metre scale) observations are hardly ever possible. Although the study of hidden structures leaves a great deal to interpretation, a good knowledge of on-land regional geology and various cross-checks from model-derived parameters can generally lead to the proposition of a unique and realistic solution. These parameters include the velocity, density and magnetic susceptibility of formations and sometimes even the dip of reflectors. There is, nevertheless, a quantum leap in reasoning between local geological data and geophysical interpretations based on mega-scale structures. The present work makes a clear distinction between small- and large-scale observations.

1.2 The Initial Continental Arrangement

Inherent to the problems of trans-Atlantic correlation, whether geological or geophysical, is the difficult question of the initial continental configuration that should be used in such studies. Various types of criteria are known to exist for the choice of continental fit between the two sides of the Atlantic. Palaeomagnetic criteria may be used (Smith et al., 1981; Scotese, 1984), but are hampered by the imprecisions of the method which can attain 500 km in latitude and even more in longitude. Bathymetric data can also be used (Bullard et al., 1965) but the 500 fathom isobath is now considered not to represent the limit between oceanic and continental crust. A continental fit can also be constructed from the pattern of magnetic anomalies mapped on the ocean floor (Pitman and Talwani, 1972; Klitgord and Behrendt, 1979) or the geometry of transform fault zones (Le Pichon and Fox, 1971; Laughton, 1975; Le Pichon et al., 1977; Klitgord and Schouten, 1982 and 1984). However, the results of these studies are not always consistent between themselves and sometimes show differences with certain regionally-based studies (Kristoffersen, 1977; Srivastava, 1978).

Several attempts have been made to bring together geological and geophysical data in the same synthesis (Sclater et al., 1977; Biju-Duval et al., 1977; Wissmann and Roeser, 1982; Olivet et al., 1984). Although some of this work could be used to establish a basis for correlation, there are unfortunately several reasons why such an approach is not satisfactory.

Firstly, none of these methods takes account of the crustal stretching produced by the opening of the North Atlantic (Montadert et al., 1979). Above all, shearing phenomena in the Hercynian basement occurred in Triassic times before the phase of opening in the Biscay-Labrador area (Lefort, 1973). This late shearing also affected the western margins of the African continent (Swanson, 1982) and is thought to have offset the ancient markers that can be traced across the Atlantic. This is the reason for a choice of continental fit which is based on the more easily correlatable markers (Lefort and Van der Voo, 1981). These markers are mostly of Upper Palaeozoic age, thus the initial configuration of the Atlantic bordering continents used in this study can also be considered of that age (see Table 1).

Table 1. Plate rotations used for continental fit in this study

Plate rotated with respect to North America	Latitude and longitude of pole of rotation		Angle of rotation
Europe	89.1°N,	150.3°E	29.0°
Iberia	68.1°N,	344.2°E	53.3°
Africa	66.0°N,	346.6°E	76.7°

The continental assembly so obtained, shown on a Mollweide projection, is further discussed in Chapter Eight; the positions of Greenland and the Galicia Banks were not calculated using the rotations listed in Table 1 but fitted by hand in order to avoid important continental overlaps. Curiously, this reconstruction is not very different from that given by Sclater et al. (1977), Scotese et al. (1979) and Tapscott and Phillips (pers. comm., 1982). As far as

Africa and America are concerned, the present reconstruction is also close to that of Klitgord and Popenoe (1983) and Klitgord and Schouten (1986). These similarities are partly explained by the orientation of the selected markers: the lack of N-S trending markers leads to an imprecision in relative longitude shared by all reconstruction models. Even though the present technique is more precise than the others in latitude estimation, it is impossible to know whether convergence with other models results from compensating errors or relatively small Permo-Triassic displacements.

1.3 Variations in the Depth of Basement Around the North Atlantic

The maps which follow show the important variations which exist in the concealed basement roof topography. Although the data are heterogenous, and sometimes greatly smoothed out, they are nonetheless sufficient to define areas where basement is unlikely to be attainable. Thus, the depth to basement is shown to be 6–7 km in Senegal, 10 km in Western Morocco (see Fig. 1), 5 km off S. Portugal and Spain, 12 km in the offshore Aquitaine Basin and 4–5 km in the Celtic Sea and Western Approaches to the English Channel (see Fig. 2). On the Canadian side (see Fig. 3), the roof of the basement can be depressed to depths of 6–12 km, whereas several basins off the coast of the U. S. A. can attain maximum depths of 8, 12 or even 14 km (see Fig. 4). In a general way, the N. American basement is deeper than that measured beneath the European and African margins. This difference is reflected in a disparity of available geological data. Furthermore, outcrops of basement are extremely rare on the deeper parts of the continental shelf and slope off North America compared with Europe and Africa, where outcrops are fairly common. Such differences result in a far greater dependence on geophysical rather than geological data on the western side of the Atlantic.

The transitional basement which immediately borders oceanic crust is too closely associated with recent intrusions and salt diapirs to enable reliable mapping by geophysical methods alone. Certain tilted blocks at the base of the continental slope show rotations which, in some cases, lead to the outcropping of submerged basement. These outcrops are the only real constraints for the tectonic interpretation of oceanic border zones. In any case, because basement rocks from deep-water areas are often highly eroded they rarely constitute a reliable element of comparison in trans-Atlantic correlations.

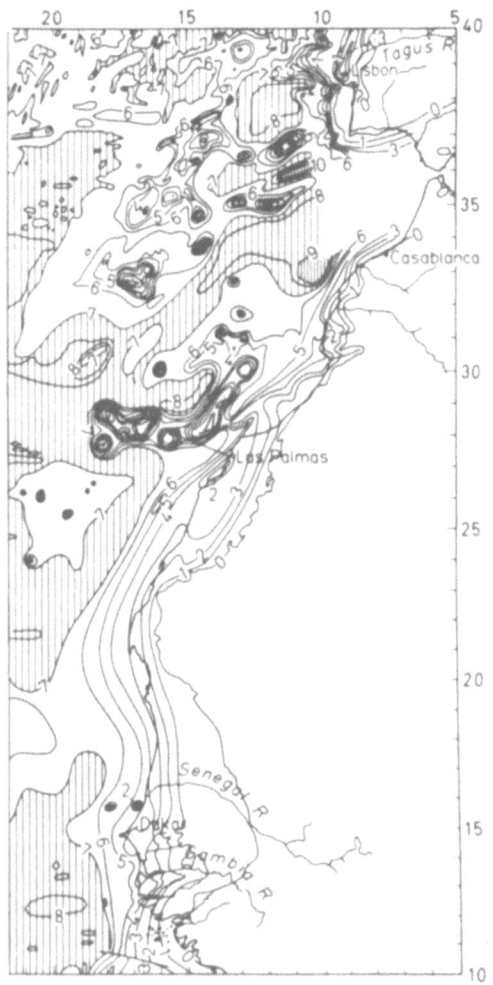

Fig. 1. Relative depth to roof of basement off N.W. Africa, expressed in seconds. (After Uchupi et al. 1976)

1.4 Major Basement Provinces Each Side of the North Atlantic

According to recent and well-accepted palaeomagnetic results, the pre-Atlantic domain was subject to at least two plate collisions during the Palaeozoic. The first collision, producing the Caledonian-Appalachian foldbelt, is linked to the approach of the Laurentian plate with Baltica and Armorica (Perrroud et al., 1984) during the Lower

Palaeozoic. The second event is the result of the northerly drift of Gondwana during Devonian and Carboniferous times leading to the formation of the Hercynian foldbelt and rejuvenation in the Northern and Southern Appalachians (see Fig. 5). If the palaeomagnetic results are correct, it should therefore be possible to find traces of at least two Palaeozoic sutures under the Atlantic margins. Otherwise, geological information from the continental areas would suggest the presence of a Grenvillian suture within the Laurentian plate, an Armorican suture within the Avalonian microplate and a Mauritanian "welt" within Gondwana.

The limits of Baltica lie, for the most part, outside the studied area, so only the structure of the concealed parts of Laurentia, Armorica and Gondwana will be considered in the present study.

Particular attention is paid to the criteria used for the identification of the sutures mentioned above. As an aid to comprehension, a clear distinction is made between Caledonian and Hercynian events. This corresponds, furthermore, to two well-defined phases of plate collision.

Most of the areas included in this study are thus restricted to the continental shelf and adjoining on-land basins. The main aim of this work is to propose correlations and examine the consequences implied by these correlations. Thus, it is considered unnecessary to provide lengthy descriptions of each orogenic belt around the North Atlantic margins. Apart from the fact that systematic descriptions would be tiresome, their inclusion would obscure the fundamental approach followed in this study. Because of this, the reader is referred to a copious bibliography, listed at the end of the book, which provides the sources used by the author as well as more general works.

The approach adopted by the author consists of a brief presentation of the main on-land units and structures (as they appear at the coast or at the edge of overlying cover successions) followed directly by a discussion of their concealed prolongations at depth. The choice of tectono-metamorphic units or zones has been made on a scale which is compatible with a simplified structural picture of the region concerned. Occasionally, it is necessary to extrapolate more widely when a major unit is correlated with distant equivalent structures onshore; such correlations can be useful for a better understanding of large-scale structures.

The areas studied in this book extend from the Rockall Bank to the Gulf of Guinea along the eastern borders of the N. Atlantic, and from Greenland to Florida in the West. This zone corresponds, therefore, to the continental crust bordering the Atlantic Ocean basin and does not include the Caribbean Sea, the Gulf of Mexico, Baffin Bay, the Norwegian Sea, the North Sea or the Mediterranean.

Fig. 2. Average depth to roof of basement off Western Europe (contour intervals are km)

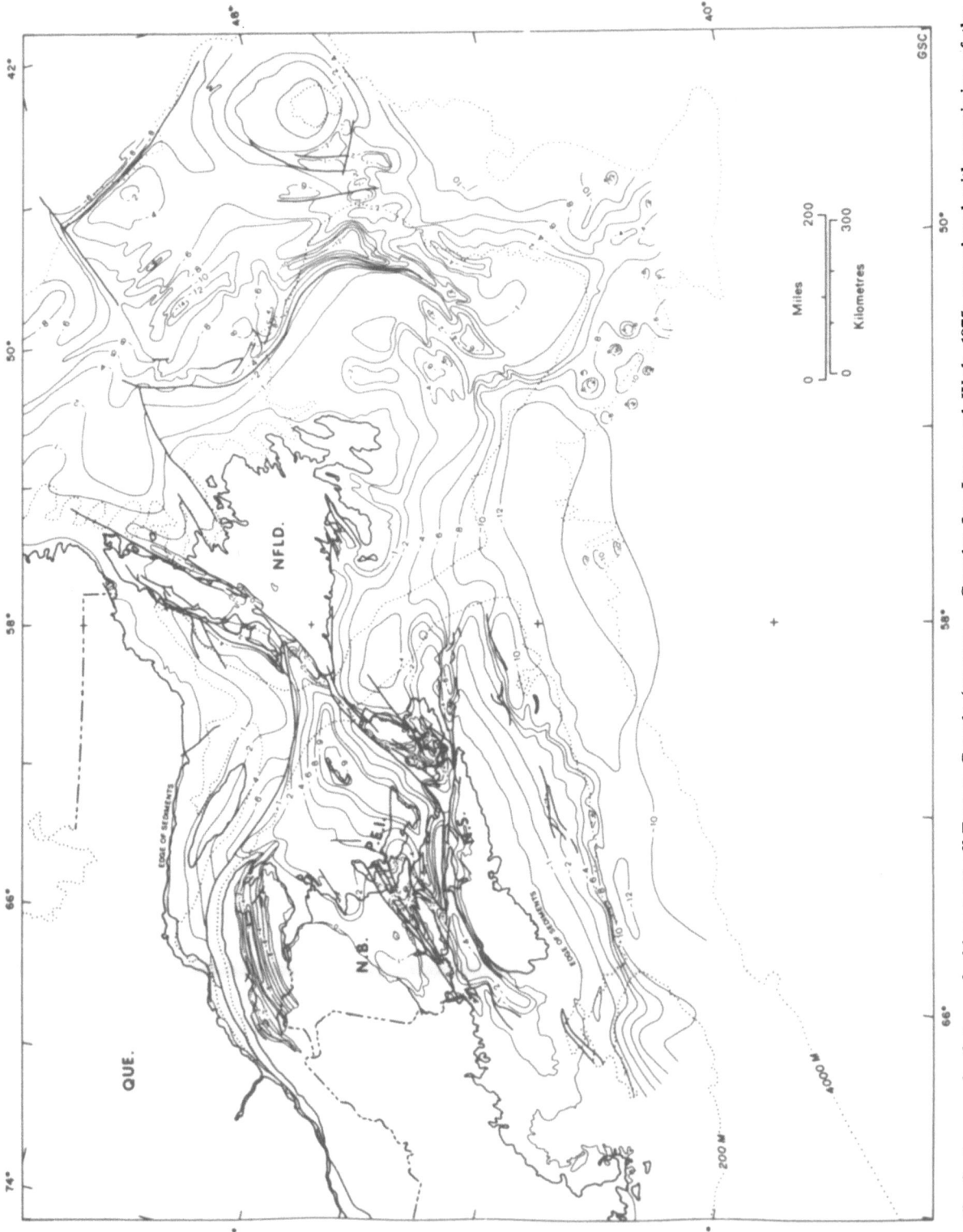

Fig. 3. Average depth to roof of basement off Eastern Canada (contour intervals are km). (After map-sheet 1400 A of the Geological Survey of Canada; In: Jansa and Wade 1975; reproduced with permission of the Minister of Supply and Services, Canada)

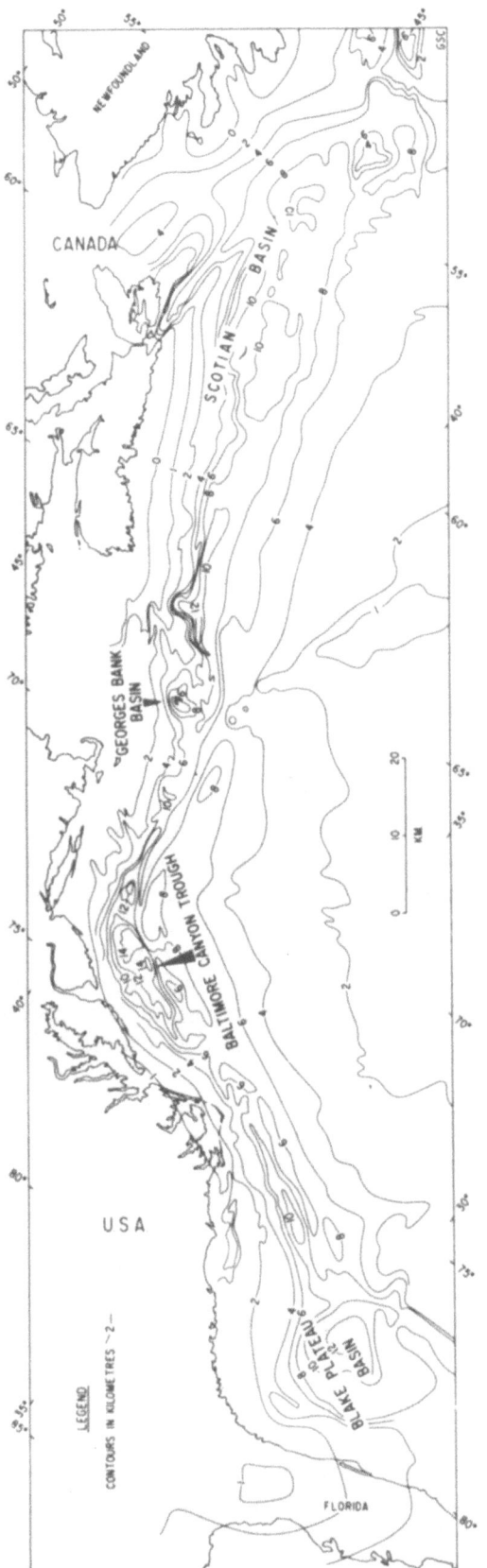

Fig. 4. Average depth to roof of basement off the Eastern U.S.A. (After Jansa and Wiedmann 1982)

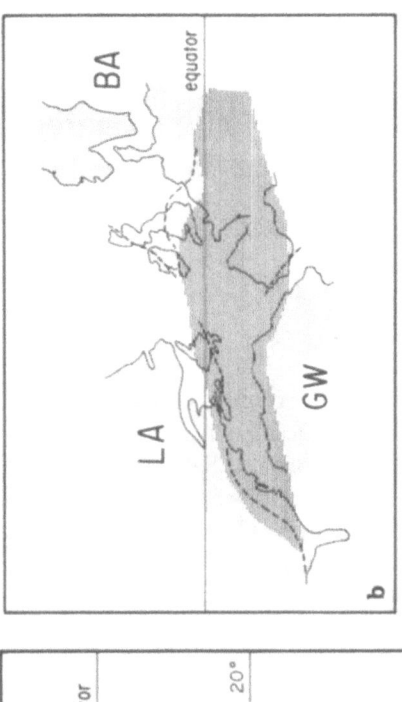

Fig. 5. Major lithospheric plates and collision zones in the pre-Atlantic domain according to palaeomagnetic results. **a** Caledonian collision; **b** Hercynian collision. A: Armorica; BA: Baltica; GW: Gondwana; LA: Laurentia. Armorica is itself made up of an Avalonian and an Iberian fragment. Horizontal ruling shows extent of foldbelts (see text for explanation). (After Perroud et al. 1984)

Chapter 2

The Submerged Part of the Laurentian Basement: Former Western Margin of the Iapetus Ocean

This basement area corresponds to the western margin of the Iapetus Ocean which extended from Scandinavia to the southern U.S.A. (and probably still further South) during Lower Palaeozoic times. It is made up of two different types of terrain; in the North, pre-Grenvillian terrains dated at between 1,370 and 2,480 Ma and, in the South, Grenvillian terrains yielding ages around 955 Ma. An ancient suture zone can be defined between these two types of terrains — this is known as the Grenvillian Suture.

2.1 Fragments of the Pre-Grenvillian and Grenvillian Basement in the North Atlantic Area

Pre-Cambrian metamorphic rocks corresponding to known on-land outcrops have been cored in five oil prospection wells to the South of the Labrador Sea (Umpleby, 1979). However, the connection of these terrains with Greenland is problematic due to the fact that Greenland would have to be displaced 300 km to the South in order to satisfy a fit with Northern Labrador. This is incompatible with the compiled magnetic data for oceanic crust in this area (Srivastava et al., 1981). Even if the displacement of Greenland occurred before the opening of the Labrador Sea, the problems of correlation are not solved because, further South, units of different age are brought together — these are the Grenville Front in Canada and the Ketilidian belt in Greenland (Bridgwater et al., 1973). Thus, there is no current solution concerning the juxtaposition of the Labrador and Greenland basements. Along the Labrador Coast, the Grenvillian basement crops out as a narrow band and its offshore extension has been mapped mainly by seismic reflection methods (Grant, 1972; Van der Linden and Srivastava, 1975; Hinz et al., 1979). Similar methods have been applied to the mapping of offshore basement between Labrador and Newfoundland (Haworth and Sanford, 1976; Haworth et al., 1976).

On land, the Grenvillian basement is known to be made up of a wide variety of schists, gneisses, granites and gabbros. This great variability in composition contrasts with the homogeneity in gravity. Grenvillian terrains are, in fact, characterized by a clear negative gravimetric anomaly (- 20 milligals) extending over a vast area from the North of the St Lawrence Gulf to the Humber Zone of N.W. Newfoundland, where basement crops out again (Haworth, 1975) (see Fig. 6). Because of this, the typical gravimetric features of the Grenvillian basement can be used here to extend offshore the geology of terrains observed on the S.E. Margin of Labrador (Douglas, 1970). Laurentian-type basement has also been found South of Lat. 57°N on the Rockall microcontinent (Scrutton, 1970), where two main localities have been sampled. The northern locality has provided an acid granulite with small amounts of pyroxene. Hornblende and whole-rock samples have yielded Laxfordian ages (1,566 Ma and 1,670 Ma respectively, Roberts et al., 1973). Further South, another granulite has been dated as Grenvillian (ca. 1,000 Ma; Miller et al., 1973).

This ancient basement is also found on the Shetlands Plateau, to the West of the Orkneys, where a series of horsts cutting the Permo-Jurassic is seen to upfault blocks of Lewisian and Moinian basement (Evans et al., 1982) (see Fig. 7).

The strong positive gravity anomaly extending from Lat. 60°–61°N is elongated parallel to the Orkney-Shetland axis and shows that the mapped ridge of Pre-Cambrian basement continues at depth for some considerable distance under cover formations. The concealed ridge is probably composed of Lewisian granulites. A positive magnetic ridge is parallel to the gravity anomaly and passes through the westernmost Lewisian horsts — this structure corresponds to an ancient basement ridge extending all the way to the continental slope.

Further South, the Lewisian basement is frequently composed of gneisses and forms extensive outcrops

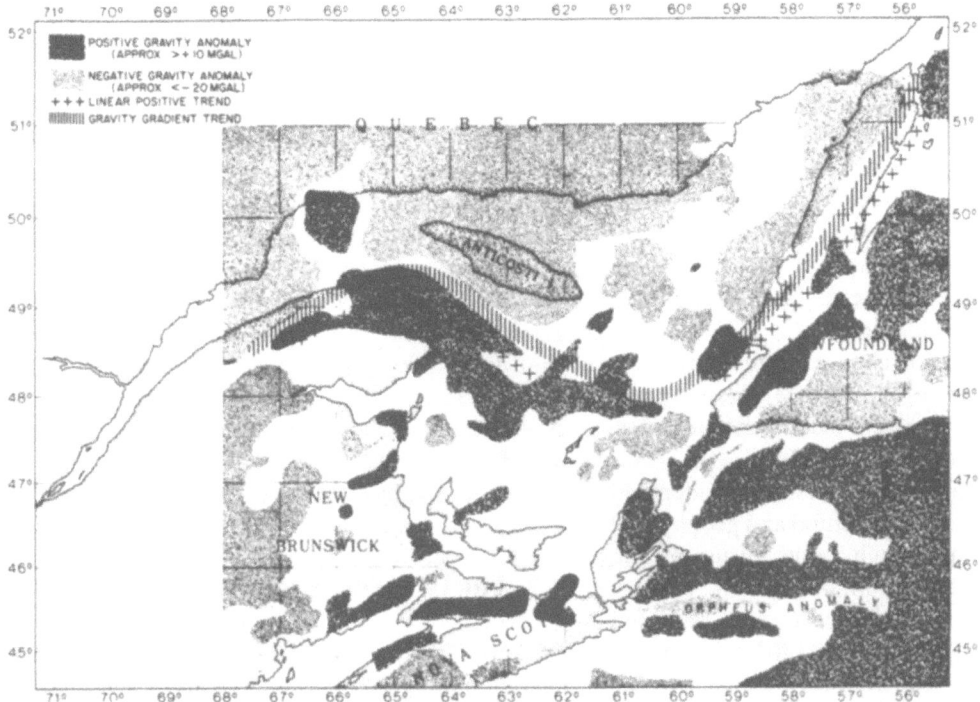

Fig. 6. Sketch map showing the extent of Grenvillian basement in the Gulf of St. Lawrence according to gravity data: areas shown as light stipple, and bounded on the South by a strong gravity gradient (zone indicated by vertical ruling), are underlain by Grenvillian basement. (After Haworth 1975; reproduced with permission of the Minister of Supply and Services, Canada)

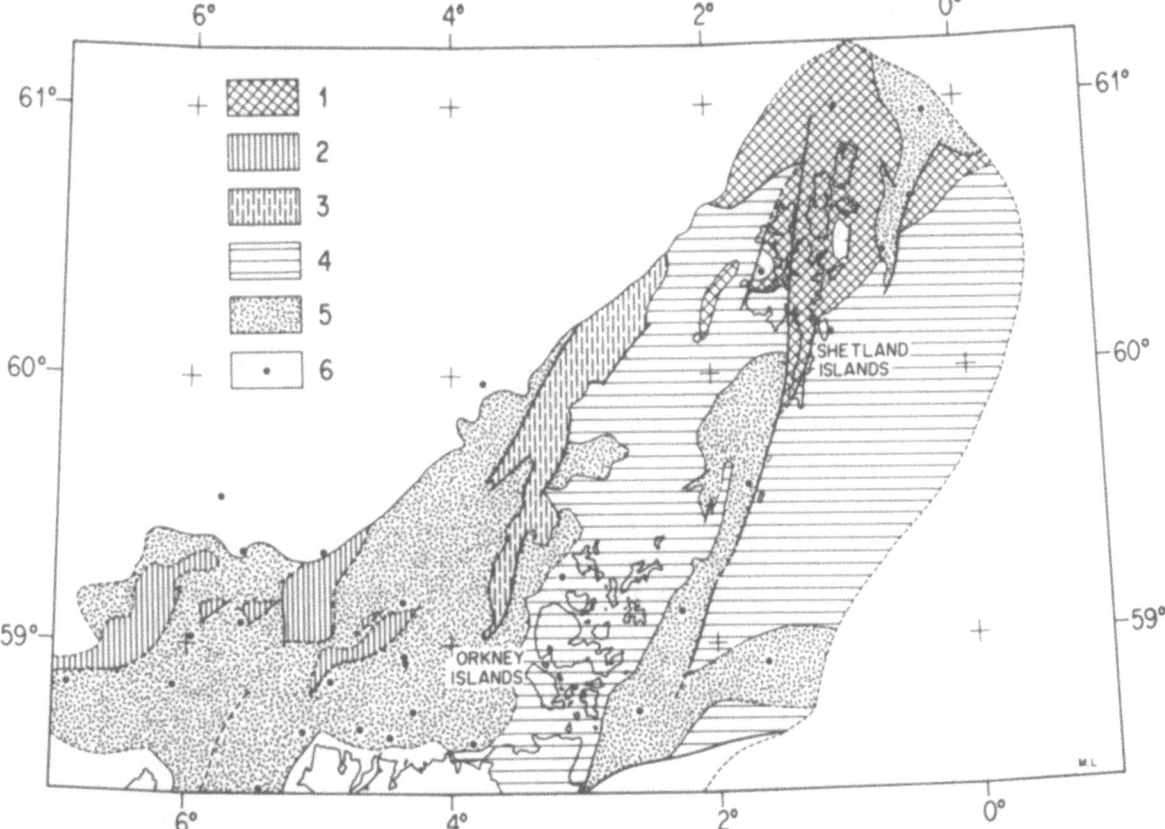

Fig. 7. Offshore basement to the North of Scotland. 1: Dalradian; 2: Lewisian; 3: Lewisian or Moinian; 4: Devonian; 5: Permo-Jurassic; 6: sampling localities. (After Evans et al. 1982)

in the Hebrides; this terrain shows initial structures following a NW-SE trend which have been sliced up into strips by major Caledonian faults. These include the Minch, Camasunary-Skerryvore and Great Glen Faults which have brought up ancient basement to outcrop (see Fig. 8). The initial structural trends are still apparent in the Hebrides basin where they form transverse steps (Binns et al., 1975); where these steps break through the cover there appear outcrops of either Lewisian or Torridonian (Chescher et al., 1983; Eden et al., 1973). On the whole, this basement is characterized by positive gravity anomalies but the presence of basic granulites, as seen on the eastern side of the Hebrides, tends to re-inforce this regional effect.

The Scottish mainland to the East shows an identical basement, with the same structural and geophysical characteristics even though basic intrusions are more abundant at outcrop. Above all, a thick development of Torridonian cover is to be noted which, furthermore, builds up in offshore areas.

Some data are also available for basement to the West of the Outer Hebrides, which appears to be almost entirely made up of Lewisian rocks. This has been verified by sampling (Evans et al., 1982) and geophysics. Otherwise, it is possible that the outer part of the margin is richer in dense granulites than areas closer to the Hebrides (Jones, 1981).

Apart from the presence of a pre-Grenvillian basement, the main interest of this margin lies in its lack of subsidence. This could be due to the existence of large quantities of low-density granitic bodies of Laxfordian age at depth (Watson, 1977).

Finally, as far as the Malin Sea to the North of Ireland is concerned (see Fig. 9), the Lewisian basement is generally characterized by high frequency magnetic structures linked probably to basic granulites. By contrast, the Moinian, Torridonian and Dalradian display very weak magnetic structure. To the North of this region, the Skerryvore Fault brings the Lewisian to surface on the Skerryvore Bank where metamorphosed limestone and basic granulites have been sampled — the rest of this Bank appears to be made up of gneisses (Evans et al., 1980). On the Stanton Banks, Lewisian amphibolites and granulite facies orthogneisses are found (Gérard, 1979; Gérard and Boillot, 1977). This material is injected with leucocratic and sometimes highly cataclased granites, which may be assigned to the Laxfordian (Binns et al., 1974).

To the North of the Stanton Banks, the Minch Fault (Gérard, 1979) is seen to form a southern limit to gneiss-dominated basement areas with probably less basic material — this is because gravity is less

positive. The outer margin situated further West has not been fully investigated, but is characterized by a gravimetric and magnetic high which may correspond to upfaulted basement (Bailey et al., 1974; Roberts, 1975) despite the presence of Upper Cretaceous and Eocene cover (Gérard, 1979). Gérard (op. cit.) has proposed that this uplifted block corresponds to a fragment of Grenvillian basement identical to that found on the Rockall Plateau. In this way, it is almost certain that the Malin Sea basement is marginal to the main pre-Grenvillian domain and that the underlying basement is more likely composed of Grenvillian. Such terrains also crop out in N.W. Eire, in County Mayo, where they have been dated by radiometric methods (Max and Sonet, 1979; Van Breeman et al., 1978).

2.2 The Grenville Front

In Labrador, this boundary can be traced with precision by geophysical methods — although it shows rather less continuity at the surface — and is considered as evidence for subduction from NW to SE (Thomas and Tanner, 1975) that took place up to about 1,000 Ma ago. The Grenville Front is a linear structure which is bordered by a long wavelength gravity anomaly and a narrow magnetic anomaly. These anomalies are both negative and of large amplitude; they are not precisely superimposed on each other. Near the Canadian coast, amplitudes fall off sharply and the anomalies become difficult to follow onto the margin (Valliant et al., 1975); also in this area, the gravimetric anomaly has even a tendency to split into two segments (Jacobi and Kristoffersen, 1981).

On the continental slope, by contrast, a clear gravimetric lineament enables recognition of the suture trace right up to the boundary between continental and oceanic crust (Van der Linden and Srivastava, 1975). It would appear that the Grenville Front separates on-land terrains which are broadly denser in the South than in the North and that this observation can be extended to the continental margin.

On the Rockall Bank, outcrops of Laxfordian and Grenvillian granulites are separated by a magnetic structure elongated ENE-WSW (Roberts, 1975) which might represent the eastern prolongation of the Grenville Front. Furthermore, it is possible that the gravimetric anomaly (Roberts, 1970) which extends to the South of the Rockall Bank between dated samples, but slightly to the North of the trace chosen by Roberts (1975), is also part of the Gren-

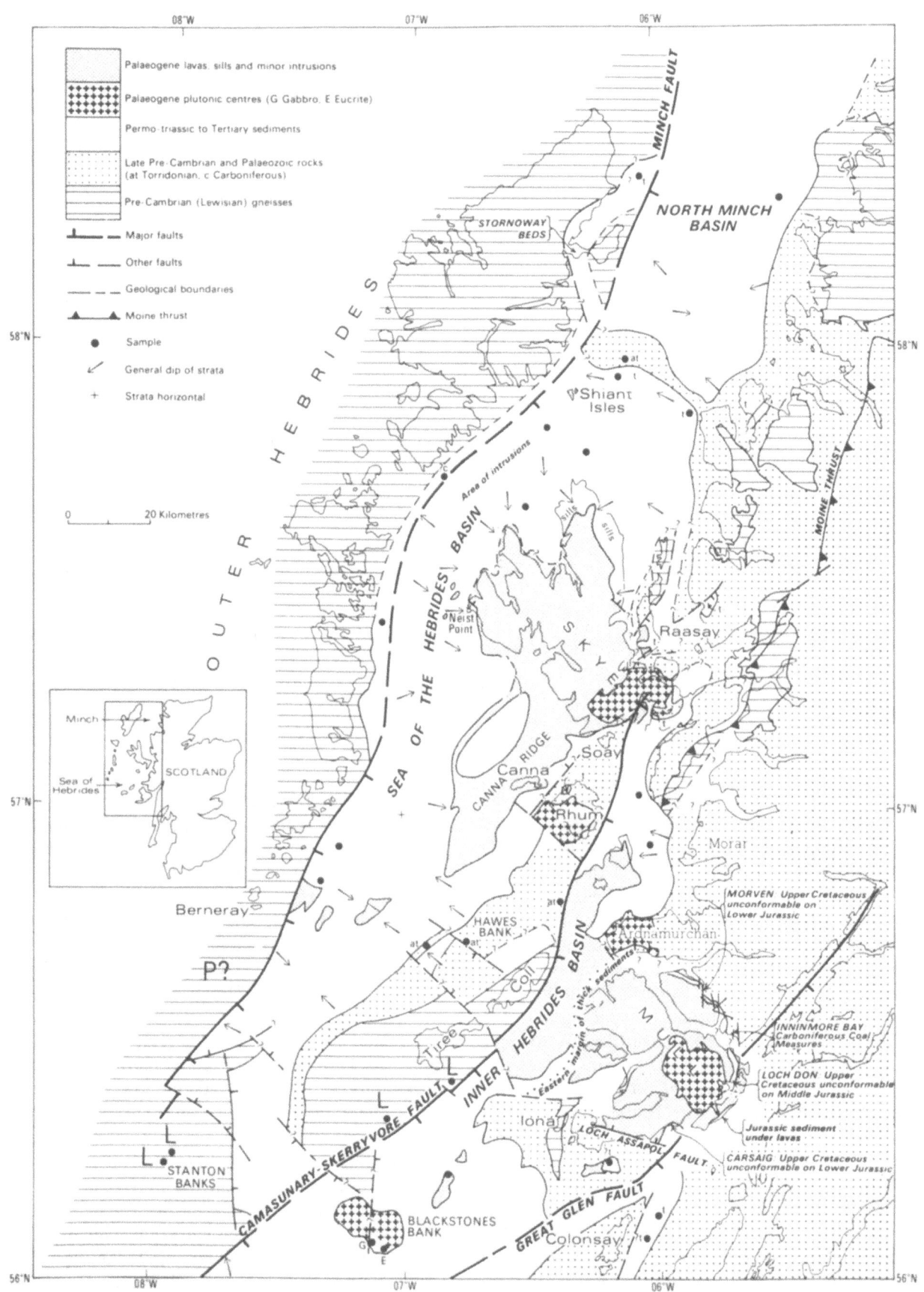

Fig. 8. Basement in the Hebrides Sea. L: Lewisian; P: Palaeozoic. (After Binns et al. 1975; reproduced with permission of Elsevier Applied Science Publishers)

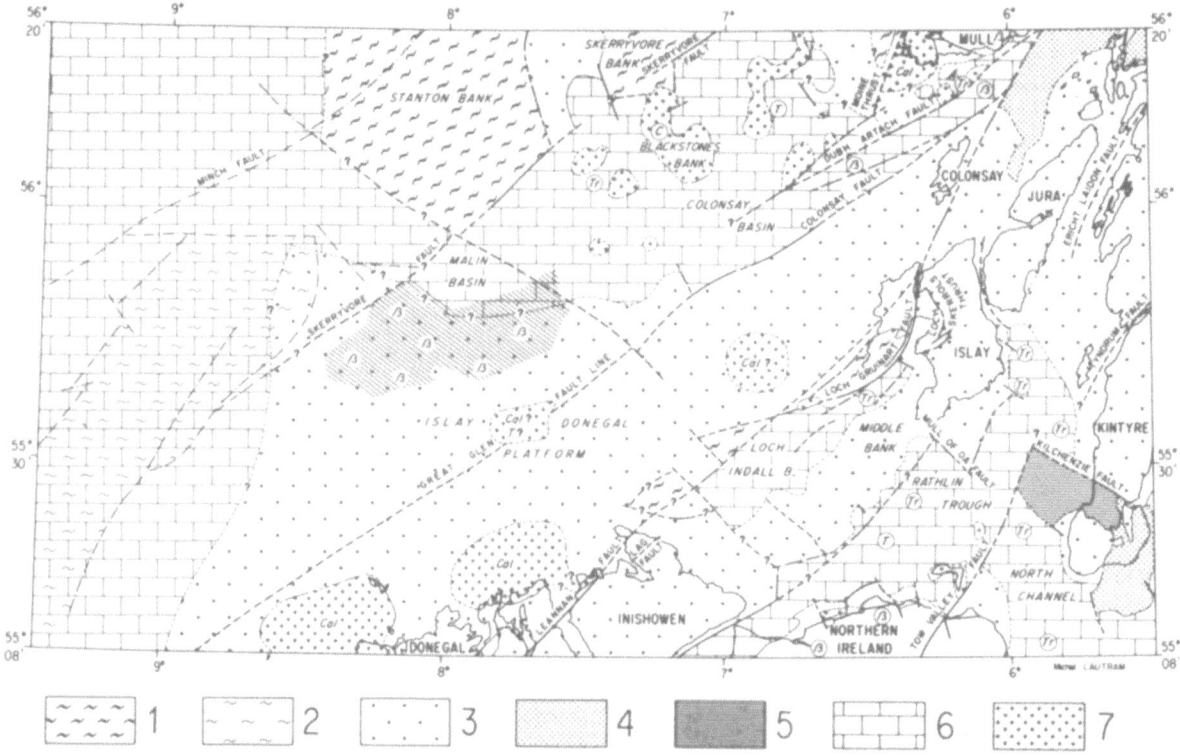

Fig. 9. Geological sketch map of the Malin Sea. β: Tertiary basalts; —·—·—: faults concealed by cover; D: Dalradian; Tr: Triassic. (Modified from Gérard 1979 and Evans et al. 1980) 1: Lewisian; 2: probable Grenvillian concealed by cover; 3: Lower Palaeozoic and Upper Proterozoic; 4: Old Red Sandstone; 5: Carboniferous; 6: Mesozoic and Tertiary; 7: intrusions (Cal: Caledonian; C: Cretaceous; T: Tertiary)

ville Front. However, the relationships between the negative gravity anomaly, the Grenville Front and the Grenvillian Suture have not been fully elucidated (Irving et al., 1974). It should also be noted that there is a change in orientation of the central graben recognized on the Bank. The southern part of the graben shows a N-S direction and the northern part a NW-SE trend; this bend is situated near 57°N and could be associated with the proximity of the Front. The same could apply to the straight transform margin seen further S.W. towards 55°50N (Roberts et al., 1979).

When the Rockall Trough is closed (Le Pichon et al., 1977; Scrutton, 1972), the Hatton-Rockall block closely follows contours on the continental slope off Northern Ireland. This fit juxtaposes the Grenville Front with an E-W-oriented fault which follows the 56th Parallel. The E-W-oriented fault is clearly marked by an elongate magnetic anomaly and has already been interpreted as the possible prolongation of the Grenville Front onto the European plate (Bailey, 1975; Bailey et al., 1975). This magnetic

anomaly has been extended on the continental shelf (Gérard, 1979) where it is superimposed onto a negative gravity discontinuity (Hydrographic Department, 1973). In any case, it is difficult to separate, from a gravimetric point of view, which part belongs to the anomalies following the Front and which part belongs to the Mesozoic basin controlled by the Front. The Grenville Front cannot be followed eastward of 8°W since it appears to be interrupted by the offshore prolongations of the Camasunary-Skerryvore Fault (McQuillin and Binns, 1973); an extension up to the Great Glen Fault, however, cannot be totally ruled out. To the South, another E-W fracture bounds the concealed horst which lies beneath cover at the edge of the northwest Irish continental margin.

It is interesting to note that the Grenville Front, defined in this way, is seen to truncate the gravimetric anomaly which marks the edge of the margin not only off Canada but also to the West of Ireland. In addition, it can be seen that this Front is bounded to the North by important morphological highs which bring up basement on the Makkovik Bank off Labrador (Van der Linden and Srivastava, 1975), Rockall Island on the Hatton-Rockall Bank (Roberts et al., 1973) and on the Stanton Banks off Northern Ireland. Further to the East, near Scotland, most of the Grenvillian has been masked by the Caledonian Moine Thrust and the Grenville Front cannot be found. In this area, the contact between pre-Grenvillian or Grenvillian and younger terrains is formed by a major Caledonian shear fault.

2.3 Southern Limit of the Laurentian Basement

This boundary is situated in mainland Canada along the Baie Verte – Brompton Line (Williams and St Julien, 1978). Further South in the American Appalachians, this same limit is marked by an important gravity gradient which probably results from a density contrast between Grenvillian Crust and the Appalachian orogenic belt (Diment et al., 1972). An identical gradient exists West of Newfoundland where the eastern limit of the Laurentian basement is observed. Furthermore, this gradient is compatible with a compilation of seismic refraction data in the region (Weaver, 1967). In fact, this more or less well-defined gravity gradient can be followed right along the border of the Appalachians; it only disappears where allochthonous structures have been recognized (Haworth, 1981). In Newfoundland, this gradient itself is perturbed by granitic intrusions, small Carboniferous basins or even klippes of obducted ophiolite (Jacobi and Kristoffersen, 1981). To the North of Newfoundland, the gravity gradient zone comes close to the Labrador Coast and is interrupted only at 51°N by a positive anomaly attributed to a failed Palaeozoic rift (Haworth, 1980). A clear linear magnetic structure follows the gravimetric gradient and is superimposed on the Cabot Fault which is known to form an eastern limit to Grenvillian terrains. The only disparity between magnetic and gravimetric information occurs much further North near 53°N. Here, the magnetic discontinuity continues northward, whereas the gravimetric marker turns eastward (Jacobi and Kristoffersen, 1981); such geometric relations can be explained by the influence of Carboniferous movements on the Cabot Fault or post-Grenvillian thrusting along a straight front.

On the East Canadian continental slope, the southern limit of the Laurentian basement causes an interruption in the gravimetric edge effects that normally follow the continental margin (Srivastava et al., 1981). This interruption is entirely comparable to that produced by the Grenville Front.

The position of the southern limit of Laurentian terrains is more speculative on the Rockall Bank. Since this limit displays an eastward bend in the Labrador Sea and must necessarily pass to the South of Mayo County in Eire (Van Breemen et al., 1978), there has to be a southerly deflection of the limit in the region of the Rockall Plateau. The best candidates for defining such a limit (Lefort, 1984) are provided by the arcuate magnetic anomalies (Vogt and Avery, 1974) that mark the SW extremity of the Rockall micro-continent (Roberts and Jones, 1978). These markers cut the N 40/50°-trending features which are characteristic of this part of the plateau. One of the consequences of this interpretation is that Grenvillian basement should exist to the NE of Porcupine Bank; this has already been proposed by Bailey (1979), who used magnetic data to support an extension of Grenvillian terrains southward to the offshore prolongation of the Great Glen Fault. The southern limit of the Laurentian Continent is difficult to locate in areas further East, but is thought to pass through NW Ireland and perhaps as far as the Highland Boundary Fault in Scotland (Phillips et al., 1976). Certain authors even extend the edge of the Laurentian Continent to the Southern Uplands Fault (Dunning and Max, 1975). It is, however, far from certain that flakes of Grenvillian basement caught up in the Caledonian orogeny could be present so far South (Piasecki et al., 1981).

Figure 10 shows, in a simplified manner, the extension of Laurentian basement on the N. Atlantic margins as well as the main boundaries of Grenvillian and pre-Grenvillian terrains. The sinistral offset which appears between the Rockall and Porcupine Banks is not entirely conjectural since it is known that the Great Glen Fault (and its satellite, the Leannan Fault) has shown late-stage sinistral movements amounting to about 220 km (Max, 1978).

The results of seismic refraction provide a remarkable contribution to the distinction between Grenvillian and pre-Grenvillian crust. Comparisons of seismic velocities are nevertheless only valid if one takes into account that velocities show a regular decrease, within the same terrain, from West to East. This variation is probably due to differences in petrographic composition.

Thus, one can correlate crust with a 6.61 km/s velocity North of the Grenville Front on the Labrador margin (Hinz et al., 1979) with the 6.41 km/s material occurring off the Hebrides (Bott et al., 1979) and also material in the 6.2–6.45 km/s range found to the North of the Great Glen Fault (Bamford et al., 1977). Further South, in the Grenvillian, velocities are always lower than those known on the other side of the Front at similar longitudes. In this way, velocities ranging from 5.18 to 6.47 km/s between Newfoundland and Labrador (Sheridan and Drake, 1968), can be correlated with the 6.33 km/s known to the south of Rockall (Scrutton, 1970) and even with the 6 km/s recorded N.E. off Ireland (Hall et al., 1984). Furthermore, these velocities are never more than 6.05 km/s South of the Great Glen Fault on land (Bamford et al., 1977).

2.4 Palaeozoic Cover of the Submerged Laurentian Basement

To the North of the Gulf of St Lawrence, the Grenvillian basement is overlain by typical margin deposits such as carbonate-bearing successions, mudstones and prograding clastic sequences; the same sediments are found to the South and East of the Labrador Coast. In the St Lawrence Gulf, sediments are mainly Ordovician and Silurian with the youngest part of the succession cropping out on Anticosti Island. Between Newfoundland and Labrador, however, Cambro-Ordovician has been found (Loring, 1974). These terrains generally show very low dips, which confirms the fact that they belong to a platform area hardly deformed by the Caledonian orogeny (Haworth and Sanford, 1976). Towards the North, terrains become progressively older until Lower Cambrian and even Upper Hadrynian rocks are encountered to the East of Labrador (Haworth et al., 1976). If the assumed position of the Laurentian margin is correct, then the Ordovician sediments sampled near the Cartwright Arch and SW of Greenland must also belong to this margin. Sediments from the Cartwright Arch are dolomitic and shaly/silty with shallow water fauna (Bousquet et al., 1977), whereas sediments off SW Greenland are further from the

ancient margin and contain breccias with Pre-Cambrian clasts as well as sandstones and limestones suggesting proximity to the coast (Johnson et al., 1975).

According to seismic data, there should be little or no Palaeozoic cover on the Pre-Cambrian basement to the North of Hamilton Bay (Grant, 1972). By contrast, the Palaeozoic appears to exist locally further South where it occurs under Mesozoic and Tertiary cover rocks. The Lower Palaeozoic successions of the former margin of Iapetus are even, in places, covered by Carboniferous (Umpleby, 1979) but only exceptionally do such rocks lie directly upon Pre-Cambrian granites (Bousquet et al., 1977). The locations of all these sampling points are indicated on Fig. 12, which also shows the undersea extent of the Caledonian-Appalachian belt.

The nature of Palaeozoic sediments covering the Rockall Plateau is unknown; if, however, samples from the slope to the East of Greenland are taken into account (Johnson et al., 1975), the northern part of the Rockall Plateau could be covered with Upper Devonian or Lower Carboniferous strata. In this case, the succession would consist of dolomicrites, sandstones and mudstones probably in direct contact with the ancient basement. North of Ireland, there are extensive outcrops of Dalradian, forming a prolongation of the on-land Donegal terrains, which are sometimes overlain by Old Red Sandstone or Carboniferous. Furthermore, seismic data appears to indicate that Carboniferous rocks underlie the Rathlin Trough (see Fig. 9). Towards the North, in the Hebrides Sea, the Palaeozoic cover of the Iapetus margin is poorly known because it is often hidden underneath the Permo-Triassic or Recent sediments and is difficult to differentiate from Upper Proterozoic formations (Binns et al., 1975) (Fig. 8). Nevertheless, Lower Palaeozoic rocks could form the cover to the South of Skye.

Throughout this region, one of the problems which remains to be solved is the existence at depth of the Old Red Sandstone and Carboniferous (Binns et al., 1974), as suggested locally by seismic reflection and the re-working of spore assemblages (Eden et al.,

▶

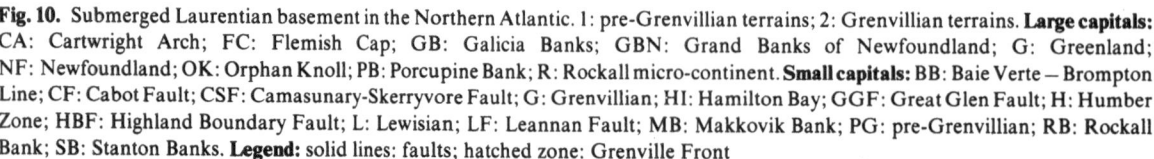

Fig. 10. Submerged Laurentian basement in the Northern Atlantic. 1: pre-Grenvillian terrains; 2: Grenvillian terrains. **Large capitals:** CA: Cartwright Arch; FC: Flemish Cap; GB: Galicia Banks; GBN: Grand Banks of Newfoundland; G: Greenland; NF: Newfoundland; OK: Orphan Knoll; PB: Porcupine Bank; R: Rockall micro-continent. **Small capitals:** BB: Baie Verte – Brompton Line; CF: Cabot Fault; CSF: Camasunary-Skerryvore Fault; G: Grenvillian; HI: Hamilton Bay; GGF: Great Glen Fault; H: Humber Zone; HBF: Highland Boundary Fault; L: Lewisian; LF: Leannan Fault; MB: Makkovik Bank; PG: pre-Grenvillian; RB: Rockall Bank; SB: Stanton Banks. **Legend:** solid lines: faults; hatched zone: Grenville Front

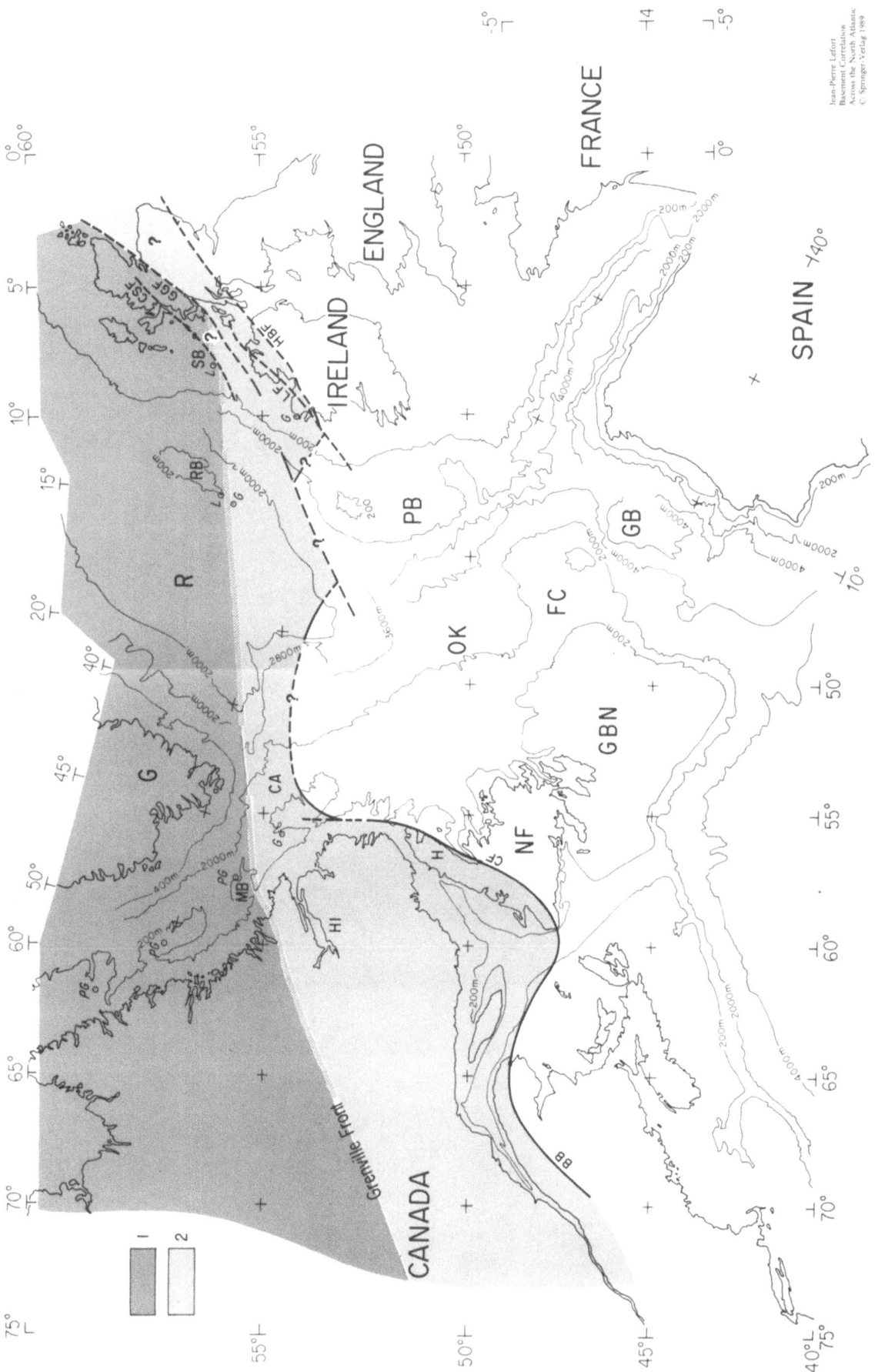

Jean-Pierre Lefort
Basement Correlation
Across the North Atlantic
© Springer-Verlag 1989

1973). In any case, the Old Red Sandstone makes up most of the area of outcrop between Scotland and the Shetlands (Evans et al., 1982). Thus, it can be seen from the geological and marine geophysics point of view that it is easier to determine the extension of the Laurentian basement across the Atlantic than to delimit the sedimentary prism which covers it. Intense re-mobilization of the European margin by transcurrent Caledonian deformation, added to numerous vertical re-activations during the Mesozoic and Tertiary, has led to great complexity in the present-day basement/cover relationships. As far as the ancient basement is concerned, it can generally be maintained that the width of Grenvillian outcrop decreases in a regular manner towards the East; this merely reflects an original oblique relationship which existed between the Grenville Front and the edge of the Laurentian continent.

Chapter 3

The Submerged Part of the Caledonian-Appalachian Mobile Belt

Because of the types of reconnaissance techniques employed and the limitations mentioned above, only certain aspects of the belt can be taken up in this chapter. Emphasis is laid, above all, on the definition of external boundaries and the large-scale structures encountered within the belt, as well as the distribution of basic material; this latter is the only convenient criterion which exists to recognize the Iapetus suture. Since the major units of the submerged mobile belt have only very rarely been sampled, there is generally a great reliance on geophysical arguments in the following presentation.

3.1 The Caledonian Front

In Eastern Canada, this Front is known as "Logan's Line" and is seen to separate highly deformed terrains in the South from slightly folded units in the North. Folded Cambro-Ordovician formations can be detected by seismic reflection to the East of the Gulf of St Lawrence but they are poorly represented at outcrop and are overlain to the South of Logan's Line by a thick Permo-Carboniferous succession which can attain 5–9 km in thickness off northern Nova Scotia and South of Newfoundland (Hobson and Overton, 1973) (Fig. 11).

Logan's Line is itself recognized by seismic reflection and is often situated to the North or to the West of the Grenvillian basement margin; this implies the local existance of important thrusts. The importance of these thrusts can be appreciated by comparing Figs. 10 and 12. The positive magnetic anomalies constrained between the Grenvillian margin and Logan's Line often correspond to ophiolitic klippes (Haworth and Keen, 1979). To the West of Newfoundland, the boundary of klippes and thrust faults has been mapped with precision using seismic methods (Haworth and Sanford, 1976); this limit seems locally to be cut by late stage shears. There appears to be no continuity between Logan's Line and the thrust contact bordering the

eastern edge of Labrador (Haworth et al., 1976). The ophiolitic and sedimentary klippes of Taconic age in the Humber Zone (which border the Dunnage zone in the West) make up an area of transition between the two main belts of thrusting.

A deep seismic reflection profile recorded North of Newfoundland shows that allochthonous units within the foldbelt could be far-travelled (Keen et al., 1986) and that Grenvillian basement extends at depth to at least the central parts of the island (see Fig. 13). In this way, a large part of the succession corresponding to the suture zone could be more or less allochtonous. To the West, as confirmed by geological arguments, superficial structures clearly show an easterly dip. By contrast, reflectors at the base of the Grenvillian crust indicate structures dipping in the opposite direction; although this phenomenon is rather different from that found in the northern Caledonides, it is similar to that observed on the WINCH profile surveyed between Ireland and Scotland (see Fig. 13). This could result from imbrication between upper and lower crust and should, therefore, be of Caledonian age. However, other explanations are possible, especially since the orientation of these reflectors is only derived from a single profile.

In any case, it is clearly seen that the topography of the roof of the Grenvillian basement shows the presence of ancient listric faults which marked the initial opening of Iapetus.

In the absence of any evidence, it may be supposed that the northern limit of the Caledonian Front follows more or less the Grenvillian margin as far as the Porcupine Bank (see Fig. 12). It is traditional to consider the Moine Thrust in Scotland as representing the Caledonian Front (Watson, 1978); this thrust can be followed further North as far as the Shetlands (Brewer and Smythe, 1984). Offshore, the seismic reflection profile MOIST suggests that the main thrust may be split into two. The Moine Thrust places Cambro-Ordovician quartzites and carbonates belonging to the Laurentian margin in contact

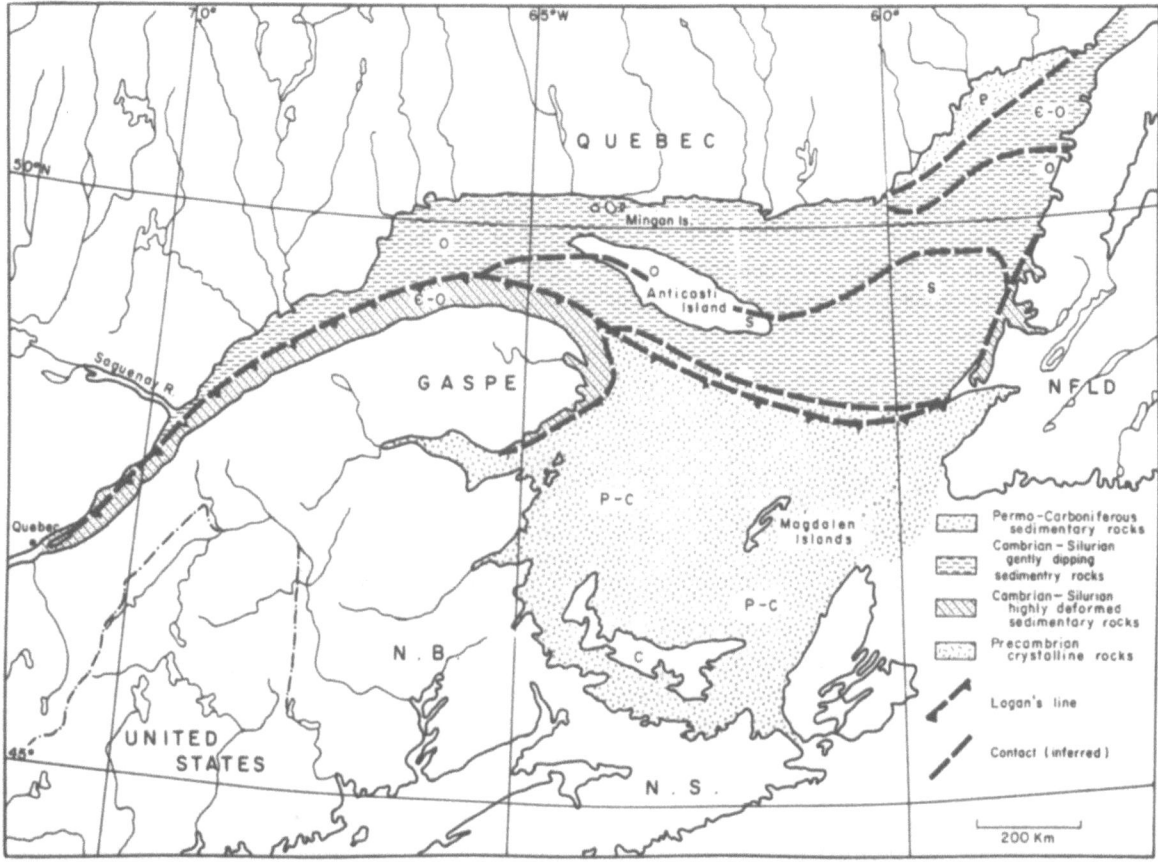

Fig. 11. Basement in the Gulf of St. Lawrence. (After Loring 1975; reproduced with permission of the Minister of Supply and Services, Canada)

with metamorphic rocks of the Moine series. The MOIST profile also shows that the Lewisian basement plunges fairly deeply beneath the Caledonian foldbelt, but that the basement dips off far more steeply and appears better rooted than the thrust planes recognized in the Southern Appalachians (Fig. 13).

The main thrust in the Outer Hebrides, also of Caledonian age, shows the same tectonic style but the environment is completely intracratonic. Such a doubling of thrust zones is also known in Canada and thus renders the definition of a Caledonian Front rather ambiguous in some places. The ancient thrust faults have been re-activated as normal faults, thus providing a straightforward explanation of the fact that sediments on the Laurentian margin are often concealed beneath Devonian, Carboniferous or even Permo-Triassic basins.

The WINCH profile (Fig. 13) shows that, for areas further South, between Ireland and Scotland, the Caledonian Front remains poorly defined and that

there are nappes and important thrusts on each side of the traditional boundary – this complicates even more the precise localization of the contact. Such difficulties are made worse by the cutting up of this zone into short segments by late shears – the Great Glen and Leannan Faults. In agreement with Hall et al. (1984), it is thought that the Caledonian Front passes beneath the Loch Indall basin; further West, the trace of the Front is totally unknown. It should also be noted that, in the Shetlands, this Front is associated with allochthonous ophiolitic units (Flinn et al., 1979) which may be considered as forming a pair with the obducted ophiolites of Newfoundland. Their situation and mode of emplacement suggest that remnants of oceanic crust may exist at depth beneath the Caledonian foldbelt. The large-scale sinuosities apparent in the Caledonian Front, as given in Fig. 12, are more important than mapping errors; thus the curved trace of the Front is not an artefact and evokes the type of sinuous structures already described in the

Appalachians (Thomas, 1977). As in America, these large bends could be related to initial irregularities in the outline of the Laurentian continent and may reflect the control of pre-existing Grenvillian structures on the ancient margin. With regard to these bends, it has been suggested that they were related to the opening of the Iapetus ocean which may have occurred from a series of triple point junctions (Rankin, 1976). Because this hypothesis implies that Grenvillian terrains should also exist East of the foldbelt, it is not favoured by the present author.

3.2 Major Structures Recognized in the Caledonian Appalachian Belt

Permo-Carboniferous sandy facies cover nearly all the orogenic belt in the Gulf of St Lawrence; there are only a few places where deformed Cambro-Ordovician rocks crop out around the Gaspé Peninsula (Loring, 1975) (Fig. 11). The zonation defined by Williams (1972) between New Brunswick and Newfoundland cannot be recognized offshore, even though the general trend of structures is clearly seen through magnetics and gravity. The limit of Acadian deformation is bounded by a strong positive gravity anomaly which is only interrupted where thick Carboniferous basins mask the roof of the Appalachian basement (Sheridan and Drake, 1968). The negative magnetic anomaly, which is generally superimposed (Haworth, 1975) on the gravity anomaly, shows clearly that a continuous basic intrusion is not present and that this phenomenon is due to an edge-effect between the foldbelt and the Grenvillian platform.

Further South, the geophysical lineaments proper to New Brunswick, which correspond to volcanic and metasedimentary basement horsts (Miller and Garland, 1953) of possible oceanic origin (Haworth, 1975), are abruptly interrupted off Prince Edward Island. As in the centre of the Gulf, this may be due to the presence of thick Carboniferous basins (Hobson and Overton, 1973). However, since this interruption also corresponds to a straight and positive magnetic ridge, it is possible that the structure has a tectonic origin. If the concave outline of the Grenvillian plate in this area is taken into account, as well as the parallelism of the magnetic ridge and the Grenvillian margin, then one is led to consider this fault as a transform structure inherited from the Appalachian Collision. The equivalents of the New Brunswick magnetic lineaments can be thus found in Newfoundland (Haworth, 1975). This interpretation is shown on the schematic map of the fold belt (Fig. 12).

In Newfoundland, the eastern limit of the Appalachian belt is marked by the Dover-Hermitage Fault (Kennedy et al., 1982); here, the contact between the foldbelt and bordering terrains is picked out by a strong positive gravity anomaly. An identical effect can be recognized off southern Newfoundland, at Cape Breton and in southern New Brunswick — this occurs wherever there is a sharp contact between the foldbelt and the Avalonian basement (Williams and Hatcher, 1983). This geophysical signature therefore duplicates what has already been observed on the northern edge of the foldbelt. However, a positive magnetic anomaly is here seen to overlap with the gravity anomaly and suggests the existence of basic rocks at depth. Some transverse local trends, notably between Southern Newfoundland and Cape Breton, can be related to Carboniferous tectonics which have upfaulted deep basement to the surface (Sheridan and Drake, 1968; Haworth, 1975). An interpretation of the seismic refraction profile recorded between southern Newfoundland and Anticosti Island reveals the deep crustal structures beneath the Gulf (see Fig. 14). The following features can be recognized:

— Pre-Carboniferous terrains folded in the South and slightly deformed in the North, separated by a thrust.

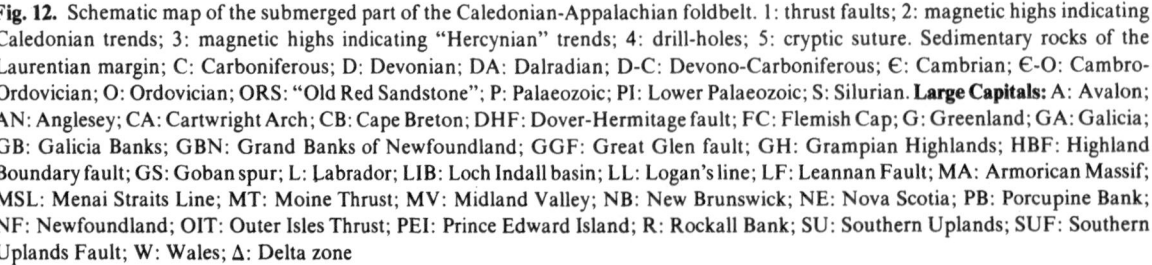

Fig. 12. Schematic map of the submerged part of the Caledonian-Appalachian foldbelt. 1: thrust faults; 2: magnetic highs indicating Caledonian trends; 3: magnetic highs indicating "Hercynian" trends; 4: drill-holes; 5: cryptic suture. Sedimentary rocks of the Laurentian margin; C: Carboniferous; D: Devonian; DA: Dalradian; D-C: Devono-Carboniferous; ϵ: Cambrian; ϵ-O: Cambro-Ordovician; O: Ordovician; ORS: "Old Red Sandstone"; P: Palaeozoic; PI: Lower Palaeozoic; S: Silurian. **Large Capitals:** A: Avalon; AN: Anglesey; CA: Cartwright Arch; CB: Cape Breton; DHF: Dover-Hermitage fault; FC: Flemish Cap; G: Greenland; GA: Galicia; GB: Galicia Banks; GBN: Grand Banks of Newfoundland; GGF: Great Glen fault; GH: Grampian Highlands; HBF: Highland Boundary fault; GS: Goban spur; L: Labrador; LIB: Loch Indall basin; LL: Logan's line; LF: Leannan Fault; MA: Armorican Massif; MSL: Menai Straits Line; MT: Moine Thrust; MV: Midland Valley; NB: New Brunswick; NE: Nova Scotia; PB: Porcupine Bank; NF: Newfoundland; OIT: Outer Isles Thrust; PEI: Prince Edward Island; R: Rockall Bank; SU: Southern Uplands; SUF: Southern Uplands Fault; W: Wales; Δ: Delta zone

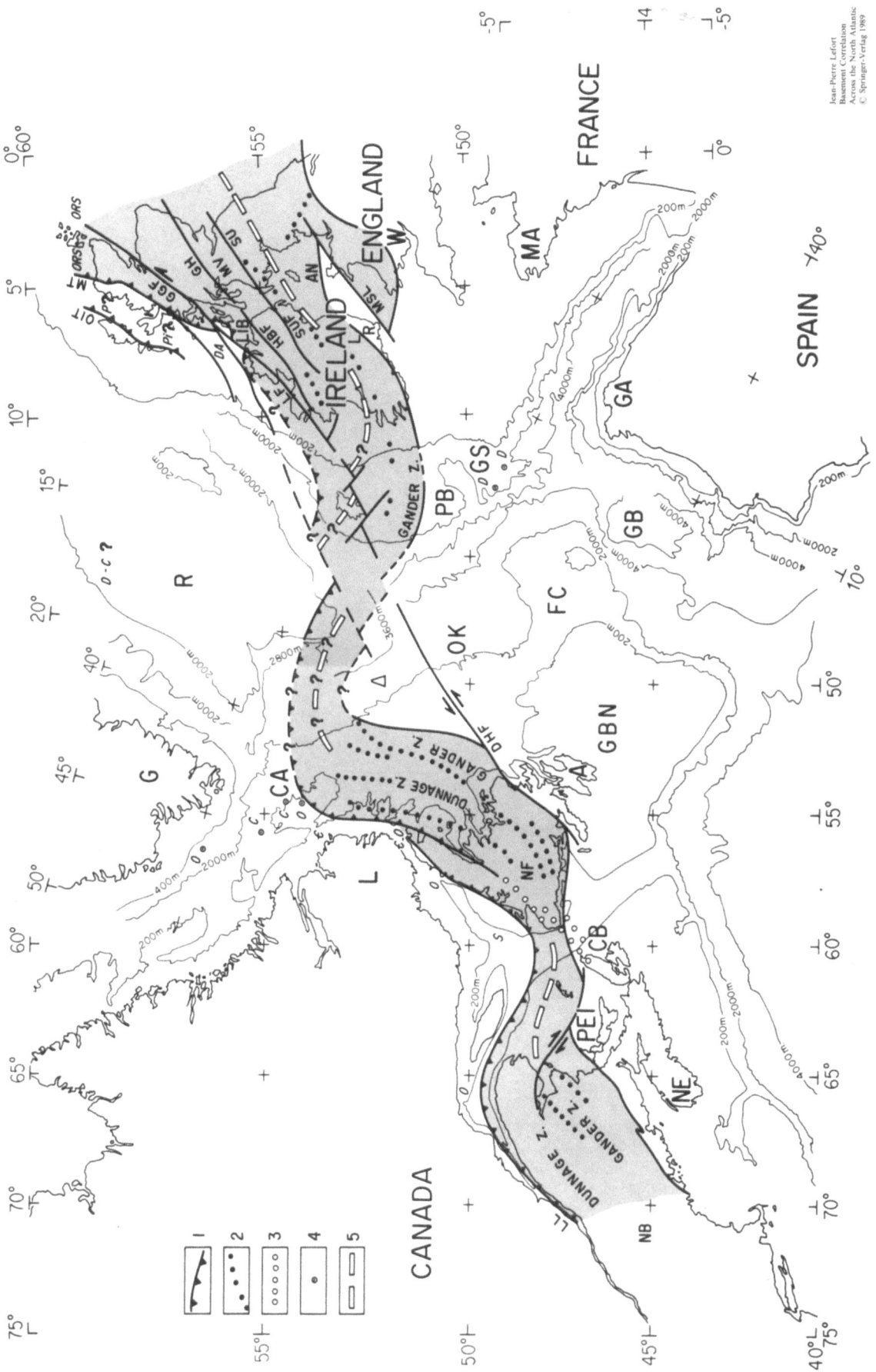

Jean-Pierre Lefort
Basement Correlation
Across the North Atlantic
© Springer-Verlag 1989

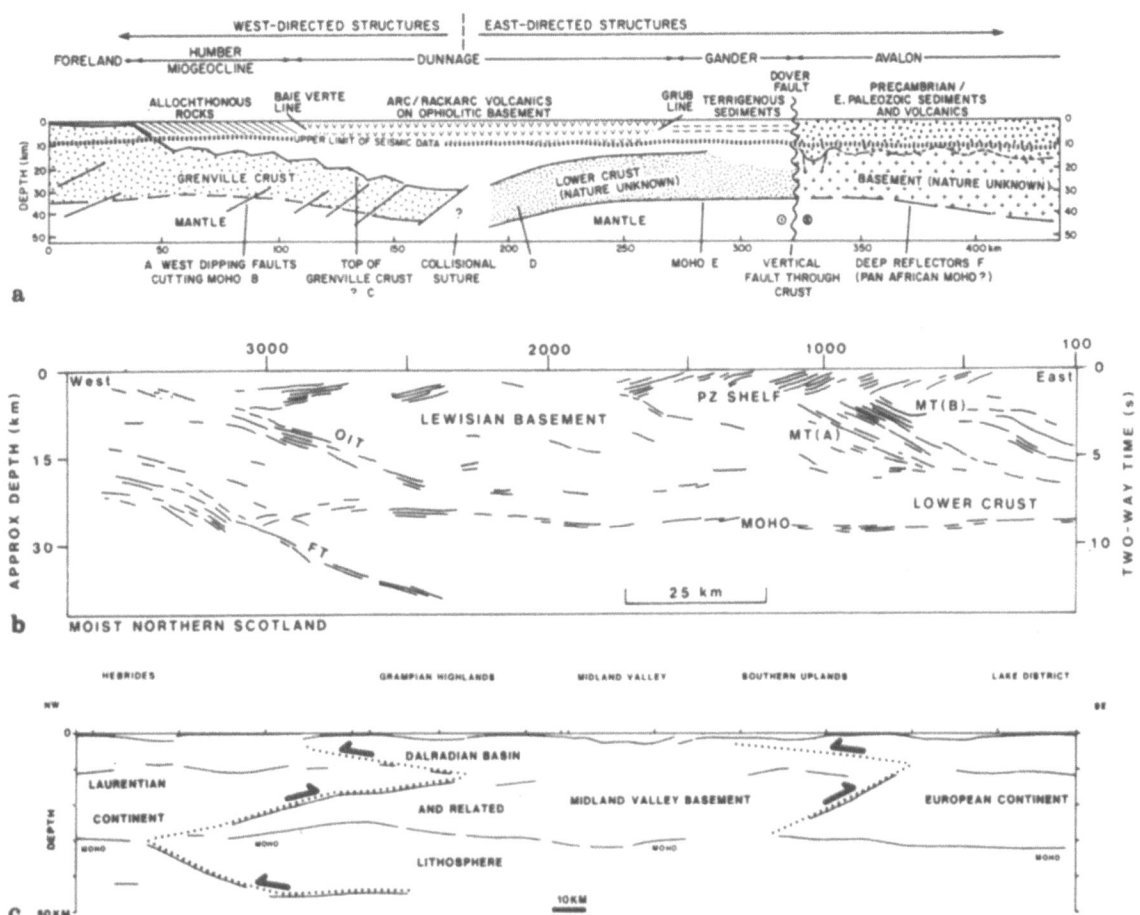

Fig. 13. Seismic profiles of the submerged mobile belt. **a** Profile surveyed off Eastern Canada. (After Keen et al. 1986); **b** MOIST profile from Northern Scotland. (After Brewer and Smythe 1984). FT: Flannan Thrust; OIT: Outer Isles Thrust; PZ shelf: Palaeozoic margin; MT(A) and MT(B): possible position of Moine Thrust; **c** WINCH profile from area between Ireland and Scotland. (After Hall et al. 1984)

— An upfaulted block of basic basement having a seismic velocity of 7.03 km/s, which foreshadows the obducted ophiolites observed in the Humber Zone on western Newfoundland.

— Thick Carboniferous basins displaying a velocity of 4.5 km/s.

— A late-stage fault responsible for the uplift of crystalline basement off Cape Breton and the truncation of uppermost Palaeozoic basins, with low seismic velocities, which conceal central areas of the foldbelt (Fig. 14).

To the North of Newfoundland, formations affected by Caledonian-Appalachian deformation have been sampled and are mapped by seismic reflection (Haworth et al., 1976) (Fig. 15). To the West, sediments of the Laurentian margin, which include Cambro-Ordovician limestones, sandstones and shales, are partly overlain by Mississippian and Pennsylvanian formations. In the East, on the other hand, the succession is dominated by acid and basic volcanics, peridotites and basalts showing rare interbeds of Ordo-Silurian sediment. Jacobi and Kristoffersen (1981) have attempted to extend the on-land zonation by magnetic and gravimetric methods; the main result of this work is to show that there are remnants of oceanic crust extending from Notre-Dame Bay (in northern Newfoundland) towards the North as two ophiolitic ridges which almost crop out (Fig. 20). The westernmost of these ridges is well rooted and is probably the root zone of the klippes now located on the Grenvillian basement (Haworth and Miller, 1982).

This area of the foldbelt corresponds broadly with the Dunnage zone. The ophiolitic ridges are bounded to the East by the Gander Zone which is gravimetrically light and poorly magnetic. All these structures turn towards the East near the southern tip of Greenland, thus following the Grenvillian re-entrant described above; in this area, both zones are covered by Carboniferous strata (Umpleby,

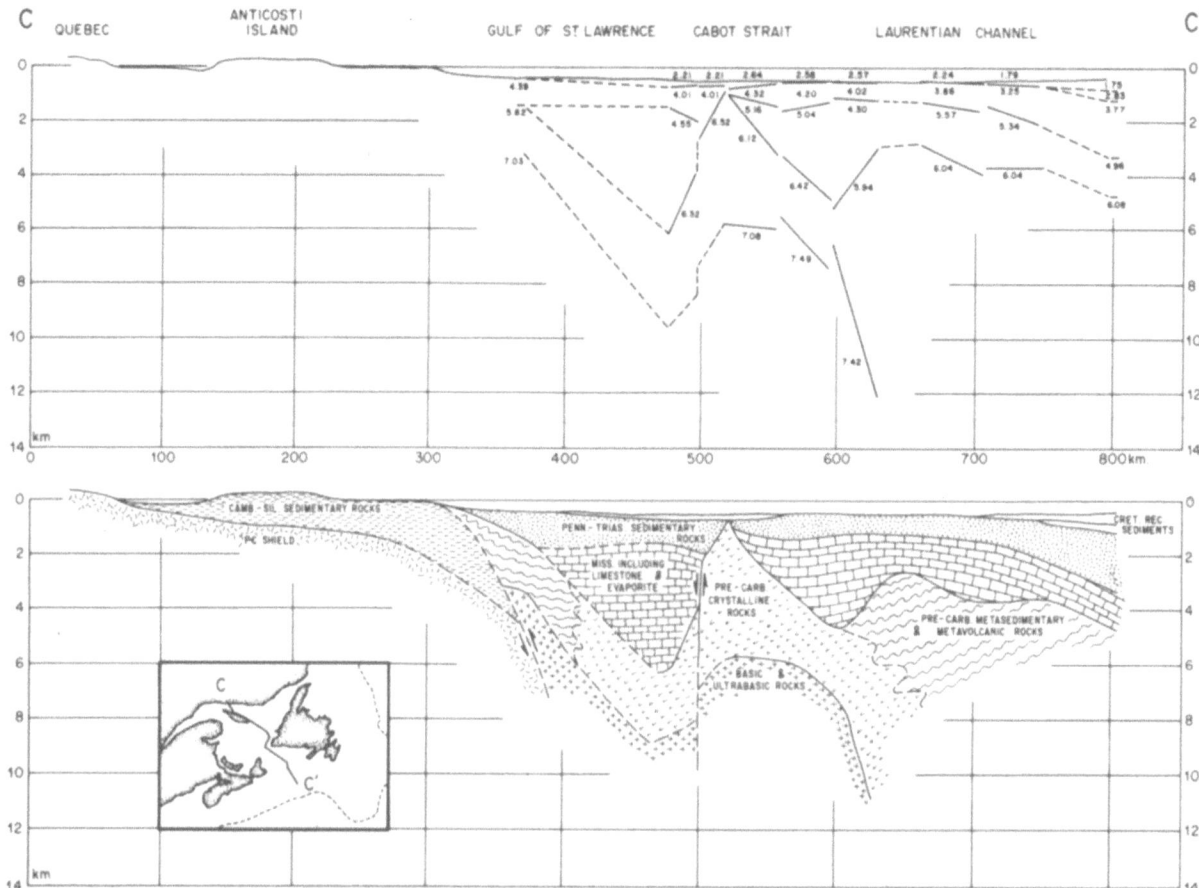

Fig. 14. Interpretation of seismic refraction data in the Gulf of St. Lawrence. (After Sheridan and Drake 1968)

1979). On land, the Dunnage Zone is considered as representing remnants of the Iapetus Ocean; it is completely sheared out near the Grenvillian spur in the Gulf of St Lawrence and only re-appears in northern New Brunswick (Williams and Hatcher, 1983). The Gander Zone has been interpreted as the southern continental margin of the Iapetus Ocean (Kennedy, 1976). The negative anomalies typical of the Gander Zone could be related to the presence of granites at depth (Haworth, 1981).

Offshore, between the Gander Zone and the Dover-Hermitage Fault (which forms an onland limit to the Avalonian terrains), there exists an enigmatic triangular area, known as the Delta Zone (near 52°N), which displays seismic velocities around 5.2 km/s.

These velocities are less than those usually measured in the foldbelt (5.8 – 6.5 Km/s; Sheridan and Drake, 1968). At the surface, the Delta Zone is probably composed of non-metamorphosed terrains which may be identified as a thick Carboniferous basin

(Cutt and Laving, 1977). Nevertheless, since the Irish equivalent of the Dover-Hermitage Fault cuts transversally across the Caledonides of North Europe, it can be assumed that the Dover Fault does not always represent the limit between the foldbelt and the Avalonian platform. Therefore, the question arises as to whether the deep basement of the Delta Zone really belongs to the fold belt. As a working hypothesis, it is proposed that the Delta Zone is external to the orogenic belt since the bend in the foldbelt to the West of Rockall is clearly marked, if rather poorly defined. This is contrary to the hypothesis of Haworth (1981).

The systematic association between positive and negative anomalies known along the contact separating the Gander Zone from the Avalonian basement is also found on the Porcupine Bank, where a coupled anomaly is picked up again on the far side of the Delta Zone, thus indicating a prolongation of the contact into Europe. As in Canada, this double geophysical marker forms the limit of a zone of high

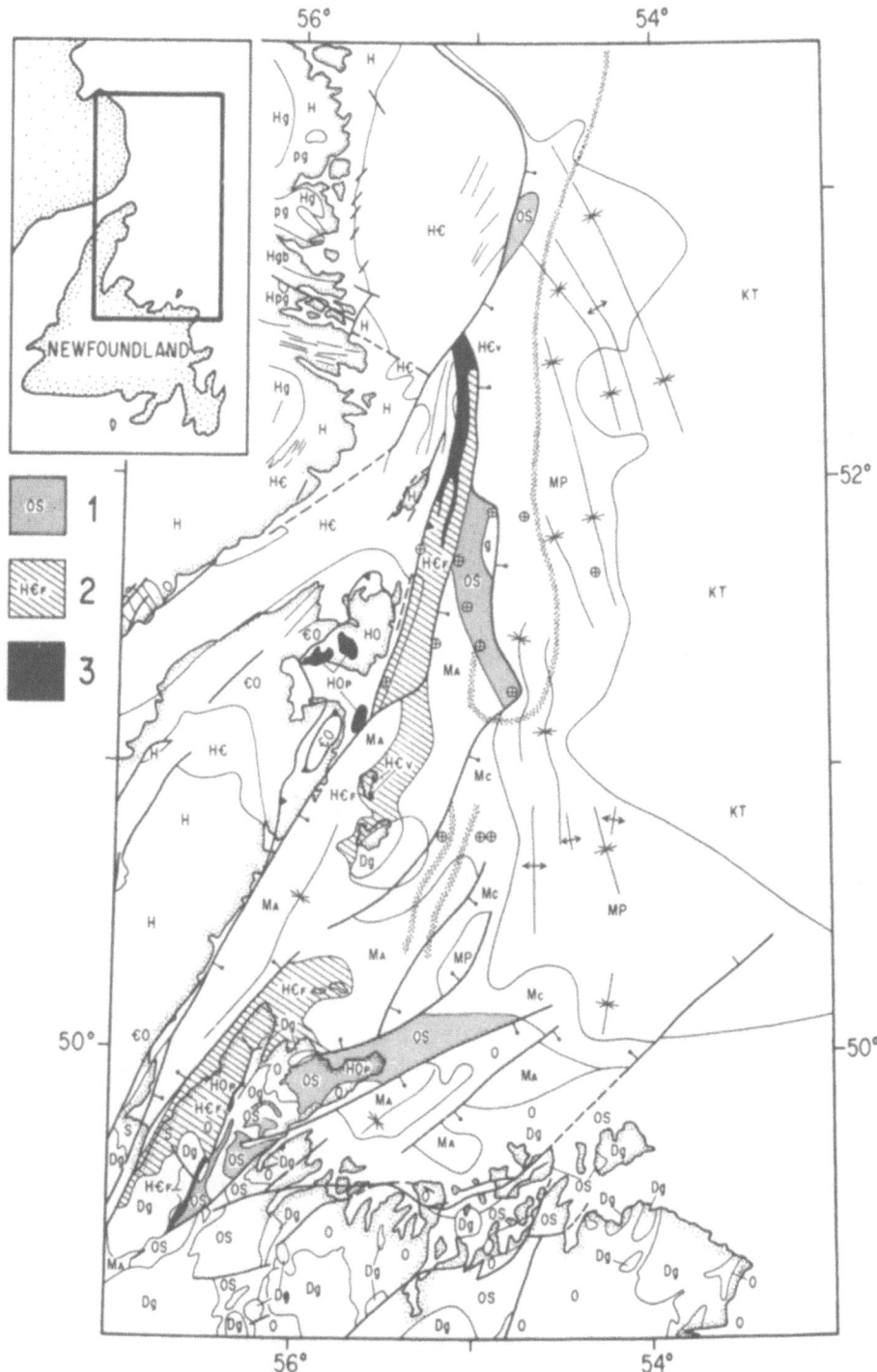

Fig. 15. Basement outcrops in the mobile belt North of Newfoundland. 1: Ordovician volcanics; 2: Hadrynian and Ordovician pelites and psammites; 3: Hadrynian and Ordovician basic and ultrabasic rocks; MA, Mc and MP: Mississippian and Pennsylvanian. (After Haworth et al. 1976; reproduced with permission of the Minister of Supply and Services, Canada)

frequency anomalies; the European equivalent of the Dunnage Zone is fairly comparable to its Canadian model as far as magnetic (Max et al., 1982) and gravimetric characteristics (Buckley and Bailey, 1975) are concerned even though some of the high frequences of the Porcupine Bank are known to be due to Tertiary intrusions (Lefort and Max, 1984). This solution is practically identical to that given by Jacobi and Kristoffersen (1981).

Even though outcrops of basement are lacking over the Porcupine Bank, an inventory of the facies present can be deduced from magnetic (Lefort and Max, 1984), seismic reflection (Bailey, 1975) and gravimetric data (Buckley and Bailey, 1975). From this, it may be concluded that indurated basement with thin Palaeozoic cover exists between the South of the Bank and $51°50' - 53°N$. By contrast, areas to the North seem to be characterized by metamorphic terrains of probable Caledonian age.

A similar zonation is found on the Western Irish margin, with an offset of about 30 km to the South. However, there are no unequivocal gravimetric, magnetic or seismic criteria in this region which allow the recognition of the Iapetus suture. There is just one negative magnetic anomaly (identical to that observed in the Solway Firth discussed below) near 14°30 W and 53°N (Haworth, 1983) which appears parallel to the Gander Zone boundary and which may represent the suture. But since this anomaly shows a trend parallel to many Permo-Triassic fractures cutting the Bank (Lefort and Max, 1984), the trace of the suture remains rather hypothetical.

The southern limit of the foldbelt, as proposed here, can be extended over southern Eire. Here, the negative gravity anomaly co-incides with the Caledonian Leinster Granite whereas, further East, the magnetic and gravimetric anomalies are superimposed on Pre-Cambrian rocks of the Rosslare Complex and bordering Caledonian volcanics (Stillman, 1981). This correlation has been accepted by many authors (Kennedy, 1979; Jacobi and Kristoffersen, 1981; Haworth and Jacobi, 1983). In central Ireland, the style of geophysical signatures is markedly different from that observed in central Newfoundland — this is mainly because the suture is cryptic in Ireland and basic/ultrabasic material from the suture zone appears practically non-existent (Max and Inamdar, 1983). There are only a few elongate, low frequency anomalies which suggest the possibility of oceanic slices at depth. To the North and South of Ireland, the presence of positive anomalies would appear to be linked rather to Caledonian volcanic belts at outcrop than deep

structures (Murphy, 1981). Thus, in this area, the localization of the Iapetus suture depends entirely on indirect stratigraphic and structural criteria (Phillips, 1978) which, furthermore, have been called into question (Williams and Max, 1979).

According to whether the Anglesey formations are assigned to Caledonian (Williams and Max, 1979) or a previous period of orogenesis, the southern limit of the Gander Zone can be drawn either to the North or the South of the Anglesey area; furthermore, an identical type of problem is posed in S.E. Eire. Since, in both cases, rocks of Upper Proterozoic — Lower Cambrian age are intrusive into an ancient gneissic basement of possible Pentevrian affinity (Piasecki et al., 1981), it is more sensible to include these terrains in the Avalonian basement. An analogous geological setting is found in North Brittany (Auvray, 1979), where Caledonian events can be totally ruled out — here, uppermost Proterozoic — Lower Cambrian acid volcanics are seen to terminate the Cadomian orogenic cycle.

In the southern Irish Sea area, geological and geophysical data can be interpreted as follows (see Fig. 16):

— The Irish Sea Geanticline, forming a horst of Pre-Cambrian and Lower Cambrian terrains, belongs without doubt to the Avalonian basement. Here, the positive geophysical markers (mainly magnetic) enable the linking together of the Rosslare Complex with the Lleyn Peninsula and Anglesey, thus confirming the geological correlations (Piasecki et al., 1981).

— A little further North, the negative magnetic anomaly superimposed on the Central Basin suggests that the metamorphic basement is at great depth.

The metamorphic basement would appear to be beneath 2,500 m of Permo-Carboniferous and Triassic strata (Dobson et al., 1973); it is perhaps the same basement which re-appears off Anglesey in N. Wales overlain by scraps of Ordovician and Carboniferous (McQuillin et al., 1969).

Even further to the North, the mid-Irish Sea uplift and the Wicklow Head shelf are characterized by a strong positive gravity anomaly of 35 milligals, which ties in with the anomalies following the S.E. seaboard of Eire. This structure represents a transitional zone with the European equivalent of the Gander Zone. Its faulted northern contact does not have the same structural significance as the Dover-Hermitage Fault in Newfoundland. From the magnetic point of view, it is possible that the core of this horst is made up of Ordovician igneous rocks (Al-Shaikh, 1970). The highly magnetic zone bor-

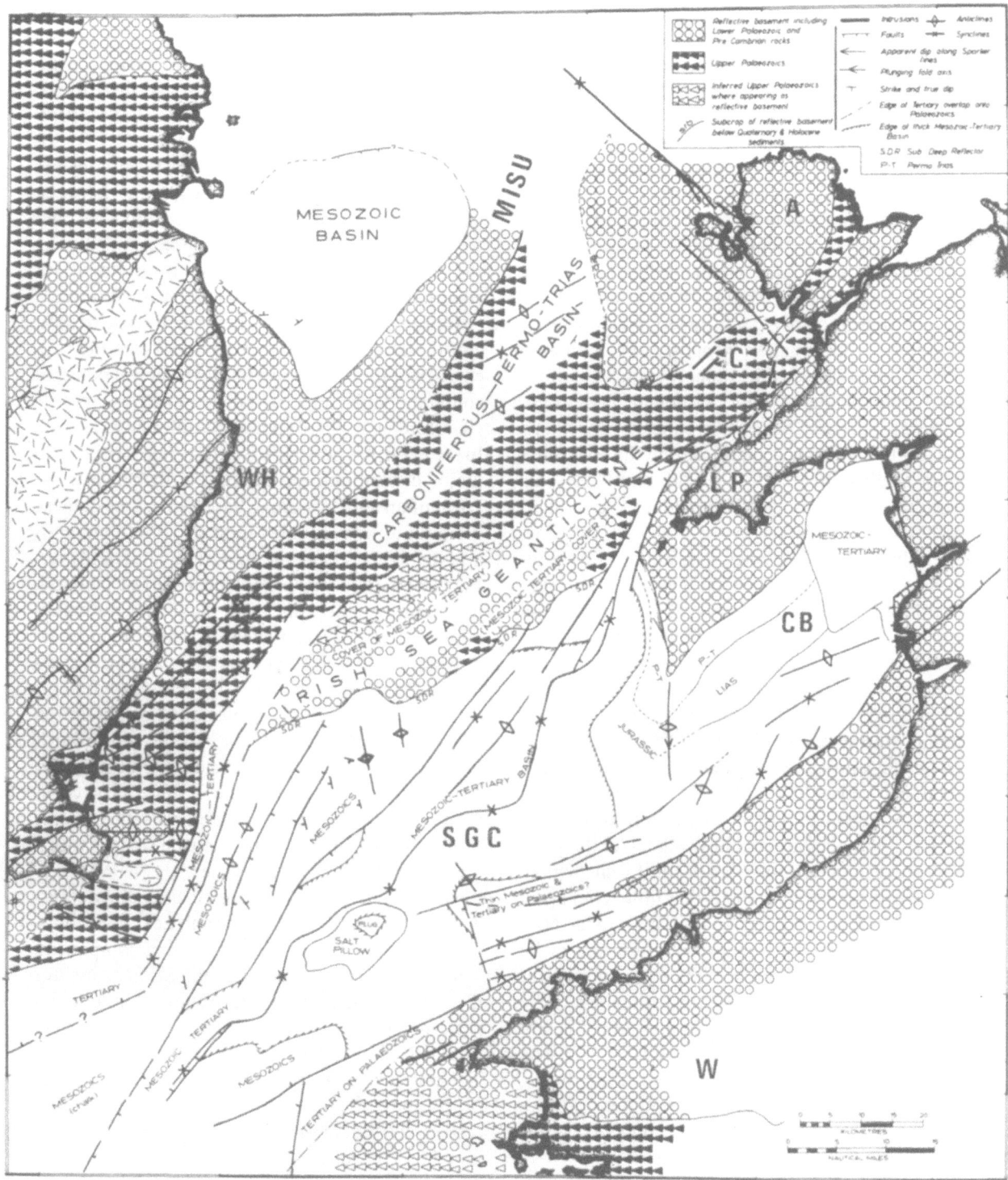

Fig. 16. Sub-Pleistocene geology of the southern Irish Sea. A: Anglesey; C: Caernarvon Bay Basin; CB: Cardigan Bay Basin; LP: Lleyn peninsula; MISU: Mid Irish Sea Uplift; SGC: Saint George's Channel; W: Wales; WH: Wicklow Head shelf. (After Dobson et al. 1973)

dering the coast could, in any case, be reasonably correlated with dolerites, diorites, andesites and ultrabasics known onland. Offshore, all this zone is considered as being composed of Cambro-Ordovician terrains (Dobson et al., 1973); it is clear that this zone lies within the orogenic belt.

The Gander Zone can be tentatively extended as far as the North of Anglesey, where a fairly wide negative magnetic anomaly follows the limit defined by Haworth and Jacobi (1983). However, the thickness of Upper Palaeozoic to the East and North (up to 3 km; Bott and Young, 1971) is too great to permit application of the same geophysical criteria. Therefore, the zonation proposed by Williams (1978) should be tested by other methods.

To include Anglesey in the Avalonian Zone does, however, present an inconvenience since this isolates the Welsh Caledonides from the rest of the main foldbelt (Coward and Siddans, 1979). This paradox can be resolved if the limit of Avalonian basement is interrupted by the Menai Straits Line. Even though this line acted as a transcurrent fault during the Lower Palaeozoic (Baker, 1971), the dextral offset indicated on Fig. 12 is perhaps only apparent − the majority of faults following this trend in Wales display a sinistral movement (Anderson and Owen, 1968). Thus, the possibility of a Cadomian crustal flake should be taken into account (Matthews and Cheadle, 1986).

The strong anomalies recorded over the Isle of Man are due to dense Pre-Cambrian rocks. Immediately to the South of the island, the observed negative gravity anomalies are probably related to the presence of granitic intrusions at depth. In this region, magnetism shows generally rather little relief apart from a N 60° oriented negative anomaly to the N.E. of the Isle of Man. This anomaly appears to be linked to an increase to the depth to basement over a line that corresponds classically to the trace of the Iapetus suture. As in Ireland, the suture is not marked by any important basic body (Fig. 17). The Peel and Solway Firth Basins were filled during Permo-Triassic and probably Carboniferous times and are evidently controlled by movements on this fault (Wright et al., 1971; Hall et al., 1984). The Ramsay-Whitehaven basement ridge, which passes through the Isle of Man and continues into Ireland (Hall et al., 1984), is probably also influenced by the existence of this plate boundary. Elsewhere, the existence of such basement uplifts near the Appalachian or Grenvillian sutures has already been discussed. A second, N-S-oriented

ridge, situated between the Isle of Man and Anglesey, probably corresponds to a structure formed prior to Caledonian closure since it is interrupted by the suture. Further South, these trends appear to delimit the Anglesey Zone (Baker, 1971) and may be of Lower Palaeozoic age.

In Scotland, the Iapetus suture has no surface expression but constitutes a sharp limit between Caledonoid magnetic trends (WSW) and Charnoid trends (NW) (Watson and Dunning, 1979). The suture is also manifested at depth by an important conductivity anomaly within the crust. A seismic reflection profile, recorded between the Isle of Man and the Malin Sea (Hall et al., 1984) (see Fig. 13), cuts across this structure and shows certain points in common with the on-land refraction profile (Bamford et al., 1977) but great differences as well. In the main, it is possible to say that there are just two major discontinuities dipping North between the Laurentian and Avalonian basements. One of these discontinuities is situated in the Southern Uplands and the other is beneath the Grampian Highlands. Between the two, in the Midland Valley of Scotland, basement would appear to have a Laurentian affinity in view of on-land seismic velocities in excess of 6.4 km/s (Bamford et al., 1977). Nearer the surface, Dalradian and Lower Palaeozoic sediments seem to mask these discontinuities. The obliteration of deep structures could be due to northward-directed thrusts, which implies a decoupling between indurated basement and the sedimentary cover (Hall et al., 1984). This explains why the seismic profile gives an impression of crustal wedges which are imbricated within each other.

Certain faults so far considered as mainly transcurrent, such as the Great Glen and Leannan Faults, could also have displayed a thrusting component. The recognition of superficial thrusts accounts for the fact that the Iapetus suture shows neither magnetic nor positive gravity anomalies − the upper crustal levels of the suture zone have been thrust out. The presence of a negative magnetic anomaly vertically above the Iapetus suture trace demonstrates, conversely, that it is really the suture zone structure that has controlled the Peel Basin. An identical association of sedimentary basins and negative magnetic anomalies exists at the limits of several Caledonian thrusts and suggests that, as in northern Scotland, post-orogenic basins often result from the reactivation of reverse faults in a normal sense (Fig. 17).

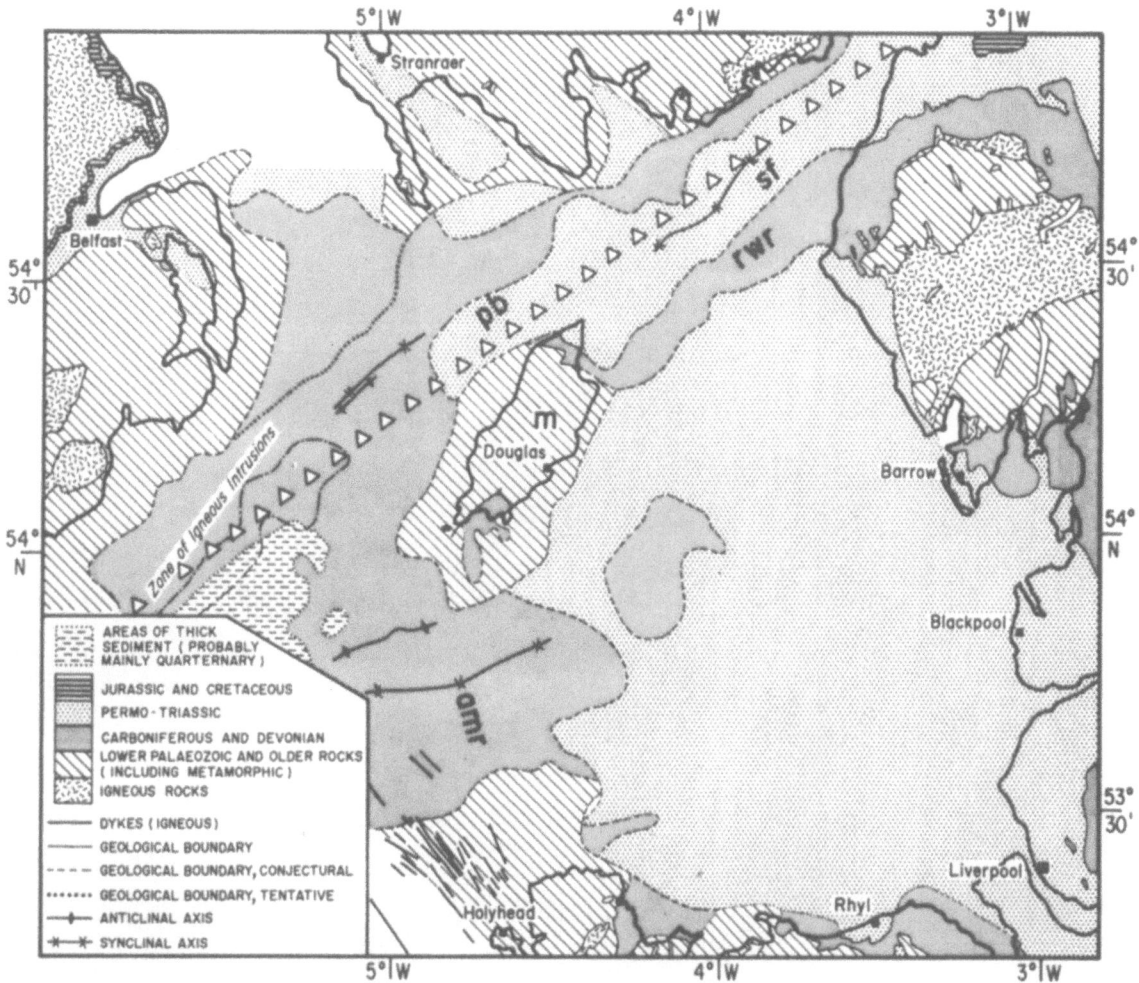

Fig. 17. Geological sketch map of the northern Irish Sea. amr: Anglesey-Man ridge; pb: Peel Basin; rwr: Ramsey-Whitehaven ridge; sf: Solway Firth. Triangles indicate theoretical position of Iapetus suture. (Modified from Wright et al. 1971)

3.3 Shearing Along the Caledonian Belt

Although there are many transcurrent faults in Wales (Anderson and Owen, 1968), these do not appear to disrupt the structure of the foldbelt so profoundly as is the case in Scotland. In this region, the major faults (Great Glen, Leannan, Highland Boundary and Southern Uplands) have juxtaposed various segments of the foldbelt which were originally far separated from each other in some cases. On land, the LISPB profile (Bamford et al., 1977) shows the dominant role played by steeply dipping faults such as the Highland Boundary Fault, the Southern Uplands Fault and, to a lesser extent, the Great Glen Fault. Mainly for technical reasons, deep seismic reflection profiles in the northern Irish Sea are unable to detect steeply dipping faults (Hall et al., 1984). Furthermore, these faults are concealed

offshore beneath thick accumulations of Carboniferous and Permo-Triassic sediments. Although the Highland Boundary Fault can certainly be traced as far as the Isle of Arran (Chesher et al., 1972) (see Fig. 18), it is difficult to follow further West using either geological (McLean and Deegan, 1978) or magnetic evidence (Dobinson, 1978). However the Highland Boundary Fault can probably be correlated with the Fair Head – Clew Bay Fault in Ireland, even though geophysical continuity is not very apparent (Max and Riddihough, 1975).

On the other hand, the submerged part of the Southern Uplands Fault can be easily traced from Scotland to Ireland owing to the existance of magnetic lineations beneath the North Channel (Dobinson, 1978; Max and Inamdar, 1983). The traces of both faults are indicated on Fig. 12. It should be pointed out that basement areas between

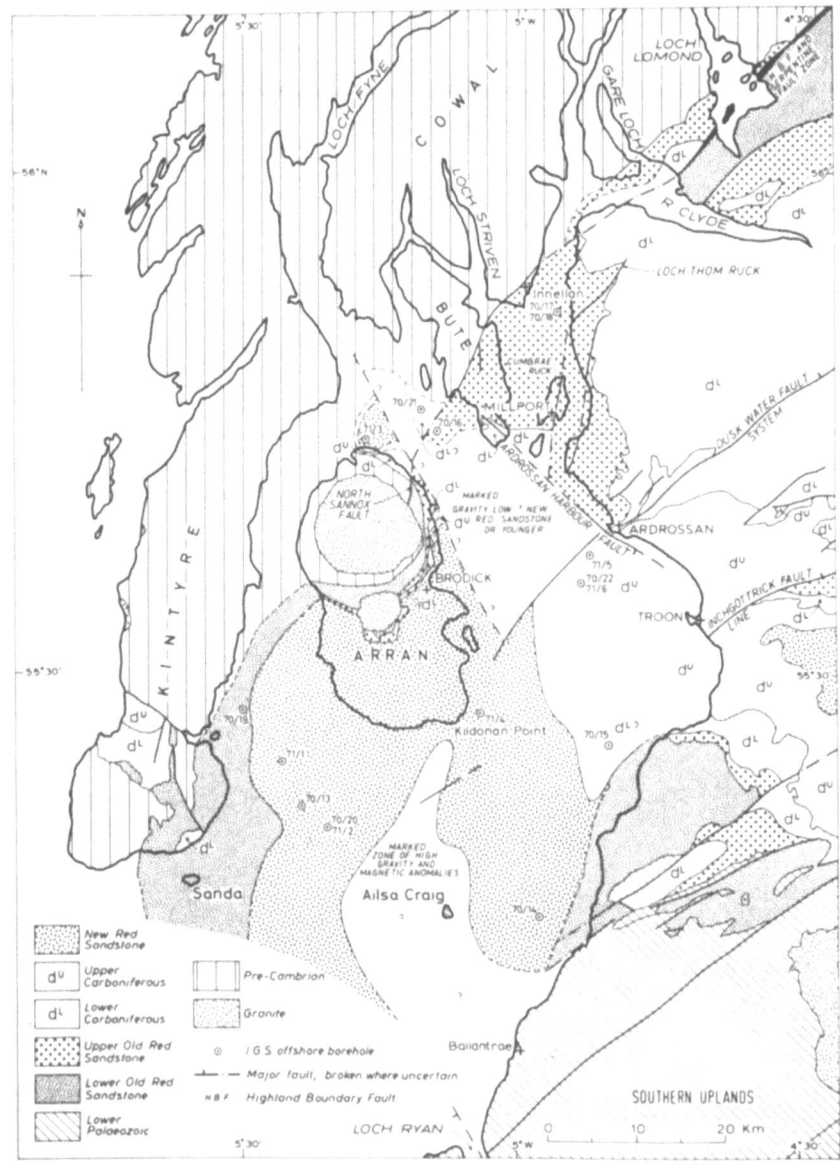

Fig. 18. Geophysical sketch map of the Firth of Clyde. (After Chesher et al. 1972)

these structures are always highly magnetic, whether on land or offshore (Dunham, 1970).

In the Malin Sea (Fig. 9), N 60°-trending anomalies are well defined in the East (Evans et al., 1980) but become far too diffuse in the West to enable recognition of the Minch, Skerryvore, Great Glen or Leannan Faults. Bailey et al. (1975) have summarized the different possible traces for these faults, basing their work mainly on magnetism and gravity. At present, there is only general agreement that the continuation of the Great Glen Fault forms a northern limit to the Porcupine Bank (Bailey, 1979). On the Bank itself, towed magnetic data (Riddihough, 1975; Riddihough and Max, 1976) show that there is a major N 60°-trending fault in the North. More recent airborne surveys demonstrate

that, furthermore, this fracture is cut up into many small segments by Permo-Triassic faults trending N 130° (Lefort and Max, 1984).

The Southern Uplands Fault (Max, 1976) and the Highland Boundary Fault (Max and Riddihough, 1975) are presently considered as "ancient" fractures which were not re-activated after Devonian times (Max, 1978). The Great Glen (Pitcher, 1969) and Leannan Faults (Max and Barber, 1978) appear to have cut the "ancient" fractures later, producing a sinistral displacement (Max, 1978); this observation is compatible with constraints on the offsets which are applicable to the foldbelt between Rockall and the Porcupine Bank (see Fig. 12). In the present reconstruction, the Dover-Hermitage fault in Canada is the prolongation of the N 60°-trending fault on the Porcupine Bank (Lefort and Max, 1984). It is also known that the eastern part of the Gander Zone was subject to sinistral shearing during the Devonian (Hanmer, 1981). At first sight this appears to agree quite well with the European data. Nevertheless, since the "ancient" Caledonian fractures in Ireland are said to turn northward (Max and Riddihough, 1975), it is really only feasible to use the Leannan Fault in trans-Atlantic correlations. This poses a problem, since the Dover Fault is cut by a Devonian-Carboniferous granite and thus is generally thought not to have been reactivated afterwards (Bell et al., 1977).

Three hypotheses are currently possible:

— Early (?) Carboniferous fault displacements observed in Europe were not propagated westward,

— the Dover-Hermitage Fault should in reality be correlated with one or other of the "ancient" Caledonian fractures of Europe, which in this case would remain straight,

— Carboniferous movements occurred in the Gander Zone according to the model of Elias and Strong (1982).

In fact, this problem does not involve correlations between Porcupine Bank and Newfoundland but rather concerns the offshore geometry of the different transcurrent faults known west of Ireland. The possibility of a correlation between the Dover Fault and the Great Glen Fault has occasionally been put forward, but this hypothesis is difficult to support on either geometrical or chronological grounds.

The idea that the Great Glen Fault could be responsible for a sinistral displacement of several thousand kilometers during the Carboniferous (Van der Voo and Scotese, 1981) is not compatible with the trans-Atlantic correlations favoured in this study. Although these correlations are rather imprecise between Rockall and the Porcupine Bank, they do not support such a major transcurrent shear. Further South, there is also probably evidence for sinistral faults of small magnitude on the edge of the Irish Sea Geanticline, as suggested by the presence of "en echelon" folds in the Carboniferous (Fig. 16). Finally, it should be noted that the large transcurrent faults recognized to the West of Scotland are often sinuous on the meso-scale; this is contrary to the argument by which all these faults are thought to have continued their transcurrent movements up to recent times. In many cases, it is clear that the straightness of ancient transcurrent faults has been disturbed by normal faulting at a later date (Parson, 1979).

3.4 Conclusions Concerning the Submerged Mobile Belt

Because of the abundance of Devono-Carboniferous and Permo-Triassic in the East and the thickness of recent cover in the West, Caledonian basement is rarely visible. Therefore, most information depends on geophysical methods (Haworth, 1983). Despite some doubtful points, a certain number of conclusions can be drawn:

— The submerged part of the foldbelt shows two major bends — to the East of the Gulf of St Lawrence and to the West of Ireland — which are separated by a re-entrant situated to the South of Greenland; these structures are a northerly prolongation of the sinuosities already observed in the Appalachians.

— There is a strong contrast in the degree of preservation of ophiolite fragments from one side to the other across the Atlantic. This is probably related to the mode of consolidation of the suture zone upon closure of the Iapetus Ocean (Barker and Gayer, 1985) — the suture is always cryptic in the East, whereas incomplete suturing or obduction of oceanic crust appears to be the rule in the West.

— In view of these observations, it is fruitless to try to systematically extend the zoning known from Canada to Europe. Only the Gander zone can be followed with any confidence from one side of the Atlantic to the other.

— The northerly Front of the Caledonian belt is also clearly continuous, and is always a thrust contact on the Laurentian side.

— The late stage transcurrent shears do not strictly follow the foldbelt but are seen to cut across the sinuosities. The multiplication of these shear faults on the European side may have led to the apparent widening of the orogenic belt.

— It is probable, furthermore, that the post-orogenic basins which partly conceal the mobile belt are mostly related to the relaxation which developed after tangential tectonic activity.

— Finally, it is apparent that the southern edge of the orogenic belt is locally concealed beneath the Hercynides. This overprint is especially important for Western Europe, where the only available control on placing the southern boundary of the Caledonides occurs South of the Porcupine Bank, on the Goban Spur. Here, detrital muscovites showing Cadomian ages are preserved in a Devonian sediment that lacks any sign of Caledonian reworking. This suggests an important distance between this site and the metamorphic front of the Caledonides (Lefort et al., 1984).

Chapter 4

The Avalon Spur; Former Southern Margin of the Iapetus Ocean and Northern Border of the Theic Ocean

During the Lower Palaeozoic, the Avalon Spur made up the southern margin of the Iapetus Ocean. Taking into account the limits of the Gander Zone defined in the previous chapter, it can be seen that the northern margin of this spur includes parts of New Brunswick, the Avalon Peninsula of Newfoundland, the S.E. fringe of Eire and most of Wales. Towards the South, the Avalon Spur was bordered by the Theic Ocean. The best criterion for defining the southern limit of the Spur is given by the sudden and systematic interruption of N-S trending structures in the Avalonian basement which abutt against a continuous belt of basic bodies. Avalonian basement terrains are also to be found in northern Nova Scotia as well as the Iberian and Armorican massifs of Western Europe.

One of the typical features of Avalonian terrains in the Variscan massifs of Europe is the occurrence of seismic velocities which are generally lower (less than 6.3/s) (Sapin and Prodehl, 1973; Bamford, 1979) than Lewisian terrains, which are often characterized by velocities greater than 6.4 km/s. This line of reasoning can be extended to Eastern Canada, although it is necessary to exclude various basic intrusions which intrude this zone (Sheridan and Drake, 1968).

4.1 The Pre-Avalonian Basement

Most of the rocks making up the basement of the Avalon spur are of Upper Proterozoic age; more ancient rocks are certainly present, but they are restricted to rare outcrops mostly located in Western Europe. Such ancient relics are unknown in Newfoundland, even though their presence has been suspected (King et al., 1974). Three dates near 1,000 Ma have been obtained from gneisses cropping out at Cape Breton (Olszewski et al., 1981) and in the Bay of Fundy area (Olszewski and Gaudette, 1982). In S.E. Eire, the Rosslare Gneiss yields an age of 2,000 Ma (Max, 1975) and thus probably constitutes an extension of the migmatitic gneisses known from Anglesey and Wales (Piasecki et al., 1981). Other gneisses, dated at between 1,800 and 2,600 Ma old, re-appear in the Channel Islands and in North Brittany. These outcrops are considered as evidence for a Pentevrian or Icartian orogenic cycle (Auvray, 1979).

To the North of Spain, on the Ortegal Spur, some rocks have been dredged which show untypical characteristics for this region (Capdevila et al., 1974). These rocks include coarse-grained, granoblastic acid granulites, fine- to medium-grained basic charnockites and medium-grained, granoblastic acid charnockites. This group of rocks makes up a metamorphic series belonging to the low pressure (hornblende-bearing) granulites facies; these rocks are considered as being of Pre-Cambrian age.

Further East on Le Danois Bank (Capdevila and Vidal, 1975; Capdevila et al., 1980), the granulites are not found in situ but only as large clasts in a conglomerate of probable Lower Cretaceous age (Fig. 19).

The presence of ancient "Pre-Cambrian" granulites on the North Iberian margin shows that the Cadomian rocks of this region, as in the English

Fig. 19. Basement of the Avalon Spur. 1: Cadomian ridges; 2: ridges of uncertain age; 3: non-magnetic Cadomian ridges; 4: inferred connexions; 5: post-Cadomian faults; 6: Cadomian rise; 7: pre-Cadomian basement outcrops; 8: acid volcanics; 9: pre-Cambrian granite; 10: conglomeratic and pebbly mudstone. **Large capitals,** CA: Cartwright Arch; G: Greenland; NB: New Brunswick; NE: Nova Scotia; NF: Newfoundland; R: Rockall Bank. **Small capitals,** A: Avalon; AN: Anglesey; AU: Audierne Bay; CB: Cape Breton; CO: Cape Ortegal; FC: Flemish Cap; GA: Galicia; GB: Galicia Banks; GBN: Grand Banks of Newfoundland; L: Lisbon; LDB: Le Danois Bank; LP: Lleyn Peninsula; MA: Armorican Massif; OK: Orphan Knoll; VR: Virgin Rocks; W: Wales; Δ: Delta zone

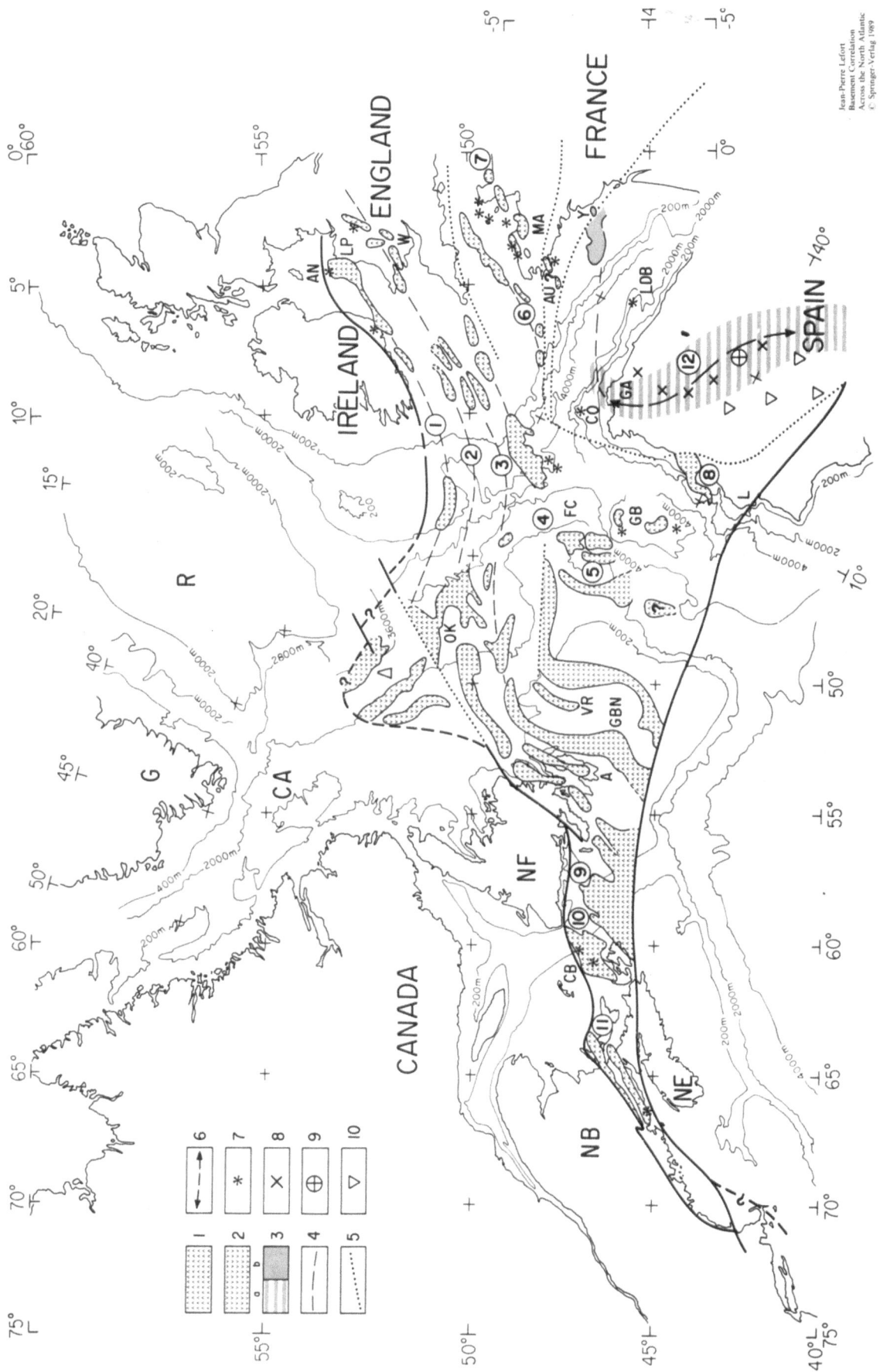

Jean-Pierre Lefort
Basement Correlation
Across the North Atlantic
© Springer-Verlag 1989

Channel, are underlain by an ancient basement. It is unlikely that the uplift of these rocks is due solely to the vertical tectonics associated with Mesozoic opening of the Bay of Biscay, or to the crustal imbrications which developed during the Tertiary subduction in this area (Boillot et al., 1979; Malod et al., 1984).

The Hercynian orogeny was undoubtedly in part responsible for this uplift, which may have already begun before deposition of the Palaeozoic. The main problem concerning the granulites of Le Danois Bank derives from the fact that biotites and phlogopites yield ages between 1,460 and 1,660 Ma and that no Hercynian metamorphism is apparent. It is thus possible to imagine that the present-day outcrops of Palaeozoic were metamorphosed in the greenschist or amphibolite facies during the Hercynian and then transported at a late stage onto ancient granulitic basement situated outside the axial zone of the metamorphic belt.

As already noted by Boillot et al. (1973), it appears that the positive gravimetric anomalies in Northern Spain coincide with the zones of granulite outcrop. It is also significant that no clear positive magnetic anomaly is seen to be superimposed on the gravity highs on the edge of the margin (Leprêtre, 1974). Taking account of the measured densities and various geophysical models applicable to this problem, it is probable that basic rocks are rare in the upper parts of the crust and that the origin of the gravimetric highs has a deeper source.

The predominance of acid rocks (acid charnockites) in the dredge hauls appears to confirm this hypothesis. Since the ancient basement now at outcrop corresponds to a deep structural level — implying the existence of major and repeated thrust slices — it is possible that various basic intrusions (? Mesozoic in age) now at temperatures above the Curie point were emplaced preferentially into zones of crustal weakness along the thrusts. This interpretation is rather similar to that proposed for regions further West (Bacon and Gray, 1970) and East (Rousseau, 1971). Granulites are probably in situ on the Galicia Banks (mainly pyriclasites) and the Vigo Mount (gneisses, acid charnockites, orthopyroxene-bearing ultrabasics and, again, pyriclasites). This offshore material ressembles rock-types found in northern Spain, although the situation at depth appears different; here the Palaeozoic is rather thin (less than 4 km; Black et al., 1964; Grau et al., 1973) and passes downward without transition into material with a seismic velocity of 7.08 km/s. This is abnormally fast for such a shallow depth. Taking into account the known ages of granulites

from northern Spain and the velocities measured from the granulitic basement of NW Scotland (Smith and Bott, 1975) — and also allowing for the fact that basic granulites are more abundant than acid granulites on the Galicia Banks — it may be supposed that a large part of the crust in this region is made up of heavy and seismically fast basement probably older than upper Proterozoic in age.

Granulites and charnockites have been sampled as almost in situ blocks off western Brittany and show identical characteristics to those dredged from the N.W. Iberian margin (Didier et al., 1977); this demonstrates the extension of ancient basement towards the North and lends support to the hypothesis of continuity between Brittany and Spain during the Proterozoic.

To summarize, Avalonian-Cadomian terrains contain remnants of an ancient gneissic basement in many different parts of the belt; however, this basement is not always of the same age on both sides of the Atlantic. In Canada, these ages are around 800 Ma, whereas 2,000 Ma ages are to be found in Europe; work in preparation even suggests the possibility of Archaean crust in the Variscan belt of Europe (Guerrot, unpublished data).

4.2 Avalonian Structures

The Avalon Spur is characterized by the widespread occurrence of Upper Proterozoic markers; these are classically termed Avalonian in Canada and Cadomian in Europe. On land, the intensity of deformation is very variable from place to place. Deformation is non-existant or very slight in the Avalon Peninsula and New Brunswick, but penetrative and polyphase in character in Nova Scotia (Keppie et al., 1983).

In Europe, a well-defined orogenic belt can be recognized in Brittany and Lower Normandy (Cogné, 1974); in Wales, Eire and also in Spain, deformation appears rather less intense away from the Armorican Massif (Piasecki et al., 1981; Capote, 1983). All tectonic activity occurring between 700 and 570 Ma is here included in the Cadomian orogeny. Due to widespread plutonism and basic volcanic activity during this period, it is fairly easy to use magnetic and gravity markers to detect the main structures of the concealed orogenic belt. These markers have also enabled the recognition of the overall structure of the foldbelt across the North Atlantic. For example, Fig. 20 illustrates the clear distinction which can be made, using magnetism, between each side of the Dover-Hermitage Fault.

To the NW of this fault, structures are associated with the Iapetus suture, whereas, to the East, structures are characteristic of the Avalonian orogeny (Haworth, 1977).

4.2.1 Tectonics of the Central Zone

In this zone, positive magnetic and gravimetric anomalies have long wavelengths and are of deep origin; these anomalies are arranged in belts which, in many places, coincide with outcrops of Pre-Cambrian basement of basic composition. This supports the interpretation of the Central Zone in terms of parallel belts which are described below from North to South (see also Fig. 19).

Belt n° 1: extends from Canada to Anglesey, passing through S.E. Eire and the Lleyn Peninsula of Wales. On land, Upper Proterozoic basic rocks are responsible for anomalies such as those observed in the Carnsore granodiorites of Ireland and in Anglesey. These basic rocks, along with their associated sediments, are considered as the remnants of an ancient subduction trench (Wright, 1969; Cogné and Wright, 1980). Subduction is thought to

have commenced around 700 Ma ago, with the downgoing slab dipping South.

Belt n° 2: extends from the Avalon Peninsula to Wales, where positive magnetic anomalies are associated with calc-alkaline volcanics of Upper Proterozoic age. These rocks form part of a magmatic arc that was generated during the subduction mentioned above (Piasecki et al., 1981). This belt can be traced into the Celtic Sea using positive geophysical anomalies which are considered as Pre-Cambrian (Jacobi and Kristoffersen, 1981); the marker can then be followed across the Avalon Peninsula where it is again superimposed on Upper Proterozoic volcanics. In detail, Haworth and Lefort (1979) have shown a good correlation between the basic volcanics of Harbour Main, Bull Arm and Love Cove and the magnetic anomalies. Because of this correlation, there is no point in discussing whether these basic volcanics are related to the initial opening of Iapetus (King, 1977) or whether they actually represent a volcanic arc (Hughes and Brükner, 1971). The horst and graben tectonics observed in the Avalon Peninsula is contemporaneous with the eruption of volcanic rocks, which

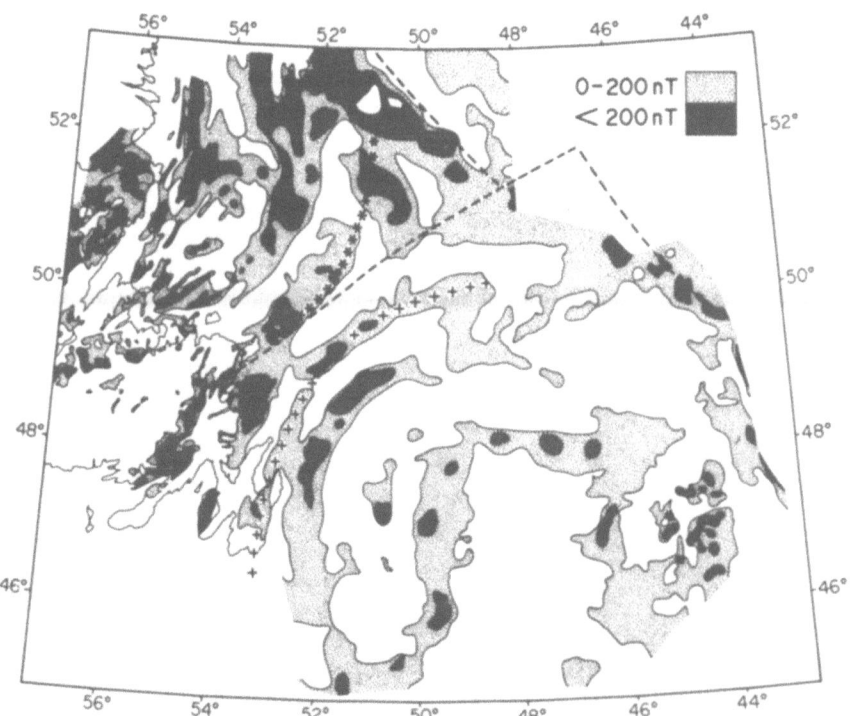

Fig. 20. Sketch map showing divergence of magnetic structures between the Iapetus suture (North of Newfoundland) and the Cadomian ridges of the Grand Banks of Newfoundland (East of Newfoundland). (After Haworth 1977)

themselves may have resulted from subduction processes.

According to this model, the Rosslare-Anglesey zone (Belt 1) has been faulted out along the Dover-Hermitage Fault.

Belt n° 3: displays identical characteristics to Belt 2 and may correspond to a splitting of the volcanic arc discussed above if traced through the geophysical lineaments in the Celtic Sea. Furthermore, this belt comes very close to Pre-Cambrian volcanic rocks in the extreme East of the Avalon Peninsula. The small, oblong anomaly, which bounds the eastern edge of this belt, is actually superimposed on Virgin Rocks where conglomerates similar to those of the Conception Group have been sampled (Lilly, 1966). The conglomerate from this small islet contains granitic and volcanic fragments which ressemble formations known on the Avalon Peninsula (Hughes and Brükner, 1971).

Ridge n° 4 is the only belt which has no outcrop control; it is considered to be Pre-Cambrian in age on the basis of geophysical comparisons (amplitude, wavelength and depth of sources). This ridge has the particular feature of being cut by the western continuation of the South Armorican Shear Zone (Lefort and Haworth, 1978).

Ridge n° 5 is made up of a group of anomalies associated with the Flemish Cap Horst. This horst is known to be composed of granites, granodiorites, dacites and volcano-sedimentary formations (Pelletier, 1971; King et al., 1985). Various samples of intrusive rock have yielded ages between 615–830 Ma, which clearly indicate that this basement ridge is Pre-Cambrian. However, ages are slightly older than those measured from on-land outcrops. Although the volcanic rocks are identical to those cropping out in Newfoundland, they have not been directly dated. Seismic refraction and reflection

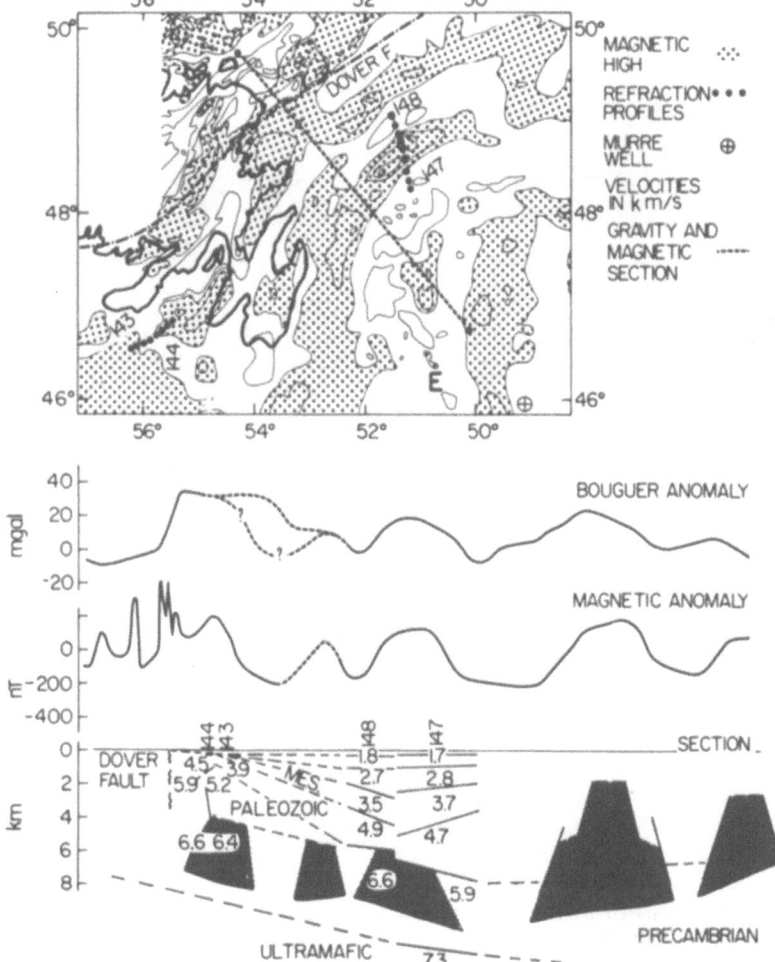

Fig. 21. Map and section showing geophysical structure of Pre-Cambrian ridges detected to the East of Newfoundland. (After Lefort and Haworth 1979)

studies have shown that the westernmost belts are truly intra-basement features, whereas towards the East, the belts have been re-activated as horsts during the opening of the North Atlantic. All these basement ridges appear to be characterized by velocities between 6.4 and 6.6 km/s; the areas in between belts seem to correspond to Hadrynian sedimentary formations (Lefort and Haworth, 1979) (Fig. 21).

Belt n° 6 is situated just at the northern boundary of the Domnonean Domain in the Armorican Massif (see Fig. 22). It corresponds to units whose magnetic susceptibility is identical to that of the basic/ultrabasic complex of the Baie d'Audierne. Modelling of this structure suggests an alignment of dyke type bodies generally dipping towards the South (Lefort and Segoufin, 1978). This alignement

is interrupted in the North by the Lizard-Start Point Thrust and in the South by the South Armorican Shear Zone. It is thus pre-Hercynian in age.

The basic/ultrabasic belt of the Baie d'Audierne shows the same orientation as the mid-Channel anomaly; in the past, these two lineaments were considered as one single belt separated by the South Armorican Shear Zone (Lefort, 1975), which would have acted in sinistral sense before Hercynian times. In fact, the Baie d'Audierne belt — which has been traced offshore as far as 5°W (Lefort and Peucat, 1974) — corresponds to a complex body in which several "ophiolitic" units may have been mixed up. It is now thought possible that the Upper Proterozoic — Lower Cambrian succession known in the Baie d'Audierne may represent back-arc spreading away from the main Channel body. This interpre-

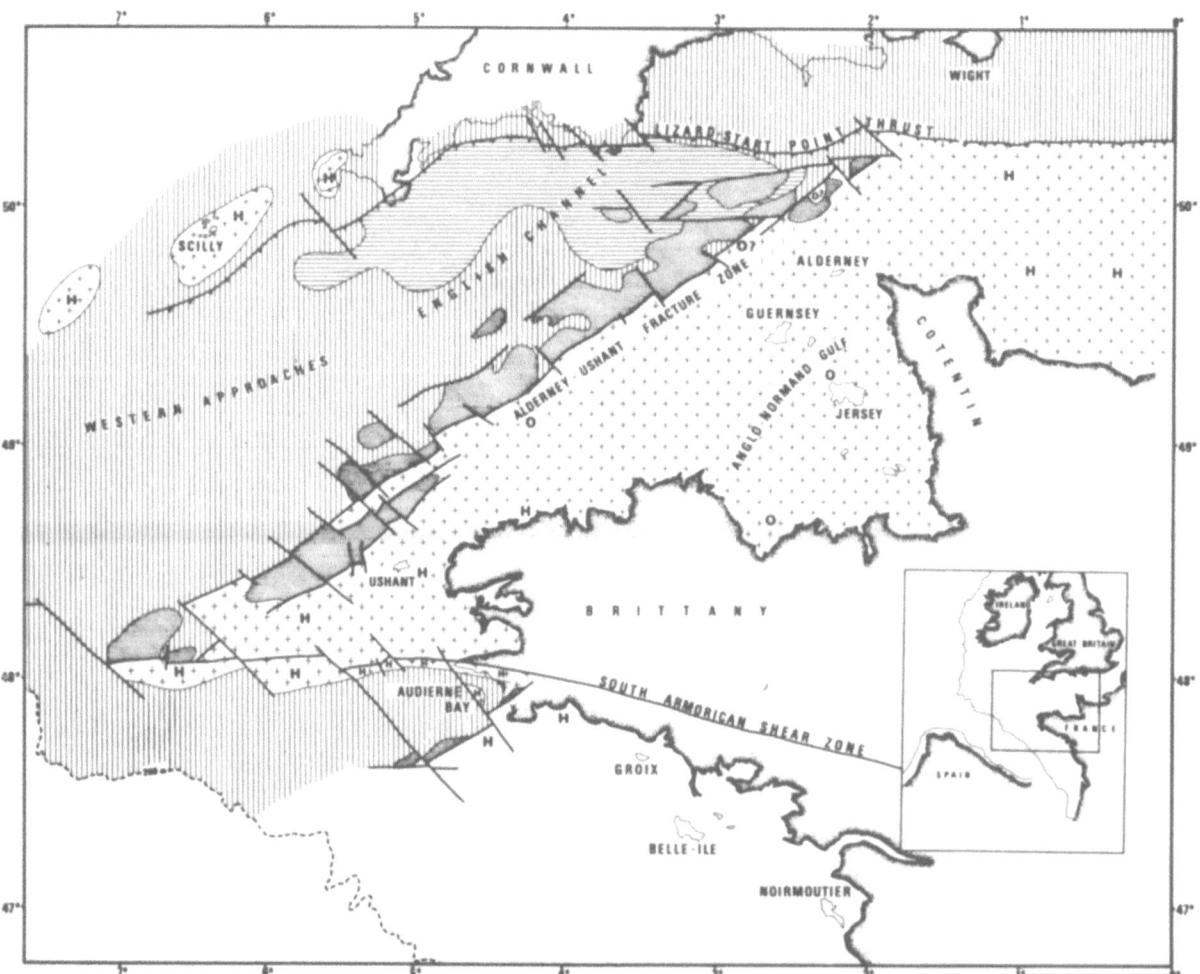

Fig. 22. Structural sketch-map showing nature of basement roof in the Western Channel, based on geophysical data. The Domnonean Domain extends from the Mid-Channel suture (Alderney-Ushant fracture zone) to the northern part of the Armorican Massif (Brittany) H: Hercynian granites; O: Ordovician granites. (After Lefort and Segoufin 1978)

tation is quite likely in view of the fact that the South Armorican Shear Zone has a dextral displacement on the Grand Banks of Newfoundland where it cuts belt n° 4 (Fig. 19). In this way, it could not have produced a sinistral movement of the Baie d'Audierne Complex, which would be the case if the two bodies had the same origin. The mid-Channel body is currently interpreted as a plate suture resulting from subduction beneath the Domnonean Domain.

Belt n° 7: this crosses the Domnonean Domain of the Armorica Massif which is characterized by the presence of Pentevrian basement relicts. It extends from the Cotentin to northern Brittany and consists of Brioverian spilites and ignimbrites (uppermost Proterozoic) as well as a band of gabbros and diorites (Auvray, 1979). At different levels of erosion, this belt corresponds to a back-arc basin and an active margin volcanic arc (Auvray and Lefort, 1980; Graviou and Auvray, 1985).

The SWAT seismic reflection profiles across the Western Approaches to the English Channel enable a more detailed picture of the geological history at depth. Profile n° 10, in particular (see Fig. 23), shows the presence of thrusts and crustal slices directed towards the North to the South of the mid-Channel magnetic anomaly. These thrust faults may be considered as synthetic with respect to southward subduction of the Channel Ocean (Bardy and Lefort, 1987; Lefort and Bardy, 1987). This thrust zone separates two crustal segments, with contrasting seismic properties, which were certainly brought together before Hercynian thrusting events. Since the terrains cropping out on the Galicia Banks

are probably either Palaeozoic or older than Upper Proterozoic, the magnetic anomaly in this area does not belong to a Cadomian basement ridge.

Ridge n° 8: in contrast, the positive ridge located North of Lisbon, which shows seismic velocities greater than 6 km/s and which is interrupted on land by an outcrop of Pre-Cambrian basement (Lefort et al., 1981), is probably a Cadomian feature. This shows that western Portugal is actually part of the Avalon Spur.

Ridge n° 9 is situated off the N.E. coast of Cape Breton Island and overlaps with the Hadrynian Fourchu Group on land (Haworth and Lefort, 1979).

Anomaly n° 10 is probably not related to volcanic formations. In fact, it coincides with the Mic-Mac gneisses which occur on Cape Breton and St Paul's Rock. These terrains are dated at 1,000 Ma and correspond to an upfaulted block of seismically fast basement (6.52 km/s, according to Sheridan and Drake, 1968) which may have been controlled by Upper Palaeozoic tectonics. Since the general orientation of this anomaly is the same as other Hadrynian (Cadomian) ridges in Canada, it is likely there has been re-activation of a horst that was originally active in the Late Proterozoic.

Finally, a double-belt (n° 11) is seen to cross New Brunswick. It is directly related to the volcanic Coldbrook Group of Pre-Cambrian age which crops out in the Kingston and Caledonian Highlands Horsts (Haworth and Lefort, 1979).

Once again, it should be remembered that, in Canada, most Cadomian ridges show velocities between 6.4 and 6.6 km/s (Haworth and Lefort,

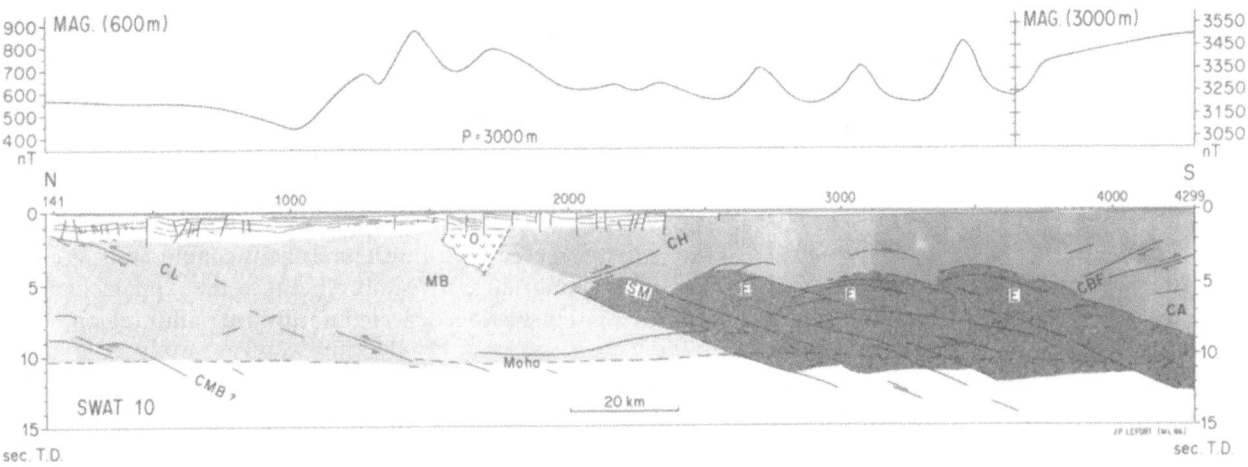

Fig. 23. Seismic reflection profile SWAT 10, drawn up between S.W. England and North Brittany. CA: Armorican crust; CFB: Baie de la Fresnaye Thrust; CH: La Hague Thrust; E: slices of lower crust; MB: Midland Block; O: "ophiolites"; MAG: magnetic profile; P: depth to magnetic bodies. (After Lefort, in press 1988c)

1979); this indicates a basic composition for the rocks. In Europe, many published seismic refraction data (Blundell and Parks, 1969) are less easy to use due to the fact that they represent average velocities over large distances. At the very most, the lower velocities around 6.1 km/s obtained from the Rosslare — Lleyn Peninsula belt would appear to reflect the abundance of sedimentary rocks in the Anglesey Suture (Cogné and Wright, 1980). This is confirmed by the known on-land geology. The 6 km/s velocities from the Porcupine Bank are derived from a structurally complex zone (Whitmarsh et al., 1974); these data lack precision and cannot be used in the present study. The 6.5 km/s velocity calculated for the Celtic Sea and across the Welsh volcanic arc (Bott et al., 1970) is in good agreement with the crustal structure inferred on the Canadian side, despite the lack of constraints at depth. Outside the magnetic bodies, between belts n° 2 and 4 South of Eire, a velocity of 6.1 km/s is indicative of seismically slow upper crust lying between the Pre-Cambrian intrusive belts.

It is thus remarkable to note the excellent continuity of basic Pre-Cambrian ridges across the Avalon Spur. Such continuity enables a reconstruction of the general organization of the Cadomian belt before the opening of the Atlantic. It can be established that the bend in the belt across the Grand Banks of Newfoundland has not been inherited from the Caledonian collision since this bend is cut by the Caledonian-Appalachian mobile belt. Furthermore, the projecting Grenvillian basement off Labrador moulds the Appalachian belt and produces a bend opposite in sense to the Cadomian ridges of the Avalon Spur. By contrast, there is no criterion for deciding whether the bend SW of Ireland is original or due to the influence of a Palaeozoic collision. The strict parallelism which exists between Pre-Cambrian ridges and the Caledonides is, in any case, due to the fact that ancient structures have controlled the southern margin of Iapetus on the European side.

4.2.2 Tectonics of the Iberian — Armorican Zone

The Cadomian basement of the Central Iberian (apart from Portugal) and South Armorican Zones is typically lacking in superimposed magnetic ridges and gravity highs. In this area, the Cadomian belt contains little or no basic intrusions. A major basement ridge (n° 12 on Fig. 19) can be distinguished across central Spain (anteclise of Végas, 1978) which corresponds to an alignment of debris flow deposits,

granites and acid volcanics of Upper Proterozoic age (Capote, 1983). An important horizontal gravity gradient borders this ridge in the west (Instituto Geographico y Cadastral, 1972); this is related not only to the presence of acid material but also to the sharp contact with generally denser Avalonian crust to the West.

It is possible to correlate this structure with a basement ridge that has been located by seismic refraction and gravimetry off southern Brittany (Lefort and Haworth, 1979). This ridge is poorly magnetic, characterized by near surface velocities of 6.3 km/s (Betz, 1965), and is found directly westward of Upper Proterozoic orthogneisses on the Ile d'Yeu (Peucat, in preparation). Continuity of this structure with the Central Iberian Zone would support the hypothesis of an arcuate Cadomian ridge to the East of the Avalon Spur, even though the ridges on the Canadian side have rather different geophysical characteristics.

It should be noted that the Iberian-Armorican Arc has an identical curvature to the bend in Cadomian basement ridges on the Grand Banks of Newfoundland. This demonstrates that the Iberian Peninsula actually belongs to the Cadomian orogenic belt even if Pre-Cambrian deformation is relatively weak. In view of this large-scale geometry, the hypothesis of a purely Hercynian Iberian-Armorican Arc (Matte and Ribeiro, 1975) can be rejected. However, this does not exclude the possibility of a late-stage intensification of the curvature of the initial Cadomian structures (Perroud, 1980).

4.2.3 Basement in the "Delta Zone"

The geophysical markers, mainly magnetic, which characterize this zone are seen to cut across trends in both the Appalachian belt (Haworth, 1981) and also the Avalon Spur. Since they are parallel to Permo-Triassic fractures on the Porcupine Bank (Lefort and Max, 1984), it is possible that they could also be of this age. Such an hypothesis provides no constraint on the age or nature of the deep crust in this area. In the absence of any better evidence, it is proposed that the Delta Zone is composed of Pre-Cambrian crust simply because the Appalachian mobile belt is deflected northward in this region.

4.3 Palaeozoic Cover of the Avalon Spur with Associated Plutonism and Volcanism

4.3.1 Formations Situated off Canada and the U.S.A.

During the Upper Palaeozoic, the western limit of the Avalon Spur corresponded to a subsiding basin known as the Fundy epi-eugeosyncline (Howie and Barss, 1975) (Fig. 24).

Here, the Mid-Upper Devonian, Carboniferous and Permian succession has a thickness which attains 9 km at Prince Edward Island and 5 km off southern Newfoundland (Hobson and Overton, 1973); these two sedimentary troughs are separated by the Cape Breton ridge. The basin fill is made up principally of clastic sediments with local development of limestones, acid and basic lava flows and coaly interbeds towards the top. According to seismic reflection, these same terrains can be found as a narrow band passing through the Bay of Maine (Ballard and Uchupi, 1975); elsewhere, notably in the Bay of Fundy, the Upper Palaeozoic is concealed beneath thick Triassic deposits (Uchupi and Austin, 1979). Older formations are seen to crop out along a coastal belt between New Brunswick and Maine. This belt is composed of either Siluro-Devonian volcanic/gabbroic formations, which can be traced as far as Cape Ann by magnetism and gravity, or Cambro-Ordovician metamorphic rocks possibly associated with Devonian intrusions. The whole succession is cut by some Hercynian granites which yield biotite and whole-rock ages between 231 and 329 Ma (Ballard and Uchupi, 1975).

The results of seismic refraction agree well with mapping based on seismic reflection (Austin and Howie, 1973); in the Gulf of Maine, the southern Gulf of St Lawrence and off southern Newfoundland, Upper Palaeozoic basins generally show velocities in the range 4.6–5.2 km/s (Ballard and Uchupi, 1975; Sheridan and Drake, 1968; Lefort and Haworth, 1979). By contrast, the surrounding terrains are nearly always seismically fast (greater than 6.2 km/s).

Drill-holes have provided additional control on the age and nature of concealed basement and appear to indicate that the roof of basement is made of formations which increase in age eastward. Thus, in the Sydney Basin and in the South of the Gulf of St Lawrence, Upper Carboniferous and even Permian strata are found beneath the Mesozoic cover (Barss et al., 1979). Red shales, anhydrite, limestone and dolomite of Visean/Namurian and Westphalian age are still encountered South of the Avalon Peninsula, whereas further East (between Pre-Cambrian ridges n° 3 and 4) only Lower Carboniferous strata are cored (Visean anhydrites overlying Upper Devonian-basal Carboniferous continental facies). Finally, only Devonian rocks have been recovered in areas between ridges n° 4 and 5.

The Murre well is located further East (between Flemish Cap and the Eastern Shoals) and shows two facies-types separated by a discordance. At the top, weakly metamorphosed sediments contain well-preserved Mid-Devonian acritarchs. Towards the base, however, the meta-greywackes and meta-quartzites cwntain a totally carbonized microflora which is undateable. The discordance in this well suggests that Acadian deformation has affected areas as far as 50°W. This should be compared with the sedimentary break corresponding to Mid-Devonian deformation and metamorphism in the Ossa Morena Zone of Spain (Chacon et al., 1983; Julivert et al., 1983). There may also be a contemporaneous, but poorly-defined break of this age in the South Portuguese Zone.

On the Grand Banks, the easternmost Palaeozoic intrusion that has been dated yields a K-Ar age of 376 Ma — this shows the extent of Acadian plutonism on the Canadian side (Jansa and Wade, 1975). Granites of similar age are to be found at Porto (Portugal) at the same palaeolatitude. The identification of an Acadian discordance in the Ossa Morena Zone and its prolongation onto the Grand Banks of Newfoundland shows that terrains to the West of the Porto-Badajoz-Cordoba Fault really belong to the Avalon Spur. This reinforces the observation already made about the Cadomian basement ridge in Portugal.

The age of 275 Ma obtained by K-Ar on red shales above the Acadian discordance on the Grand Banks indicates that there has been a late Hercynian overprint on Acadian metamorphism (Jansa and Wade, 1975). This result is close to that obtained from the Bay of Maine (Ballard and Uchupi, 1975) and is reminiscent of the late Carboniferous tectono-metamorphic event recorded in Nova-Scotia (Reynolds et al., 1980). In addition, a 272-Ma age is found for the Berlinguas Islands Granite (Portugal) (Fig. 33), which is situated opposite the Murre well site when allowance is made for opening of the North Atlantic. Thus, relatively young metamorphism and plutonism appears to be a feature of the southern border of the Avalon Spur. Even though metamorphic events are high grade in New England (Skehan and Murray, 1980) and weaker in New Brunswick (Rast, 1984), these episodes appear syntectonic in North America and late-stage, post-tectonic in Spain (Gil-Ibarguchi, 1983).

The Flemish Cap granodiorites yield whole rock ages in the range 750–830 Ma (King et al., 1985), whereas biotite and hornblende ages are slightly younger, lying between 615 and 657 Ma. This would suggest a very slight metamorphic overprint, either of Hercynian or Acadian age. Since this later metamorphism was too weak to rework the ancient whole-rock systems, it can be supposed that post-Cadomian metamorphism did not affect areas to the North of Flemish Cap; such an hypothesis is discussed in the context of the Goban Spur on the European side, for which evidence is presented in a later section.

At the latitude of the Avalon Peninsula, seismic surveys and recovered samples show that the roof of the basement is no longer Upper Palaeozoic, but rather composed of generally undeformed Cambro-Silurian (King, 1982). These sediments correspond to shallow marine deposits on a stable platform — facies which are found in Lower Palaeozoic successions from Brittany and Spain. However, the presence of middle Ordovician black shales suggests the existence of temporarily more confined environments on the Canadian side than in Europe. During the Silurian, by contrast, confined conditions seem to have been generally developed on both sides of the North Atlantic. Apart from the black shale facies, most of the succession is made up of sandstones and siltstones. Although it is tempting to place the Caradoc-Ashgill sample from Orphan Knoll as belonging to this succession (Legault, 1982), it should be remembered that there are doubts about the in situ nature of the material at this locality (Parson et al., 1983).

The presence of a detrital facies with ferruginous ooliths in the Lower Ordovician of Bell Island (Dean and Martin, 1978) and on the Eastern Shoals (Jansa and Wade, 1975), as well as their occurrence in Upper Ordovician deposits off Avalon (King, 1982), would indicate that climatic conditions and sedimentary environments were similar over an area extending from the Grand Banks to the Armorican Massif (Chauvel, 1958; Chauvel and Robardet, 1971) and Spain (Guillou, 1976).

The Red Beds of probable Devonian age which overly this succession are of continental origin and thus contrast greatly with the possibly Devonian reef deposits sampled from Orphan Knoll (Parson et al., 1983).

Putting together the observations from the Grand Banks, where deposits are generally marine (Jansa and Wade, 1975), with data from offshore Avalon which indicate continental Devonian facies (King, 1982), it is possible to trace an approximate limit between the Old Red Continent and Devonian marine facies to the South (Dineley, 1975). This boundary is certainly imprecise, due to the lack of exact dating for the different Devonian samples. Nevertheless, the proposed limit is better than speculation.

The Carboniferous of the northern part of the Avalon Spur is poorly known; according to seismic refraction studies in this area, a generally marine character is suggested by velocities of 5.2 km/s which resemble those obtained from carbonate facies. The same suggestion is also made by Cutt and Laving (1977), who base their interpretation on the presence of salt tectonic structures in the Belle-Isle basin as well as comparisons with Carboniferous evaporite-bearing successions in the North of Newfoundland (Haworth et al., 1976). These latter authors think there may be as much as 4,600 m of Visean/Namurian in this northern part of the Avalon Spur. Velocities of 3.6 km/s have been measured to the West of Belle-Isle by seismic re-

▶

Fig. 24. Palaeozoic cover of the Avalon Spur and associated intrusions. 1a: proven or inferred Upper Palaeozoic basins, 1b: thick Upper Palaeozoic successions; 2: volcanic belts of Mid-Palaeozoic age; 3: Lower and Mid-Palaeozoic of variable metamorphic grade; 4: Carboniferous granites in outcrop or proven by geophysics; 5: pre-Carboniferous granites in outcrop or proven by geophysics; 6: axes of Carboniferous basins of syntectonic origin; 7: possible limit of Old Red Continent; 8: trace of Cadomian ridges; 9: sampling and core localities not shown on small scale maps; 10: blueschists; 11: undifferentiated gneisses. Є-S: Cambro-Silurian; O: Ordovician; D: Devonian; D-C: Devono-Carboniferous; C: Carboniferous; P: Palaeozoic. **Large capitals,** CA: Cartwright Arch; CPA: Cape Ann; FB: Bay of Fundy; G: Greenland; NB: New Brunswick; NDB: Notre-Dame Bay; NE: Nova Scotia; NF: Newfoundland; PB: Porcupine Bank; R: Rockall Bank; ZPZ: South Portuguese Zone. **Small capitals,** A: Avalon; AB: Aquitaine Basin; BIB: Belle-Isle Basin; BIL: Bell Island; CAN: Cantabrian Zone; CAB: Cardigan Bay; CB: Cape Breton; CSL: Culm Basin; DOM: Domnonea; ED: Eddystone Rock; ESH: Eastern Shoals; FA: Fastnet Basin; FC: Flemish Cap; GA: Galicia; GB: Galicia Banks; GS: Goban Spur; LDB: Le Danois Bank; MB: Bay of Maine; MUB: Munster Basin; MVG: Vasco-da-Gama Mount; OM: Ossa-Morena Zone; OK: Orphan Knoll; P: Porto; OP: Ortegal Spur; RB: Rochebonne Bank; SB: Sydney Basin; SGB: Saint-George's Basin; SS: Seven Stones; SWEB: South-West English Basin; V: Vendée; VI: Vigo Mount; W: Wales; WR: Wolf Rock

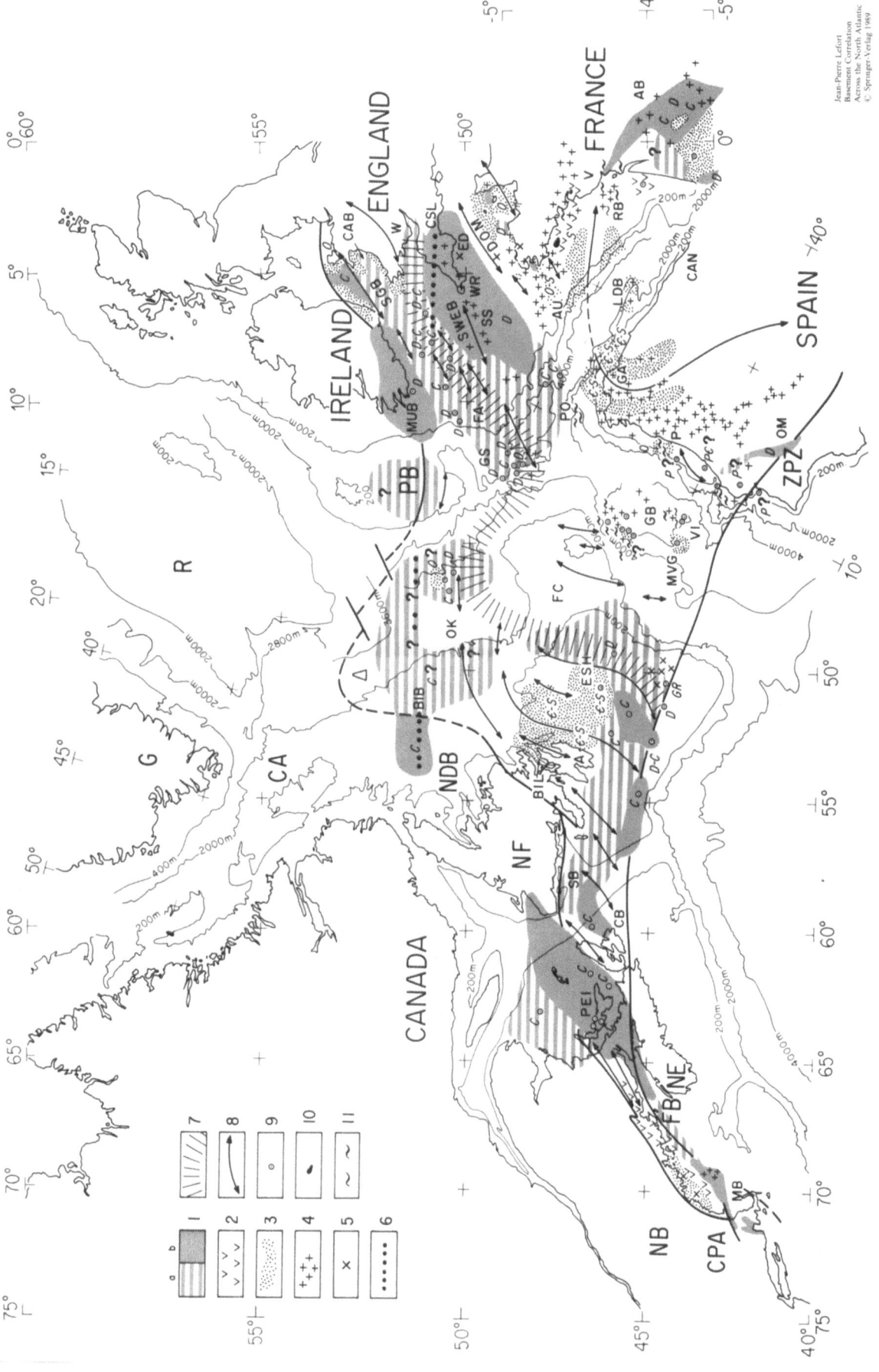

Jean-Pierre Lefort
Basement Correlation
Across the North Atlantic
© Springer-Verlag 1989

fraction (Sheridan and Drake, 1968), thus suggesting that Upper Palaeozoic formations become less calcareous towards Notre Dame Bay and are associated, as in North Newfoundland, with sandstones, siltstones and shales. This Carboniferous basin to the West of Orphan Knoll corresponds, in fact, with part of the St Anthony Basin already described by Wade et al. (1975).

To the North of Orphan Knoll there is an important negative gravity anomaly (Shih, In: Keen and Hyndman, 1979) which is aligned with the axis of the Belle-Isle Basin; even though the origin of this anomaly is unknown, it should be noted that there is a perfect continuity with structures known further West. In addition, it is at DSDP site 111, on the flank of Orphan Knoll, that anthracite clasts are found in the Bajocian (Ruffman and Van Hinte, 1973). Although the germanium content of the clasts is similar to that measured in Welsh coals, a local origin for the clastic material is very likely. Otherwise, Jansa and Mamet (1984) have invoked the possibility of a deep Carboniferous basin in this area.

Thus, there is a strong possibility that a Carboniferous basin is located over the northern part of the Grand Banks. This basin would be unusual, not only in the nature of its sedimentary fill but also in the E-W orientation of its axis. This implies that the clear relationship between Palaeozoic sedimentation and Pre-Cambrian topography that exists further South does not hold in this northerly part of the Spur. Such a disruption probably has a tectonic origin, which will be discussed further in the chapter on Hercynian orogenesis.

4.3.2 Offshore Formations near the British Isles

It would appear that there are very similar variations in axial trends of basins observed on the European side. In this way, the Permo-Carboniferous basin in the southern Irish Sea (Dobson et al., 1973) extends between a median horst composed of Pre-Cambrian – Lower Palaeozoic rocks (Fig. 16) and the Irish Sea Geanticline (or Pre-Cambrian ridge n° 1 on Fig. 19). Although the basin is broadly controlled by these structures, the lower part of the succession (fairly similar to the folded Carboniferous of Ireland), as well as the Upper Carboniferous to Permo-Triassic, shows Caledonian fold trends in the North and Hercynian in the South.

The on-land Munster Basin in southern Eire displays a certain number of resemblances with this offshore basin, notably with respect to sediment type. In fact, the Munster Basin is composed of Devonian continental facies, showing slight folding and metamorphism, overlain by Carboniferous marginal facies. The resemblance with the southern Irish Sea basin extends to its location North of a gravimetric and magnetic ridge (Cadomian Ridge n° 1 on Fig. 19) which is a continuation of the mid-Irish Sea Geanticline of Pre-Cambrian age (see Fig. 16).

South of this composite zone of sedimentary troughs, the Celtic Platform extends from Cardigan Bay to the Goban Spur, passing through St George's Channel and the Fastnet Basin (Figs. 16 and 24) (Gardiner and Sheridan, 1981). This structure behaved throughout the Upper Palaeozoic as a topographic high made up of a framework of Cadomian ridges (Belts n°s 2 and 3 on Fig. 19). The oldest crust on the Celtic Platform crops out to the SW on Goban Spur (Fig. 25).

The basement of the Goban Spur area is quite well known due to numerous dredge-hauls (Auffret et al., 1979) and two IPOD holes (Lefort et al., 1984). Granodiorites and tonalites have been sampled which yield whole rock ages of 275 Ma; these rocks resemble certain facies known from Spain which would be closely adjacent if the Bay of Biscay were closed. There are various associated granitic rocks of alkaline syenitic tendancy, two-mica granites dated at 274 Ma, cataclased potassic granites, leucogranites (cf. Brittany leucogranites dated at 310 Ma) and monzonitic granites. The importance of underlying granulites and charnockites has already been discussed in a previous section. It is obvious that nothing in this zone resembles any of the material sampled from Flemish Cap (King et al., 1985) – this is surprising since the two areas are very close together when the North Atlantic is closed.

The Goban Spur is also characterized by the presence of sedimentary and metasedimentary formations (Guennoc, 1978); they are made up of silty wackes, feldspathic sandstones and low grade schists with well developed flow schistosity. There are also some micaschists and almost in situ Visean limestones. The limestone facies are reminiscent of certain facies seen further East.

Since the red sandstones, quartzites and shales of mid-Devonian age sampled in this area show no sign of major Hercynian metamorphism (although there is a crude Hercynian fracture cleavage, the detrital biotites show a Cadomian age), the micaschist samples from the same locality must belong to a Pre-Cambrian basement (Lefort et al., 1984). This can be added to the information concerning granodiorites at Flemish Cap and re-inforces the idea

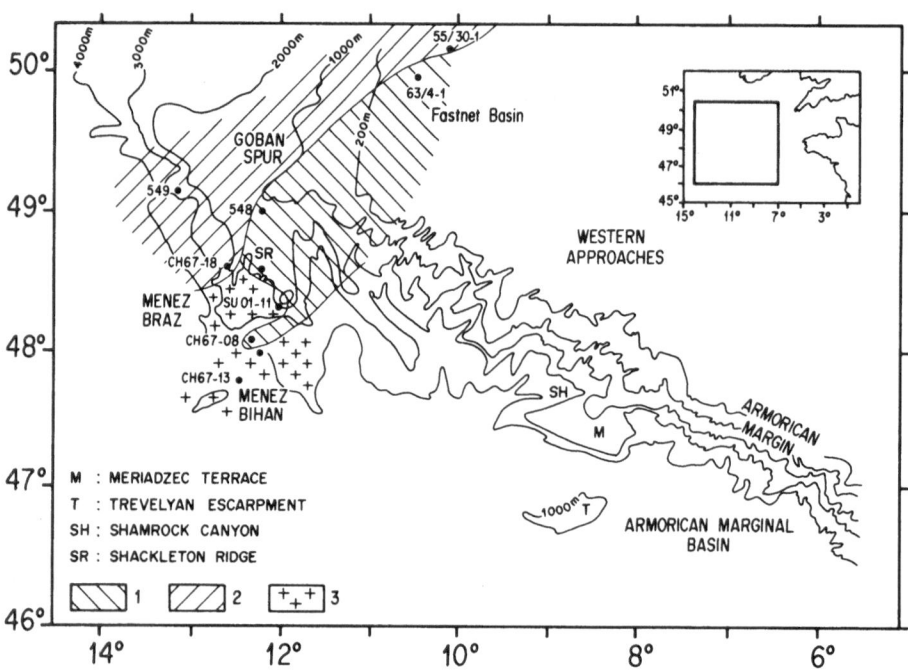

Fig. 25. Geological sketch map of the Goban Spur. 1: metamorphic schists and shales; 2: sandstone ridge; 3: plutonic-metamorphic complex. (After Lefort et al. 1984; reproduced with permission of the United States Government Printing Office)

that there was no significant Hercynian metamorphism at this latitude.

The roof of the basement in the Celtic Sea, and in its western extremity – the Fastnet Basin – is made up of low grade or unmetamorphosed Upper Palaeozoic. Figure 24 shows some of the stratigraphic information available from oil company studies.

Three wells into the Celtic Sea grabens have reached Carboniferous shales which indicate a marine environment lacking carbonates. Further North, Tournaisian and mid-Tournaisian strata have been cored. Taken together, these samples are typical of shallow water deposits. Furthermore, it is known that the southern part of the northernmost Mesozoic graben is underlain by the Stephanian and Westphalian. This part of the succession is particularly thick in Wales, so it is probable that such rocks extend over all the north-eastern part of the Celtic Sea (Naylor and Shannon, 1982). At the other end of the Celtic Sea area, in the Fastnet Basin, Frasnian continental facies are overlain by Tournaisian limestones and Namurian shallow water deposits (Gardiner and Sheridan, 1981) (Fig. 25).

Towards the South, there is a separate basin in SW England and offshore (Bunce et al., 1964) which shows a thick accumulation of Devonian and Carboniferous sediments of mostly marine facies

(Freshney and Taylor, 1980; Bigg et al., 1981). Although the limit between the Celtic Platform and the SW English Basin has not yet been precisely defined, it can be maintained (Gardiner and Sheridan, 1981) that the southern boundary of this platform corresponds approximately to the margin of the Old Red Continent. A few unpublished data from the Devonian of the Western English Channel suggest that the basin extends as far as the Alderney Ushant axis – the surface expression of the Channel Ocean Suture (Belt n° 6, Fig. 19). This zone was subsequently very unequally filled in with Carboniferous sedimentation, due to Hercynian tectonics, and the deposits are often concentrated in pockets of variable depth. These "pockets" include the black biomicrites of Visean age South of Goban Spur (Auffret et al., 1979) and the Devono-Carboniferous shales and sandstones cored South of Haig Fras.

The Munster Basin, Celtic Platform, Channel Basin and S.W. England Basin all display an undoubted control of sedimentation by N 50°- and N 60°-trending magnetic and gravimetric ridges. Only one structure is seen to cut across this framework, thus evoking once more the problem of interaction between Pre-Cambrian ridges and Hercynian tectonics. This structure is the Culm Basin of North Devon and Cornwall (Fig. 24), which is curiously

Jean-Pierre Lefort
Basement Correlation
Across the North Atlantic
© Springer-Verlag 1989

associated with an E-W trending positive magnetic anomaly that can be traced as far as Longitude 8°W. The intersection of the Culm Basin structure with Cadomian ridges has a deep crustal origin which is discussed in Chapter Seven. The proposed explanation concerning the formation of the Culm Basin is probably the same as that adopted for the northernmost Canadian Carboniferous basin.

Cadomian trends appear to have controlled not only the location of various Palaeozoic basins but also the emplacement of Hercynian batholiths as classically suggested for the Cornish granites.

In the Celtic Sea, the gravimetric map clearly shows the influence of deep basement structures and the general orientation of ridges (Blundell, 1979). From South to North, one can pick out the Cornish Platform (A and B in Fig. 26), then the Haig Fras Ridge (E, C, D in Fig. 26) and finally the Pembrokeshire — Labadie Bank (F and G in Fig. 26). These negative gravimetric axes correspond to ridges of granite.

Modelling of these anomalies shows that the Cornish granites form a bell-shaped mass which is rooted at about 10 km depth (Edwards, 1984); seismic refraction profiles across this axis (Holder and Bott, 1971) confirm such an interpretation, whereas seismic reflection profiles in the same area suggest the possibility of a causal link between certain thrust faults and batholith formation.

The Cornubian granite axis links up with a zone where various types of granitic rock are exposed on the continental slope (Goban Zone). Despite the excellent geophysical correlation with this zone, very few of the offshore granites can be assigned to the same two-mica alkali granite type as found in Cornwall.

The age of the submerged part of the Cornubian granite ridge is known from dating at Haig Fras (277 Ma, Sabine, 1965) and the Seven Stones (281 Ma; Sabine and Snelling, 1969). These granites are leucocratic and tourmaline-bearing, sometimes

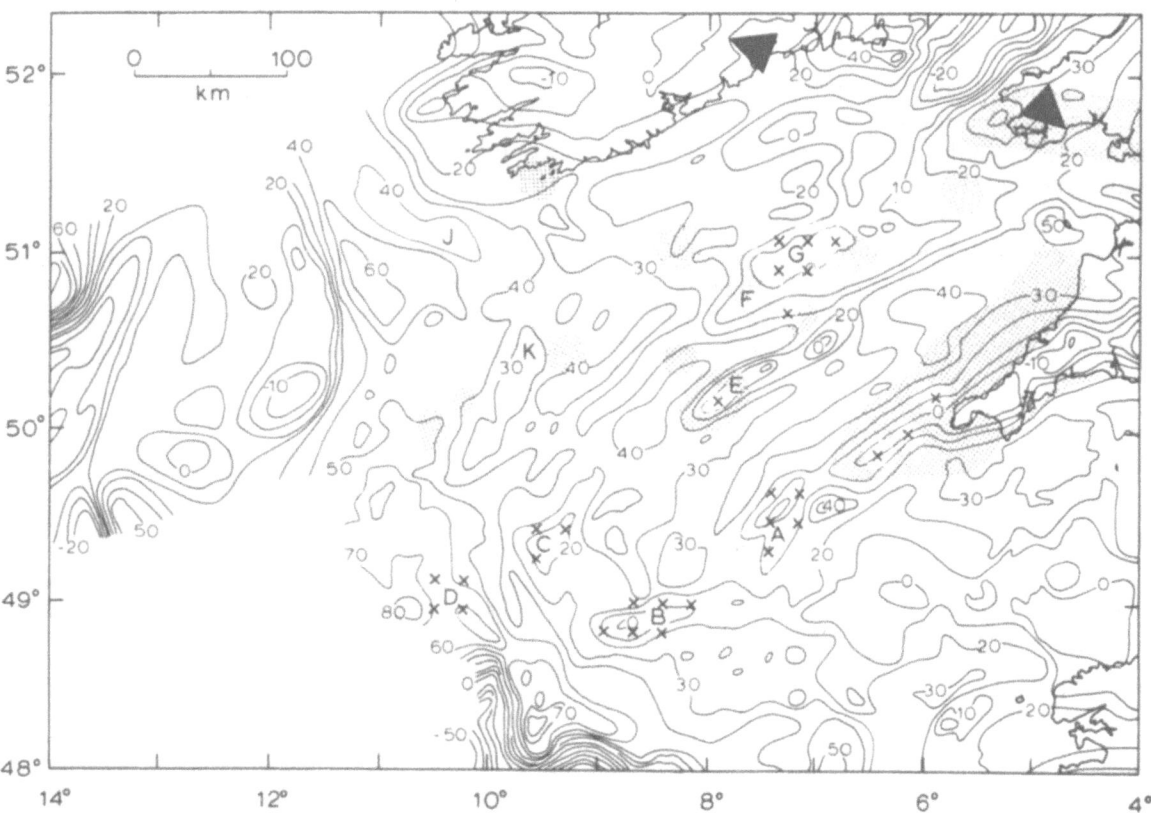

Fig. 26. Gravimetric map of the Celtic Sea. Capitals refer to granite ridges discussed in text; black triangles indicate trace of the Hercynian Front; diagonal crosses indicate zones of probable granite intrusion, whereas stippled areas show occurence of Upper Palaeozoic (mainly Devono-Carboniferous) sediments. (Modified from Blundell 1979; reproduced with permission of Elsevier Science Publishers)

showing a foliation. Two small occurrences of phonolite have been dated at 113 and 132 Ma (Phillips, In: Whittard, 1962; Sabine and Watson, 1965) but subsequent work on nepheline from the Wolf Rock locality suggests that the phonolites belong, in fact, to the Hercynian basement and were emplaced around 262 Ma ago (Snelling, 1968). In view of the discussion concerning intrusive rocks on the Goban Spur given above, there is no possible prolongation of the Cornubian granites onto the Grand Banks of Newfoundland.

4.3.3 Offshore Formations to the West of France

The submerged Armorican basement is the most intensively studied concealed basement in the world, with 1,500 sample points and a dense coverage of magnetic, seismic and gravimetric measurements which have enabled the recognition of the main structures present. In this study, it is out of question to give a detailed description of the database that has led to publication of the 4 maps showing submerged geology (1:250,000 sheets) and the 1:1,000,000 sheet showing the sub-Mesozoic basement; only the major tectonic units will be discussed in this section.

A compilation of sampling work undertaken in the Western Channel area is presented in Figs. 27 and 28 (Lefort, 1975). In the northern part of Brittany, the main structural unit is represented by the Domnonean Domain (see Fig. 22); it is made up of dioritic gneisses, epidiorites, micaschists, orthogneisses, granitic gneisses and leptynitic gneisses which constitute an indurated Pre-Cambrian block. This block was not uplifted immediately after sub-

duction of the Channel Ocean during the Cadomian orogeny, since the final phases of rhyolitic volcanism – occurring after the initial Brioverian magmatic activity along the coast – were almost contemporaneous with Lower Cambrian marine sedimentation (Lefort and Deunff, 1974; Auvray, 1979). Following on rapidly, the uplift of the region caused a marine regression which was reversed only much later. The Brioverian and Lower Cambrian volcanic belt (n° 7 on Fig. 19) is composed of porphyries, rhyolites, tuffs, albitophyres and spilites and is bordered on the S.E. and on the N.W. by Red Bed formations which are the continental Domnonean equivalents of the Arenig marine sandstones found further South in Brittany. These Red Beds directly overlie low grade Brioverian formations and leptynitic gneisses, and are overlain locally by grey shales assigned to the Ludlovian (Deunff et al., 1971); in this way, the beginning of the Palaeozoic transgression onto the Domnonean Domain is marked by the occurrence of Silurian to the East of Cadomian ridge n° 7 (cf. Lefort and Deunff, 1971). In the Normano-Breton Gulf, there are two synclines bordering the Pre-Cambrian basement horsts of the Channnel Islands which preserve evidence for such a transgression. To the North of the Brioverian volcanic arc, Lower Cambrian sediments are found in the Jersey syncline as well as black or bluish slates and Upper Ordovician spilites (Lefort and Deunff, 1974). These same sediments are overlain by grey or bluish limestones and grey shales (containing abundant acritarchs, chitinozoa, bryozoa and crinoids) of Devonian age in the Siouville Syncline (see Fig. 27). From Devonian times onwards, there was probably communication between northern and

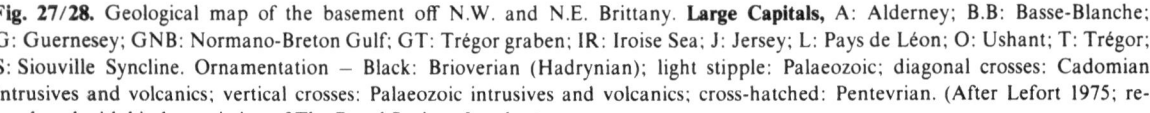

Fig. 27/28. Geological map of the basement off N.W. and N.E. Brittany. **Large Capitals,** A: Alderney; B.B: Basse-Blanche; G: Guernesey; GNB: Normano-Breton Gulf; GT: Trégor graben; IR: Iroise Sea; J: Jersey; L: Pays de Léon; O: Ushant; T: Trégor; S: Siouville Syncline. Ornamentation – Black: Brioverian (Hadrynian); light stipple: Palaeozoic; diagonal crosses: Cadomian intrusives and volcanics; vertical crosses: Palaeozoic intrusives and volcanics; cross-hatched: Pentevrian. (After Lefort 1975; reproduced with kind permission of The Royal Society, London)

Sedimentary rocks:
A^{1-4}: Dunes and alluvium; h^{1-3}:Carboniferous; d^{1-3}: Devonian; S^{2-3}/v: Silurian and Mid-Upper Ordovician and Ordovician volcanics; ^{1}: Lower Ordovician (Grès Armoricain); d^{GR}: "Grès Armoricain" (continental facies); b:Cambrian;V^{1-2}: Mid-Upper Brioverian; V^e: Brioverian volcanics
Crystalline basement (Cadomian):
Micaschists: ξ^1; Phyllites: ξ^2; Amphibolites: ξV^e; Various gneisses: ζ^2; $\zeta\eta$; $\zeta Y_{,,}$; $\zeta\eta^2$.
Crystalline basement (Pentevrian):
Gneisses and amphibolites: ζ^1; Dioritic gneisses and diorites: $\zeta\eta^1$.

Migmatites:
Mγ: Migmatites and associated granites
Intrusive rocks (Hercynian):
Microgranites: γ^1; Muscovite and tourmaline-bearing granites: γ^1t: Two-mica granites: γ^1; Muscovite granites: $\gamma,^m$; Calc alkaline biotite granites: $\gamma,$; Biotite-bearing pink granite: $\gamma,^b$; Saint-Renan granite: γ,γ^1.
Intrusive and extrusive rocks (Cadomian):
Ancient granites: $\gamma,,$; Gabbros and epidiorites: θ; Granodioritic and dioritic differentiates: $\gamma,,\delta/\eta$; Porphyrites: ν; Rhyolites: ρ; Andesites and trachytes: $\alpha\tau$.

southern parts of the Western Channel as witness the Devonian facies found in the Trégor graben (Lefort, 1970; Lefort and Deunff, 1970).

The basement inlier surrounded by Eocene cover 45 km to the North of the Trégor is made up of red sandstones which were first assigned to the "Old Red Sandstone" (Lefort, 1975). They should now be considered as an island or shoal composed of Ordovician continental facies.

The southern limit of the original Domnonean Block has been found West of the Trégor, where Arenig (?) marine facies occur (Auvray and Lefort, 1971) and to the West of Ushant, where a Llandeilian fauna is observed above Brioverian micaschists (Andreieff et al., 1973). The systematic lack of marine Arenig sandstones ("Grès Armoricain Formation") from Ushant into the Normano-Breton Gulf constitutes a characteristic feature of Domnonea during the Palaeozoic.

The intrusion of Hercynian granites is rare in the East and becomes more and more important towards the West. In the North and East of the Pays de Léon (NW Brittany) there are Devonian and Carboniferous granites, locally crushed gneisses and migmatites occurring as on land outcrops. Only two granites are specifically restricted to offshore outcrops to the South of Ushant; the Pierres Noires Granite is very cataclased, whereas the Iroise Granite is undeformed and probably of late Hercynian age. It is interesting to note that late-stage Hercynian granites off North Brittany are limited at the coastline and form a NE trending belt which is parallel to the mid-Channel magnetic anomaly. This belt is also parallel to the Cornubian granite axis. Such lineaments are thus expressions of an ancient "grain" which was preferentially reactivated at the end of the Hercynian orogeny. There are two isolated offshore outcrops which fall outside the Hercynian granite ridge; these are on the Basse Blanche (Lefort et al., 1978), situated North of the Pays de Léon at 4°10′W and 49°5′N, which is made of orthogneisses and a highly cataclased pinkish granite dated at 460 ± 10 Ma. Even though this granite is 35 km from the coast, it shows little affinity with the Léon region and rather more closely resembles various late-stage Cadomian granites in the South of the Normano-Breton Gulf (Bonnemain, Jersey, Yffiniac). This implies the possibility of important sinistral movements after 460 Ma. Finally, there is an isolated occurrence of granitic gneiss on the Eddystone Rocks in Cornwall which yields on age of 375 Ma (Sabine and Watson, 1965).

The Central Armorican Zone, made up of synclines with Devonian cores and anticlines bringing up the Brioverian, can be followed westward off Brittany (see Fig. 28). Here, the sedimentary part of the Iberian-Armorican Arc can be said to start. The affinity between West Brittany and the NW part of the Iberian Peninsular has long been recognized (McPherson, 1886) — it has recently been re-evaluated due to the stratigraphic and palaeontological work of Paris and Robardet (1977). Such an affinity has led many authors to envisage a perfect continuity between the different Palaeozoic terrains from Cambrian times up to the opening of the Bay of Biscay (Bard et al., 1971); this has now been contradicted by geological and marine geophysical studies. Three main interruptions, in fact, prevent the possibility of a simple continuum between West Brittany and NW Iberia; viz:

— **Firstly,** the South Armorican Shear Zone, with its associated suite of leucogranites dated at 330 Ma (Vidal, 1973), forms a linear discontinuity which extends to 7°W and probably as far as the continental slope (according to magnetic data; Lefort, 1975). The action of this major transcurrent fault, which does not bend southwards as often presumed, will be discussed further in Chapter Seven.

— The **second** discontinuity is constituted by the Baie d'Audierne Synclinorium, which traces out a triangular zone between granites of the South Armorican Shear Zone (Fig. 29) and a polyphase metamorphic complex in the South composed of micaschists, albitic gneisses, chloritic schists, amphibolites and serpentinites. This southerly complex seems to represent the superposition of "oceanic" crust and associated sediments, dated at 500 Ma (age of metamorphism: 370 Ma), onto ancient granulite protoliths dated at 1,300 Ma which underwent high pressure metamorphism at 385 Ma (Peucat, 1983; Peucat et al., 1982). Whatever the tectonic interpretation given to this basic/ultrabasic complex, it is obvious that it breaks the continuity of the Iberian-Armorican Arc across the Bay of Biscay.

Within the Synclinorium, there are mono-metamorphic terrains which become lower grade and younger towards the West. These terrains are made up of coarse meta-greywackes showing greenschist to amphibolite facies assemblages and fine-grained greywackes associated with low grade black shales. These shales, which are somewhat reminiscent of Calymene-bearing facies, contain microplankton fragments of probable Ordovician age. A volcano-sedimentary formation has been recognized above the metasediments. In the West, the whole succession is overlain by well-developed coal-bearing sediments (Lefort and Doubinger, unpublished data) which are assigned to the Namurian B.

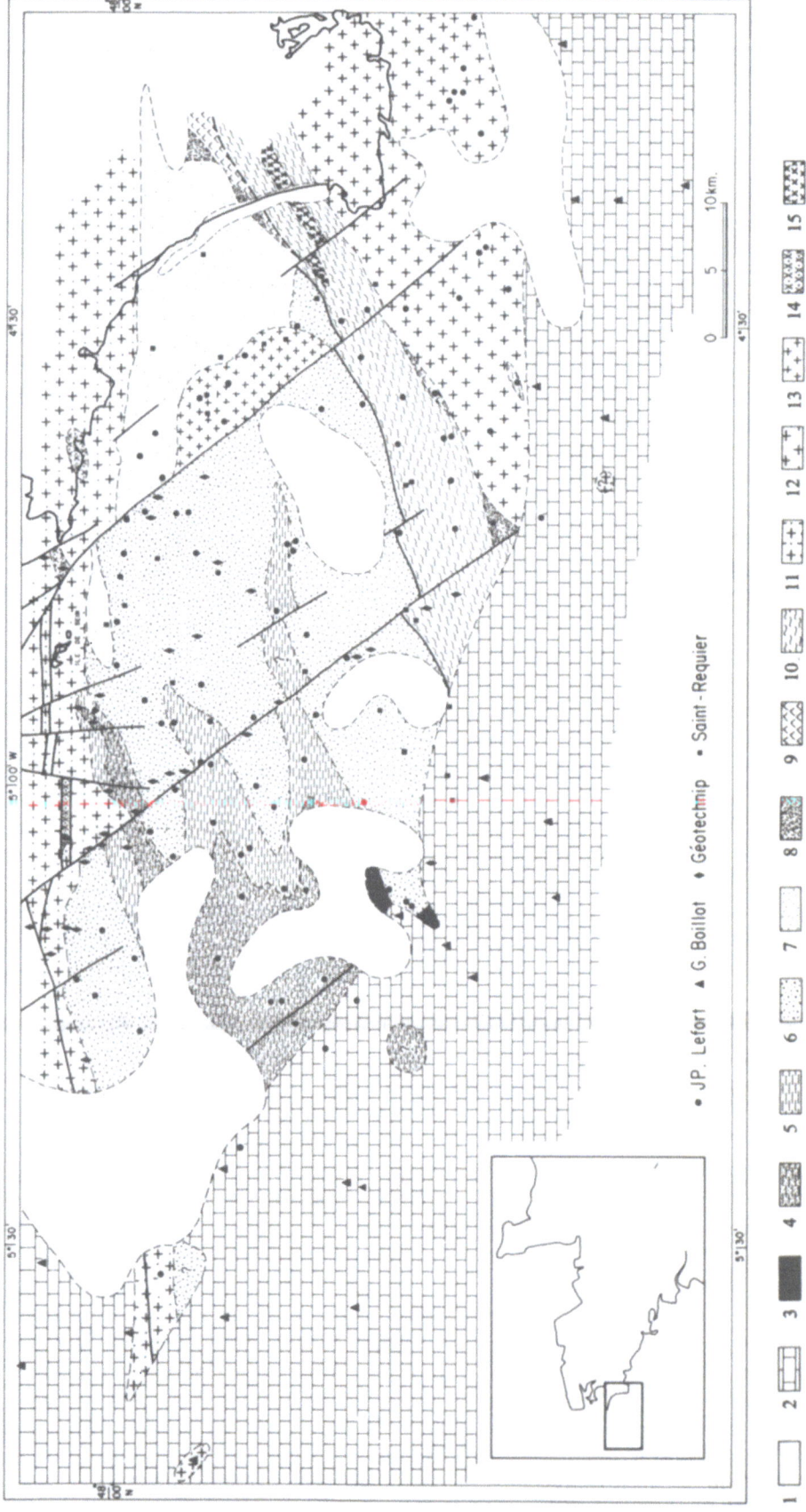

Fig. 29. Offshore basement to the West of the Baie d'Audierne. (After Lefort and Peucat 1974). **Unmetamorphosed sediments; 1**: hydraulic dunes; **2**: Mesozoic and Tertiary cover; **3**: coal bearing shales. **Group showing single phase of metamorphism; 4**: volcano-sedimentary formation; **5**: black shales; **6**: greywackes; **7**: meta-greywackes.

Group showing polyphase metamorphism; 8: serpentinites; **9**: amphibolites; **10**: micaschists and greenschist amphibolites. **Granitic group; 11**: Pointe du Raz granite; **12**: Pont-L'Abbé granite; **13**: Baie d'Audierne granite; **14**: Cap-Sizun granite – gneiss; **15**: Ploneour-Lanvern granite – gneiss

• JP. Lefort ▲ G. Boillot ♦ Géotechnip • Saint-Requier

This group of mono-metamorphic formations appears to define an extensive N 70°-trending sedimentary basin which has been cut by the South Armorican Shear Zone. These formations are partly or wholly of Palaeozoic age and are seen to lie unconformably on the polyphase metamorphosed complex on land. At sea, the same contact appears rather tectonic and thrusted (Lefort and Peucat, 1974). The description of this narrow offshore zone is interesting in that it includes greywackes of probable Ordovician age — a facies which is unknown in the Central Armorican Zone. Thus, the interruption of the Iberian-Armorican Arc is further confirmed. The presence of this sedimentary succession also shows that the age of Hercynian metamorphism, as seen throughout the South Armorican Domain, must be older than Namurian B.
— The **third** discontinuity cutting the Iberian-Armorican Arc is a geophysical marker whose nature will be discussed in the chapter on the Ligerian (eo-Hercynian)—Acadian mobilebelt. Eastwards from the Baie d'Audierne, terrains become greatly influenced by Hercynian and pre-Hercynian orogenies.

4.3.4 The Bay of Biscay Domain

This metamorphic belt is bounded on the North by an almost continuous band of leucogranites which forms a remarkably homogenous and extensive unit. These two-mica leucogranites, dated at around 300 Ma in Southern Brittany (Vidal, 1973), constitute a major arcuate structure which is far more compressed than the Iberian-Armorican Arc (see Fig. 24). The shape of this granite band can be picked out under Mesozoic and Tertiary cover rocks using offshore gravimetry (Sibuet, 1972; Lefort, 1975). Calculations show that granite massifs bordering the coast are always rooted at very shallow depth and that thicknesses are never more than 1 km (Vigneresse, 1978). The lithologies are generally undeformed, with only local enrichment in biotite or development of porphyritic texture (Audren and Lefort, 1977). There are some older granites associated with migmatization within the belt between Lorient and Vannes (Fig. 30). One can recognize four groups of metamorphic terrain within the Hercynian leucogranitic arc, viz:
— **The Ile de Groix Group** makes up an offshore blueschist belt 40 km long and 8–10 km wide trending NNW-SSE. It is mostly composed of glaucophane-bearing rocks, sometimes containing garnet, which have been more or less retrogressed into blue-green amphibole, chlorite and albite.

There are also greenschist amphibolites (prasinites of French authors) and metasedimentary rocks including albitic micaschists. This terrain has suffered a complex metamorphic history which can broadly be summarized as prograde HP/LT metamorphism followed by greenschist facies retrogression.
— **The Belle-Ile-en-Mer Group** is characterized by the presence of volcano-sedimentary horizons (porphyroids of French authors) interbedded with low grade schists and arkosic sandstones. This terrain is seen to almost encircle the blueschist belt and is abundantly developed on Belle-Ile-en-Mer.
— **The Vilaine Group** includes those rare offshore outcrops which can be assigned to the metamorphic formations seen around the Estuary of the Vilaine (Audren, 1974). It is principally composed of garnet micaschists.
— **The Port Navalo — Noirmoutier Gneisses** constitute a group which crops out as a discontinuous band to the NW of Noirmoutier. These gneisses are now included as enclaves within on-land granitic massifs. Since these formations are typically pre-Carboniferous in age, their stratigraphic position and tectonic history will be discussed in the light of orogenesis in the Ligerian (eo-Hercynian) — Acadian belt (see Chapter 5). Nevertheless, it should be noted here that the Groix blueschists and Belle-Ile-en-Mer formations have only be preserved at outcrop due to down-faulting of the Groix graben — this structure is limited by the Kerforne and Quiberon Faults, which were probably active at the beginning of the Mesozoic (Lefort, 1975).
The offshore Rochebonne Plateau is isolated in the middle of the South Armorican margin; here, the depth to basement increases only slightly seawards since magnetic data (Vaillant, 1972) indicate a depth of 400 m at 100 km from the coast. It is relatively well characterized due to some recovered samples and numerous seismic reflection profiles, and appears to be mainly composed of a granodioritic ridge (Callame, 1965). The granodioritic rocks are probably Hercynian in age and show a prophyritic, sometimes mylonitized texture which suggests the close proximity of a major fracture zone. Here and there, metasediments are found which are often represented by pelitic gneisses — however, amphibolitic gneisses and metaquartzites are also observed. Some of these lithologies are seen to be locally affected by contact metamorphism near the calc-alkaline granites. This group of rocks displays certain similarities with on land zones in Southern Vendée and forms a transition with the Aquitaine Basin basement discussed below. If these undated orthogneisses were identical to the 600-Ma-old

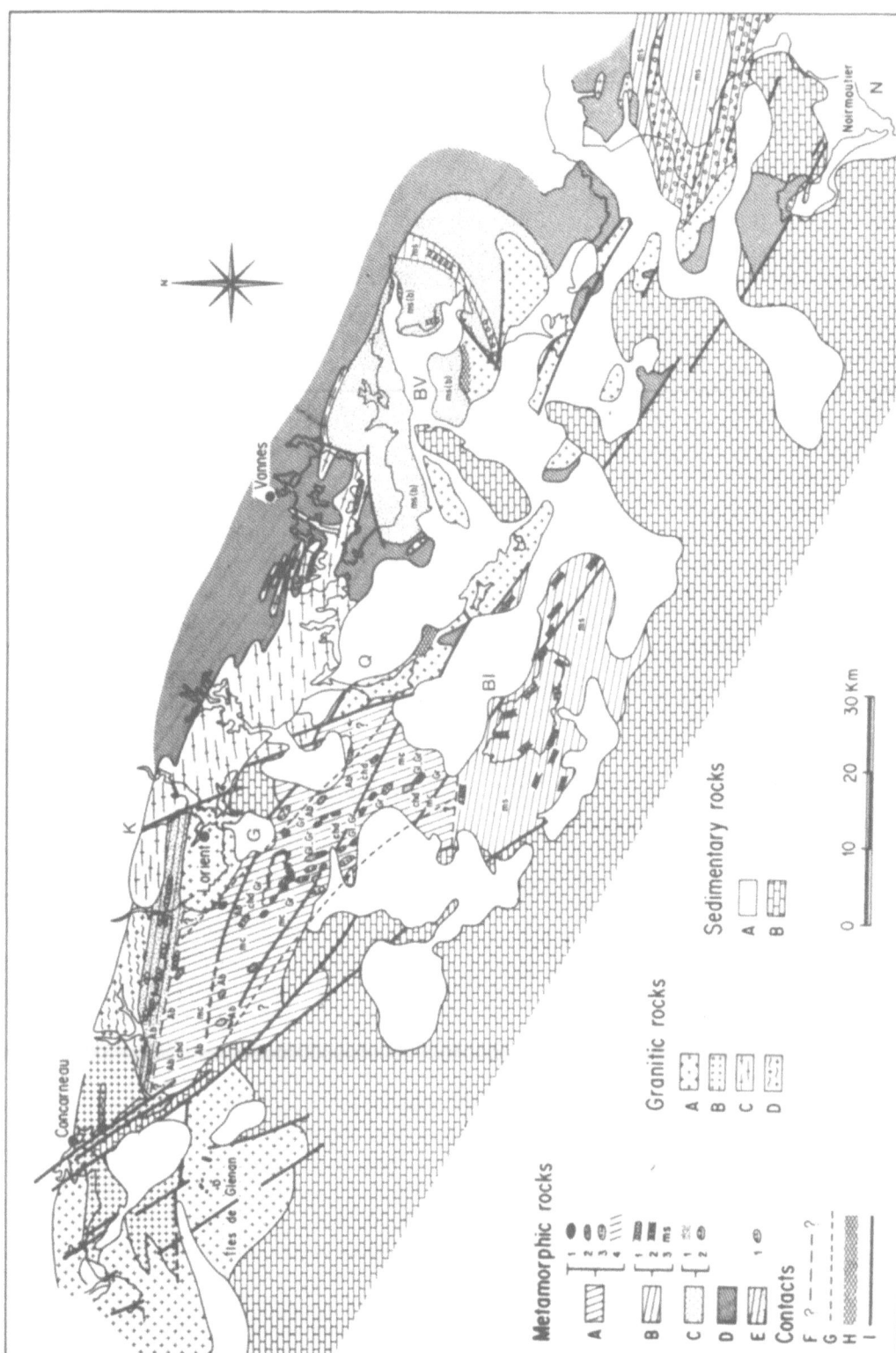

Fig. 30. Geological sketch-map of metamorphic basement outcrops off southern Brittany. (After Audren and Lefort 1977). **Large Capitals,** BI: Belle-Ile Island; BV: Vilaine Estuary; G: Groix Island; K: Kerforne Fault; N: Noirmoutier Island; Q: Qiuberon Peninsula. **Metamorphic rocks,** Ile de Groix Group; A1: glaucophane-bearing rocks; A2: amphibolites derived from glaucophane-bearing rocks; A3: greenschist amphibolites; A4: micaschists (albite + garnet + chloritoid). Belle-Ile-en-Mer Group; B1: volcano-sedimentary formations ("porphyroids"); B2: meta-arkosic sandstones; B3: undifferentiated micaschists.

Vilaine Estuary Group; C1: albite-chlorite micaschists with relict biotites; C2: pyroxenites and amphibolites. Port-Navalo Group; D: plagioclase-bearing biotite gneisses. Le Pouldu Group; E1: albite-chlorite micaschists with interbedded greenschist amphibolites. Ab: albite; chd: chloritoid; Gl: glaucophane; Gr: garnet; mc: muscovite. **Granitic rocks,** A: two-mica granites; B: biotite granites; C: cordierite granites; D: granite-gneisses. **Sedimentary rocks,** A: hydraulic dunes; B: Eocene. **Contacts,** F: presumed offshore discontinuities; G: maximum extent of glaucophane schist belt; H: mylonitized granites; I: faults.

orthogneisses on the Ile d'Yeu, we should at last have some evidence (long awaited) for the presence of Pre-Cambrian basement on the seaward side of the Aquitaine Basin. For the moment, the possibility of a relation between the Rochebonne Bank and northern Aquitaine is based solely on magnetic lineations (Vaillant, 1972).

Further West, beneath the Eocene, basement has been detected by gravimetry and seismic refraction (Fig. 31). In fact, there are two basinal structures on either side of a Pre-Cambrian ridge showing a seismic velocity of 6.3 km/s. This ridge is probably made up of orthogneiss and is non-magnetic and gravimetrically light (see Fig. 19). In comparison with known on-land seismic velocities (Lefort and Haworth, 1979), it appears that the northernmost gravimetric "basin" — filled with material showing velocities between 5.1 and 5.8 km/s (Betz, 1965) — is probably made up of slightly metamorphosed Palaeozoic formations. Even if these formations are at present in places lying on top of magnetic intrusions, showing velocities of 6.5 km/s, which are located at 1.5 km depth, they are not necessarily younger than the magnetic basement. This is because Hercynian tangential tectonics are known

to have been important in this zone (Lefort, in press 1988c). Furthermore, it is possible that the linear positive gravity anomaly situated further North (Fig. 31) owes more to the influence of the underlying magnetic intrusion than to the small "Palaeozoic" basin which is superimposed. The production of a gravimetrically heavy "basin" in the South could be due not only to the existence of a small Palaeozoic basin but also to the uplift of fast basement (6.9 km/s) (Fig.31).

It can be seen from Fig. 34 that the overall geometry of the Iberian-Armorican Arc is broadly preserved in the Bay of Biscay Domain; the presence of residual basins and basement ridges leads to a zonation that can be correlated with NW Spain where identical structures are known.

Beneath the Aquitaine Basin, it has been possible to draw up a provisional stratigraphic column and a map of concealed basement from drill-hole data (Le Pochat, 1984). These data show the presence of Upper Cambrian (with echinoderms) overlain by Lower Ordovician shales interbedded with fine-grained sandstones, then Silurian sandstones and black shales. The Lower and Middle Devonian is dolomitic and contains coral, some echinoderm

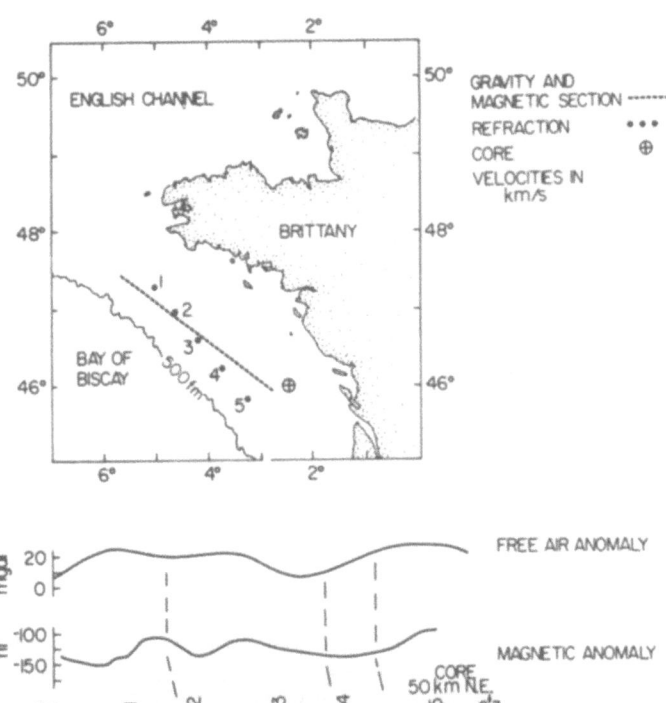

Fig. 31. Map and section showing correlation between gravimetry, magnetics and seismic refraction to the SW of the Armorican Massif. Velocities are given in km/s. (After Lefort and Haworth 1979)

debris, Bryozoans and Brachiopods. The Lower Carboniferous is composed of an alternation of shales and fine sandstones whereas the Lower Permian is conglomeratic. Samples from these drill-holes have recently been the object of renewed studies (Paris and Robardet, 1985) that have refined or modified the previous stratigraphic assignments. This explains why the map of the sub-Mesozoic basement of the Aquitaine Basin (Fig. 32) has been complemented in places with new age data. Furthermore, sedimentary formations in this region are characterized by relatively frequent re-working phenomena.

To the NW, terrains previously considered as metamorphic (Winnock, 1971) are only just in the greenschist facies — it is only at the northernmost extremity of the region that rocks have a metamorphic appearance. The Palaeozoic of the northern Aquitaine Basin bears hardly any resemblance to terrains of the same age in the Armorican Massif. From the sedimentary point of view, the thick Tremadocian succession found in drill-holes has no equivalent in the Lower Ordovician of the Armorican Massif which nearly always begins with Arenig sandstones and quartzites (Grès Armoricain Formation). Furthermore, certain typical formations of the Armorican Ordovician — such as oolitic iron-ores, grey limestones of Caradoc-Ashgill age or even the glacio-marine deposits of the uppermost Ordovician — have not been observed in the Aquitaine drill-holes. However, certain analogies can be drawn for the basal Devonian succession in both regions. The Carboniferous of the Aquitaine Basin lacks coal or thick conglomeratic members, thus providing further contrast with Brittany and showing that the Aquitaine Basin is quite separate.

Taking account of the basement geometry of SW Europe before opening of the Bay of Biscay, the Aquitaine basement should probably be correlated with the Cantabrian zone of North Spain (Paris and Robardet, 1985).

Apart from some granites pierced in drill-holes, the granites shown in Fig. 32 were detected by gravimetric methods (Rousseau, 1980). It appears that these granites only rarely cut the roof of the

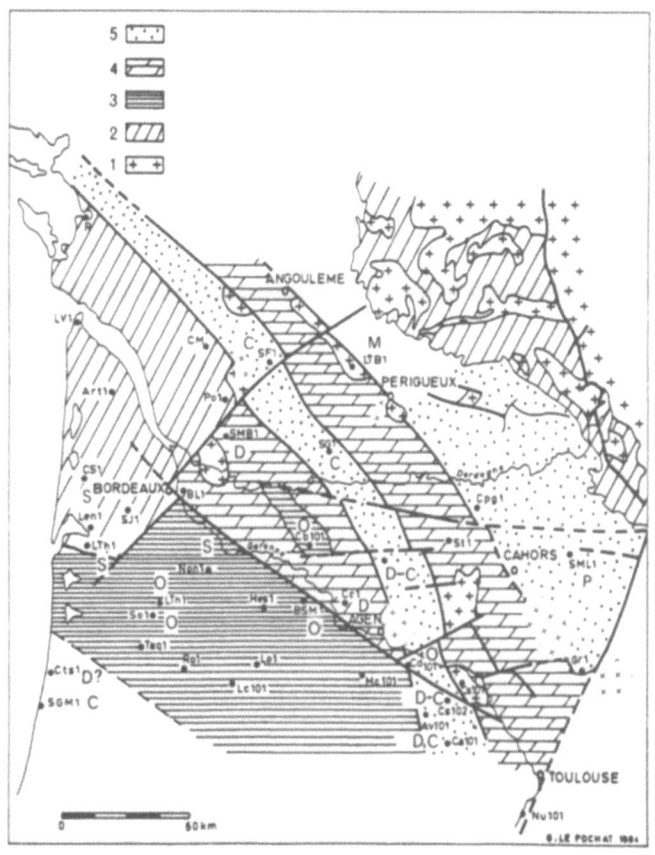

Fig. 32. Map showing possible structure of basement concealed beneath the Aquitaine Basin. 1: granite; 2: low-grade metamorphic terrains; 3: Ordo-Silurian; 4: Devonian; 5: Carboniferous. **Capitals.** O: Ordovician; S: Silurian; D: Devonian; C: Carboniferous; P: Permian (fossil dating of drill-holes). Uncontoured granites have been detected by gravimetry. (After Le Pochat 1984 and Rousseau 1980)

basement. The sub-circular form of these intrusions is different from the shape of syn-tectonic granites in the South Armorican Domain. Thus, the Aquitanian granites are either ancient massifs undeformed during the Hercynian or, more likely, post-tectonic plutons (even though they follow a typical South Armorican trend).

4.3.5 Offshore Formations near Spain and Portugal

The metamorphic basement ridge and "Palaeozoic" basins detected beneath the South Armorican margin can be correlated with identical structures mapped in NW Spain. In this region, the gravimetric signal of two heavy "basins" on either side of a less dense "ridge" can be followed as far as the continental margin if the free-air gravimetry is corrected for edge effects (Sibuet, 1972). This fit between structures on either side of the Bay of Biscay demonstrates that basins in NW Iberia were in continuity with offshore basins near Armorica throughout the Palaeozoic (Fig. 34).

The morphology of Pre-Cambrian basement ridges (Vegas, 1978) would thus appear, as in Canada, to have controlled Palaeozoic sedimentation in this area (Fig. 24).

Towards the NW of the Spanish margin, basement has been sampled in many places close to the shore and around the Ortegal Spur (Lamboy, 1976). Near the coast, samples can easily be assigned to the major tectonic units recognized on land; there are mostly sandstones, quartzites and phyllites of Silurian, Ordovician and Cambrian age belonging to the West Asturian-Leonese Zone (Fig. 24). The location of these samples confirms the purely gravimetric determination of offshore Palaeozoic basins (Fig. 34). Basic rocks belonging to the ophiolitic klippes of the Mid-Galician Zone do not extend offshore since only one serpentinite sample has been recovered at sea. Otherwise, this point of view is confirmed by detailed magnetic surveys in this area which show no detectable influence from the Cabo Ortegal Klippe away from the coastline.

In addition, amphibolites, black cherts, leucogranites, post-tectonic granodiorites and low grade schists are found associated with the sedimentary rocks. Definitely in situ rocks are rare on the Ortegal promontory — only one orthogneiss and some leucogranites have been found. If dredged angular blocks are included in the group of in situ samples, the clearest correlation is with Galicia (especially when the large number of leucogranite samples is taken into account). Otherwise, the abundance of orthogneisses and granitoids as well as the relative paucity of paragneisses (common on land) suggests a deeper erosion of the basement at the outer margin of the shelf. Studies from submersibles on the Ortegal Spur have revealed the presence of material identical to that known in eastern Galicia; however, there are granulites and high grade gneisses which show that the level of erosion is at a deeper structural level than on land.

Variscan re-crystallization in the amphibolite or greenschist facies has not been observed on the Ortegal Spur itself, even though it has been recorded on the continental shelf. Such petrographic observations have a certain importance since they show the considerable differences which exist with Flemish Cap — an area which was, however, very close before the opening of the Atlantic. This point will be further discussed in relation to the samples recovered from the Galicia Banks. On Le Danois Bank, situated between the Ortegal promontory and the French coast, it is known that the basement contains metamorphic rocks of varied composition showing lower greenschist to amphibolite facies. In addition, there are some rare late-stage granitic bodies of a type found intruding relatively high levels of the Hercynian orogenic belt (Capdevila et al., 1974). Sampling from submersibles has yielded slightly feldspathic quartzites in the lower amphibolite facies which resemble certain parts of the Lower Cambrian or Cambro-Ordovician known on land (Capdevila et al., 1980). Altogether, the margin off NW Spain fits well with what is known of the Lower Palaeozoic succession in the Armorican Massif. By contrast, the formations off NW Spain are totally lacking in Upper Palaeozoic strata — this is distinctly different from the concealed basement beneath the Aquitaine Basin.

As far as trans-Atlantic correlations are concerned, the most interesting seaboard is represented by the Portuguese margin and the Galicia Banks. There have been few successful attempts to sample basement on the margin itself. North of Lat. 41°30' (see Fig. 24), basement is composed of pelitic or arkosic gneisses metamorphosed in the amphibolite facies. There are also shales with occasional ferruginous ooliths (slightly metamorphosed) and feldspathic quartzites. All these formations are highly deformed. Certain shales have been assigned to the Llandeilo or Llanvirn and could represent the offshore prolongation of outcrops North of Porto. The greenschist facies samples may belong to the "chlorite zone" defined in Portugal (Portugal Ferreira, 1972) whereas the amphibolite facies correspond to on land outcrops (Musellec, 1974).

On the southern part of the Portuguese margin, two-mica pelitic gneisses (with well-developed flow cleavage) and highly deformed, two-mica leucogranites showing affinities with the Hercynian granites of the Evora region (NW Ossa Morena Zone, crustal re-melt complex; Capdevila et al., 1973) are found in the Berlenguas and Farilhoes archipelago (Freire de Andrade, 1937). If it is reasonable to compare the samples of Lower Palaeozoic in this region with rocks already described from the East of the Avalon Peninsula (particularly the oolitic members, and ignoring differences in metamorphism), there is no basis, however, for the prolongation of Evora-type granites westwards towards Canada.

A map of the deep basement structures between the Porto – Tomar – Badajoz Fault and the 200 m isobath has been proposed by Lefort et al. (1981). This map is based on an examination of magnetic data (Uchupi et al., 1976; provisional map of the total magnetic field of the Portuguese continental shelf, 1979) and gravimetry (Gravimetry map of Portugal, 1960) and is linked to a study of confidential geophysical information from oil and mining companies. This interpretation of deep structures has been confronted with the results of drilling on the margin (Fig. 33).

The studied area contains two major structural zones bounded by important faults – only the northerly zone will be described in this chapter. This zone includes the regions of the Estremadure and the Douro, as well as the continental margin between 38° and 41°30′ N, and is characterized by a strong irregularity in the roof of the magnetic basement. In places, there is a relief of more than 5,000 m over a few kilometres. Also to be noted is a series of morphological ridges associated with positive magnetic and gravimetric anomalies which alternate with basins responsible for negative magnetic and gravimetric markers. This alternation is clearly defined on the continental shelf but is disturbed by important edge effects on the slope and certain light diapiric structures on the inner part of the margin (Musellec, 1974). Information provided by seismic refraction profiles on the Portuguese margin is rather imprecise. At the most, it can be suggested that the basement under shot-point R.A. (Fig. 33), where velocities are 5.66 km/s, is probably Palaeozoic, whereas under R.C. and R.D. (where V = 6.07 and 6.00 km/s respectively) the basement is made up of seismically slow metamorphic or plutonic terrains. There is, in fact, no direct relationship between the measured velocity distributions and the location of ridges and basins. Furthermore, it is likely that the magnetic ridge situated under R.A. could be linked to a structure which is deeper than the level reached by seismic refraction (the depth to magnetic basement being 2,500 m and the seismic basement being at 1,000 m). Otherwise, it is known that basement was not reached under shot-point R.B. and that at R.C., situated over a basin, the seismic basement is at 3,890 m – this is less than the calculated depth to magnetic basement: 5,500 m. It is only at R.D., a zone characterized by abundant high frequency magnetic anomalies, that seismic refraction suggests a link between seismic and magnetic basement (both at around 2,300 m). Nevertheless, it is by no means certain that sub-Mesozoic basement was reached because the seismic refraction shots may have intercepted a recent lava flow. Thus, it appears that the geophysical "ridges" in this area have an origin which is generally deeper than the roof of the sub-Mesozoic basement – these deep structures have seismic velocities greater than 6 km/s. Furthermore, there is a close connection between the orientation of deep structures and the geometry of superficial basins which evokes certain observations already made on the South Armorican shelf and on the Grand Banks of Newfoundland (Lefort and Haworth, 1979). This suggests again that the West Portugal basement behaved as part of the Avalon Spur.

The basement of the Galicia Banks and the Vigo and Vasco da Gama Mounts is known to be continental in character. Seismic velocities near the surface of 4.8 km/s have been recorded by Black et al. (1964) and subsequently by Grau et al. (1973), suggesting that the upper layers of the Galicia Banks are composed of basal Mesozoic or uppermost Palaeozoic strata and that intrusions are rare or absent. However, this is not confirmed by the numerous dredge-hauls and core samples available for this region (Groupe Galice, 1979; Grimaud, 1981). In fact, the northern Banks yield gneisses, micaschists, granodiorites, porphyritic granites and microtonalites, leucogranites, granodiorites, paragneisses, micaschists and grey shales. To the South, the Vasco da Gama Mount is composed of shales and arkoses, whereas the Vigo Mount contains outcrops of orthogneiss and granodiorite (equivalent to the Galicia Banks samples). Although the granodioritic orthogneisses are of probable Cambro-Ordovician age, the late to post-tectonic granites and leucogranites are very likely Hercynian. Metamorphism varies from the greenschist to the low-pressure amphibolite facies and deformation is generally intense. Some rock types resemble certain facies found on land – in particular, the

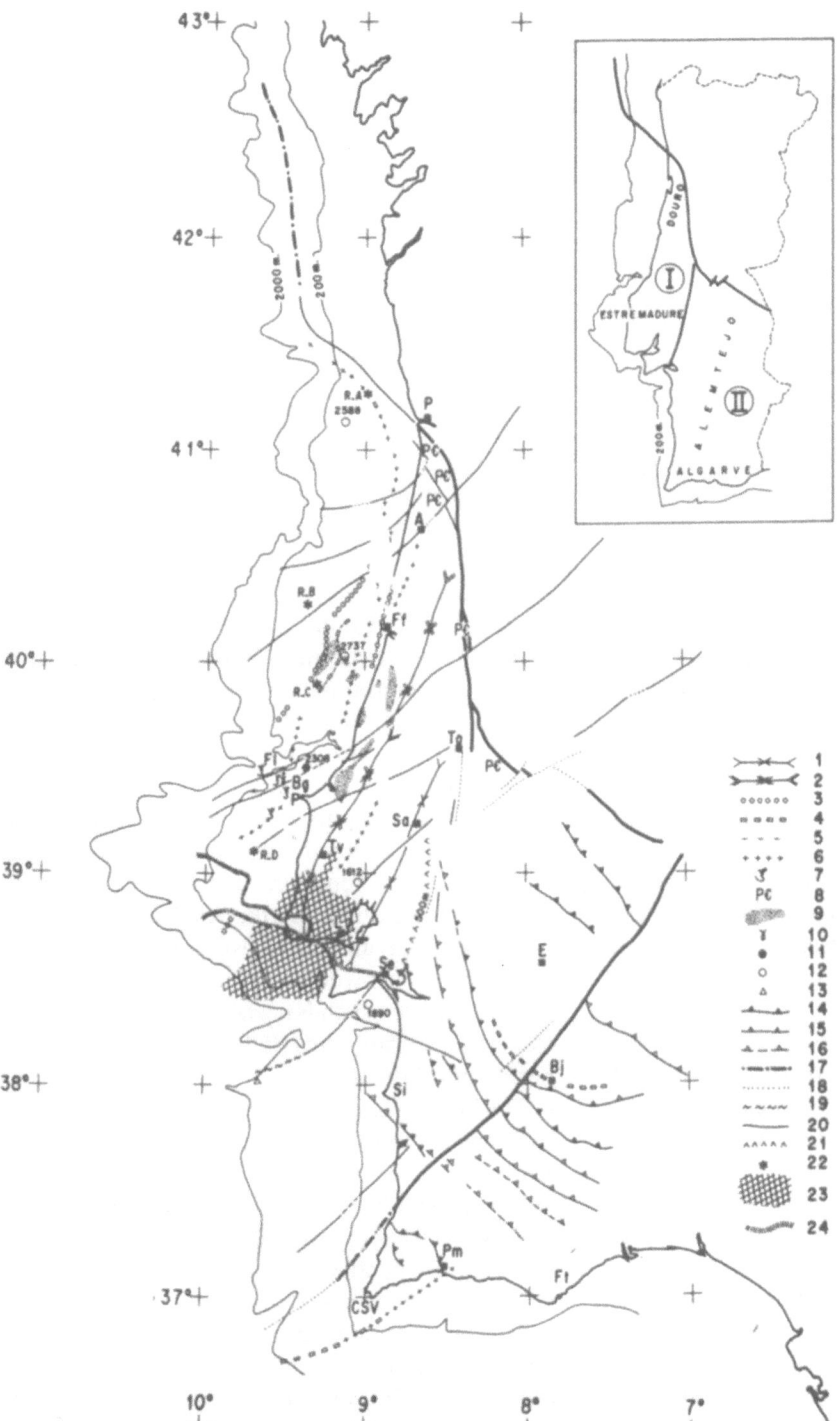

Fig. 33. Sketch map of deep structures in the West Portuguese basement. 1: axis of isopachs for onland Tertiary basin; 2: gravimetric axis of Lusitanian Basin; 3: negative gravimetric axes; 4: positive gravimetric axes; 5: negative magnetic axes; 6: positive magnetic axes; 7: Berlinguas and Farilhoes basement; 8: Pre-Cambrian outcrops; 9: diapirs recognized from drill-holes or seismic reflection; 10: granite outcrops; 11: drill-hole piercing granite; 12: drill-hole into Palaeozoic or Pre-Cambrian (numbers indicate depth of basement roof); 13: dredge-haul; 14: known thrusts (on land); 15: thrusts recognized from gravimetry (on land); 16: probable thrusts inferred from gravimetry (on land); 17: faults recognized from seismic reflection (offshore); 18: faults recognized from gravimetry (on land); 19: morphological feature (offshore); 20: fault recognized from magnetism; 21: eastern limit of Tertiary basin (500m isopach); 22: seismic refraction shot-points; 23: zone of high frequency magnetic structures; 24: Cape Roca anomaly. Key on inset map – I: Northern zone; II: Southern zone. (After Lefort et al. 1981)

lustrous shales which show two phases of deformation may correspond to the Ordovician and Silurian formations already described.

The amphibolite facies rocks correspond to deep structural levels of the Hercynian belt but, with few exceptions, are different from outcrops found on the European continent.

It is certainly possible to assume that the 7.08 km/s material detected beneath the base of the Galicia Banks corresponds, as already suggested, to the ultrabasic pyribolites (low to intermediate pressure amphibolite-bearing granulite facies) found on Vigo Mount. However, it is difficult to accept that the 4.8 km/s unit could be intruded by granites and granodiorites as found on the Western flank of the Banks. One solution is to propose that, due to the relative position of shot-points, seismic waves traversing the eastern part of the Banks have only intercepted sedimentary formations. Such rocks would show little or no metamorphism and are probably lacking in intrusions. They would be identical, in fact, to probable Stephanian — Autunian samples recently recovered from the western flank (Mougenot et al., 1985).

The abundance of high-grade rock-types recovered from the western flank of the Banks suggests that, as far as these facies are concerned, this zone corresponds to a vast tilted block with the highest grades in the West and the thicker part of the Palaeozoic cover in the East. Seismic reflection profiles across the region are rather in favour of such an hypothesis (Auxietre and Dunand, 1978). Another interesting point concerns certain granitic rocks recovered from the western flank; these deformed peraluminous granites (two-mica or muscovite-bearing) enable correlation of the Galicia Banks with the Ossa Morena Zone, and not with the Central Iberian Zone in Galicia (Capdevila, in preparation). Furthermore, there is a clear petrographic similarity between the western Galicia Banks, the Berlinguas archipelago and the Ossa Morena Zone which strongly supports a southerly origin for the Galicia Banks. The general lack of offshore Upper Palaeozoic similar to that found in the South Portuguese Zones rules out, in any case, the large-scale bend in the foldbelt between England and S. Portugal envisaged by some authors (e.g. Franke and Engel, 1986).

To summarize, the nature of the intrusive massifs and the type of metamorphism are two arguments against the correlation of the Galicia Banks with either Flemish Cap or the Western English Channel.

4.4 Conclusions Concerning the Avalon Spur

The basement of the Avalon Spur shows different ages on either side of the North Atlantic — it is approximately 800 Ma old on the American and Canadian margin and at least 2,000 Ma on the European side. These ages do not correspond to well-defined orogenic events but partly reflect the imprecision of dating methods on very ancient rocks. It may be that the Cadomian orogeny of the North Atlantic is linked to the collision of these two ancient basement blocks, but it is not certain whether present-day plate tectonic processes were operating in the distant past. A simple collision model, furthermore, does not easily explain the repetition of Avalonian structures discussed in this chapter.

— The different Cadomian ridges are more easily explained in terms of a model sometimes put forward for the Archaean (Windley, 1977) which is similar to that proposed by Strong et al. (1978). This suggests the existence of multiple crustal distensions, not necessarily synchronous, which would have produced either oceanic-type zones or basic volcanic belts according to the degree of spreading. If this model is adopted, only two areas of Cadomian basement really represent a degree of spreading beyond rifting; these are the Anglesey — Rosslare ridge (Belt n° 1) and the Mid-Channel Suture Zone (Belt n° 6). This would explain why these are the only structures which have led to the production of associated volcanic arcs (Belts 2 and 7) after a short period of subduction.

— The bend in Cadomian ridges suggested by basic rocks gives an idea of the geometry of this orogenic belt at the end of Proterozoic times. The acid igneous and sedimentary rocks which follow the framework of these markers also show the same general tectonic trend and are parallel to the orientation of the belt in northern Brittany and Iberia.

— The deep basement of the "Delta Zone" remains unknown. It never comes to outcrop and appears everywhere to be covered by Carboniferous marine facies. Only geometrical considerations suggest that it really belongs to the Avalon Spur.

— It is obvious that Pre-Cambrian ridges have controlled the location of Carboniferous intrusions and Palaeozoic basins, whether these are of Lower Palaeozoic age (as in Spain) or Upper Palaeozoic (as in Canada and the Celtic Sea). Some Carboniferous basins remain which cannot be placed in this context; in a later chapter, it is shown that such basins were probably controlled by Hercynian tectonics.

— It should be borne in mind that it is practically impossible to directly correlate the W. European margin with the Grand Banks of Newfoundland using either the nature of intrusions or type of metamorphism. From the sedimentary point of view, however, various comparisons are possible following the trace of certain Cadomian ridges; for example, this is demonstrated by the approximate border between the Old Red Continent and southerly marine facies. In any case, there is a considerable discontinuity in Hercynian metamorphism between Europe and Canada which implies either the existence of major crustal faults or a different arrangement of Hercynian tectono-metamorphic zones offshore and onshore.

— Although there is generally deeper erosion of the basement as one approaches the edge of the continental shelf, this could not have completely obliterated traces of a sedimentary belt that certain authors have taken as evidence for a continuous basin connecting S.W. England with southern Portugal. No bend of any sort can be invoked to link up these two separate domains.

— On the other hand, the major lithostratigraphic contrast which exists between the Armorican Massif and the Aquitaine Basin basement is fundamental in character and will be discussed in the next chapter. The abrupt interruption of Cadomian ridges in the area to the South of the Grand Banks of Newfoundland will also form part of this discussion.

Chapter 5

The Submerged Part of the Ligerian (Eo-Hercynian)-Acadian Mobile Belt

This segment of the mobile belt has only recently been discovered, partly because it is largely offshore but also because it is almost completely obliterated by the Hercynian orogeny.

Furthermore, the development of this belt was diachronous from East to West leading to much confusion in the literature. In common with the Caledonian-Appalachian belt, which is in part contemporaneous, the Ligerian-Acadian belt displays an easily mappable axial zone characterized by the presence of basic igneous rocks than can be detected by magnetism or gravimetry.

On the whole, this belt can be defined as the result of a collision that took place between Gondwana and the Avalon Spur during Siluro-Devonian times.

5.1 The Ligerian (Eo-Hercynian) Suture South of the Armorican Massif

Many plate-tectonic interpretations have been proposed for the Variscan foldbelt of Middle and Western Europe. But few of these models have taken into account the actual age of the main tectono-metamorphic event which occurred sometime between 540 and 360 Ma ago. This has led Autran (1978) to invoke a "Caledono-Variscan" event, whereas Cogné (1976) and Autran and Cogné (1980) have subsequently defined the concept of a "Ligerian Cordillera". The geological history of the Ligerian Domain should under no condition be confused with Hercynian folding sensu stricto, which was not related to a Cordilleran-type orogeny and was developed over a much wider area — the onset of Hercynian events post-dates the Ligerian orogeny by 50 Ma (Cogné and Lefort, 1985).

It is out of the question to discuss here all the evidence in favour of a Ligerian orogeny in Brittany; the detailed presentation of on-land geological and geophysical data lies outside the purpose of this study. Only the more salient features will be summarized.

From the stratigraphic point of view, recent on-land studies have shown the presence of Cambrian, Ordovician and Silurian fossils in low grade terrains involved in regional deformation — the oldest post-orogenic deposits are of Givetian-Frasnian (Mid-Upper Devonian) age.

The orthogneiss massifs along the South coast of Brittany (intrusive into older metasedimentary formations) all yield radiometric ages in the range 590–430 Ma. Those massifs emplaced during the Ordovician appear to be the most abundant. Major element analyses on these rocks suggest a calc-alkaline affinity (Jegouzo et al., 1986), but a review of granitic terrains of this age in Europe indicates many different origins. Certain geochemical studies show the presence of island arc volcanics, whereas others imply back-arc spreading. In places, there are even rocks of ophiolitic affinity. It would appear that the Ordovician alkali granite massifs of West Iberia are difficult to integrate into present ideas of the tectonic context.

During the Ligerian orogeny, various high pressure metamorphic facies were developed producing blueschists, eclogites and granulites that have been dated as crystallizing between 420 and 375 Ma (Gebauer et al. 1978; Peucat et al., 1979). This early event was followed by Barrovian-type regional metamorphism (Cogné, 1957; Audren and Le Métour, 1976) which culminated with the emplacement of anatectic granites in South Brittany around 375 Ma ago (Vidal, 1976). The range of eo-Hercynian ages in the Ligerian Domain covers a period from the end-Silurian to the Mid-Devonian; this corresponds precisely with a period of non-deposition in the region (Pruvost, 1949).

As far as deformation is concerned, there is an important phase of tangential tectonics involving the formation of crystalline nappe units (considered as an "Acadian" phase in the Vendée where two superimposed schistosities pre-date the Mid-Upper Devonian transgression) as well as the development of major transcurrent, sinistral shears (Audren,

1986). Some of these shear movements were responsible for the opening of basins over a time-span between the Ordovician and Devonian.

These details are useful to bear in mind since only a small part of the Ligerian belt is on-land and such features cannot be detected at sea. In addition, they also explain why so much emphasis is placed on problems concerning the suture — subduction is thought by many authors to explain the types of phenomena observed in the Ligerian Domain.

At first sight, the occurrence of blueschist eclogites and glaucophane schists in the centre of the Groix graben might suggest that the Ligerian suture passes through the island itself (Fig. 30). However, it can now be shown from detailed petrographic studies (Felix, 1972; Triboulet, 1974; Carpenter, 1976; Audren and Lefort, 1977), structural observations (Quinquis and Choukroune, 1981; Lefort et al., 1982; Cannat, 1983), geophysical surveys (Lefort and Segoufin, 1978) and radiometric dating (Carpenter and Civetta, 1976; Peucat and Cogné, 1977) that the Ligerian suture must be elsewhere. The offshore Ile de Groix Group forms a blueschist belt that has been affected by two main phases of deformation (the oldest around 400 Ma ago) and which probably corresponds to a klippe structurally above the Silurian "porphyroids" (Paris et al., 1985). This allochtonous unit, according to the studies cited above, was displaced from the South towards the North. The Ile de Groix Group, even if rather thin (Lefort and Segoufin, 1978), would appear not to form the sole of a thrust unit since there are rocks of oceanic affinity at greater depth beneath the exposed blueschists (Audrain and Lefort, 1986). Whatever the process invoked for the emplacement of this Group, there is a general consensus about the intensity of shearing deformation which appears to have been very great (Quinquis, 1980; Cobbold and Quinquis, 1980).

At sea, the band of blueschists is almost completely surrounded by "porphyroids"; these formations are considered as metasediments of volcano-sedimentary origin (Cogné, 1960; Chauvel et al., 1975). They are often associated with arkosic sandstones and locally develop a tuffaceous facies. "Porphyroids" can be followed in cores as far as the Vendée (Fig. 30), over a distance of 150 km (Audren and Lefort, 1977). Beneath the Tertiary cover, the "porphyroids" are characterized by absolutely flat magnetic anomalies (Lefort and Segoufin, 1978). It is thought that these prophyraceous tuffs are derived from the erosion of a nearby volcanic arc. On land, these terrains have been dated by radiometric and biostratigraphic methods as having an end-Silurian age (Paris et al., 1985). Furthermore, the porphyroids show a clear calc-alkaline tendancy (Peucat, work in progress).

The Ligerian belt incorporates high-grade metamorphic terrains that are hardly ever detected at sea — the migmatitic gneisses of the Port Navalo-Noirmoutier Group and high-grade micaschists of the Vilaine Estuary Group. These two Groups make up part of a high-temperature/medium-high pressure metamorphic belt which follows the southern coast of Brittany and which is parallel to the offshore Belle-Ile-en-Mer Group containing "porphyroids". The gneiss domes found near the coast are now considered (Audren, 1986) as the result of diapirism which was brought about by crustal thickening after the end of subduction. Taking account of the arrangement of metamorphic zones and the occurrence of a plutonic "hot-belt" to the North of the "porphyroids", it would appear that the Silurian volcanics in the Synclinorium of St Georges-sur-Loire (Maillet, 1977; Carpenter et al., 1982) correspond to back-arc spreading behind an active margin. This implies that the trace of oceanic crust subducted during the Ligerian orogeny must be sought further South of the active margin and, therefore, offshore.

In fact, there are several geophysical arguments which show that the Iberian-Armorican Arc discussed in the previous chapter is cut by a major crustal discontinuity. Three geophysical lineaments can be detected, more or less superimposed on each other, under the Mesozoic and Tertiary cover of the South Armorican margin; these lineaments are located at about 50 km from the coast and are broadly parallel to the coast line (Lefort, 1979).

The first lineament (see Fig. 34) is gravimetric in nature and is defined as the boundary between a western zone, where a "light gravimetric ridge" is seen separating two "heavy gravimetric basins" (characteristic of the Iberian-Armorican Arc; Lefort and Haworth, 1979), and an eastern zone, where anomalies show no preferred orientation. The imaginary line separating these two zones is known as the "G line".

The second lineament corresponds to an alignment of earthquake epicentres (Veinante and Santoire, 1980) — magnitudes are mostly comprised between 3.0 and 4.0 (Fig. 34b). The western part of this alignment runs slightly South of the "G line", possibly because of a lack of precision in locating epicentres far offshore. A study of the focal plane mechanisms demonstrates that there is no single fault plane but instead a large rupture zone dipping 70° towards the N.E. and extending to 40 km depth

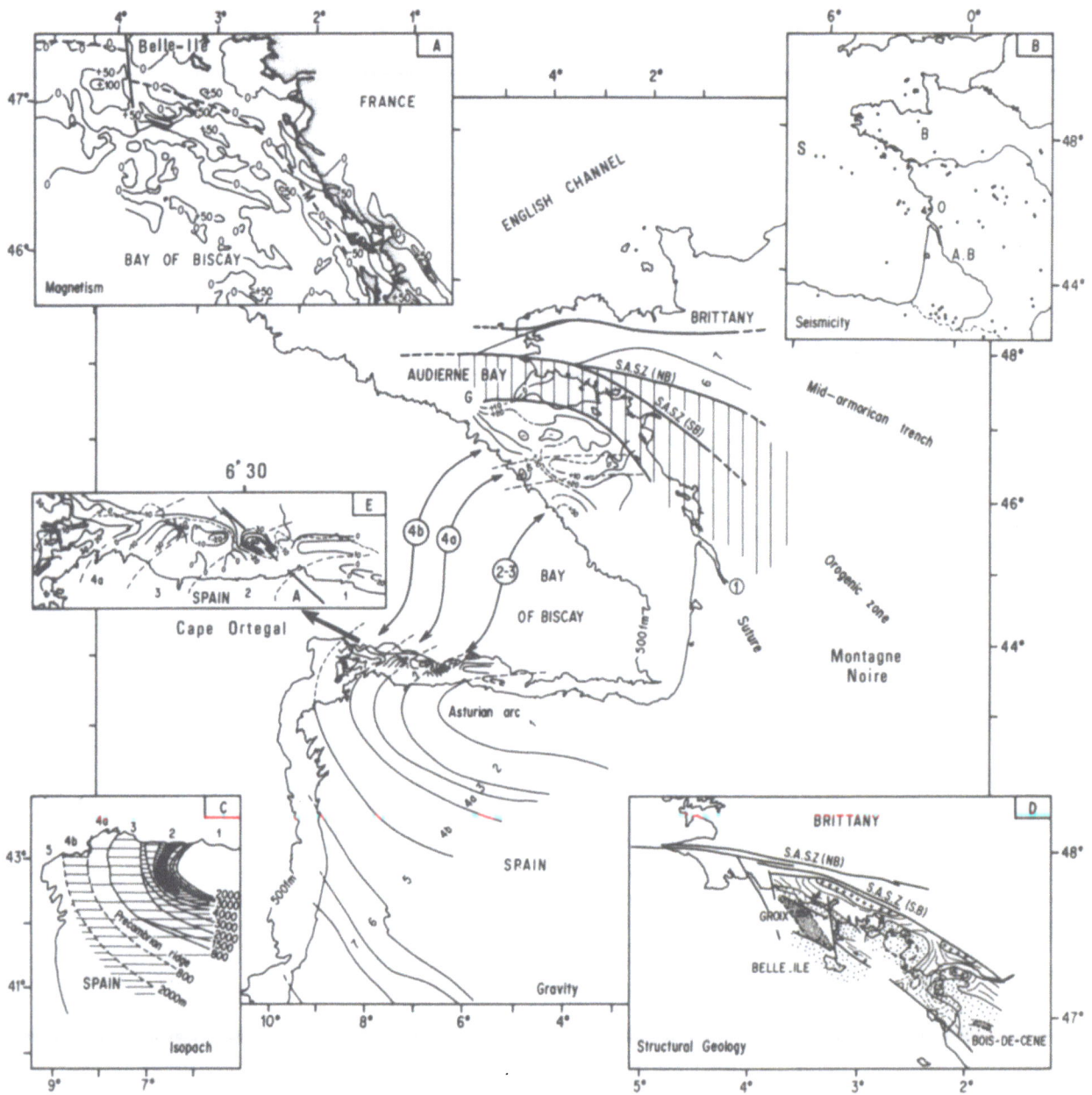

Fig. 34. Summary of geological and geophysical data concerning basement across the Bay of Biscay. Main map — gravimetry (free-air anomalies). A-E: Inset maps discussed in text. A: vertical gradient reduced to the actual pole; B: seismicity for 1977; C: isopachs for the Lower Palaeozoic in N.W. Spain; D: Ile de Groix Group blueschists (heavy stipple); E: free-air gravimetric data corrected for edge effect (N.W. Spain). (After Lefort 1979)

(Delahaye, 1976). Figure 34b shows that seismicity for 1977 is in excellent agreement with the "G-line" (L.D.G., 1977).

The third lineament is visible on aeromagnetic maps (Le Mouel and Le Borgne, 1971) and sea-towed magnetometer surveys (Lefort and Segoufin, 1978). It corresponds to an alignment of magnetic bodies with high susceptibility, whose location has been

precisely defined using the calculation of vertical gradients reduced to the actual pole (Fig. 34a) (Horn et al., 1974). To the West of 4°45′W, this body can be traced using the map of magnetic anomalies on the Atlantic Continental shelf (Segoufin, 1975).

Attempts to model this anomaly (de Poulpiquet, 1985) suggest that the magnetic body corresponds to a northward-dipping dyke over its entire length, except near Belle-Ile-en-Mer, where the same dyke is covered with a cupola of even higher magnetic susceptibility. The dips calculated from magnetism are everywhere equal to those measured from natural seismicity. These three geophysical discontinuities are currently regarded as representing the trace of the South Armorican Ocean suture (Lefort, 1979). This suture separates a northerly belt, greatly affected by deformation and metamorphism, from a southerly domain where the different zones of the Iberian-Armorican Arc are still recognizable.

The ophiolitic klippes of NW Spain, which show metamorphic ages near 460 Ma (Bernard-Griffiths et al., 1985) and which nowadays overlie the sedimentary formations of the Iberian-Armorican Arc, are derived from the NW (Iglesias et al., 1983) and represent fragments of obducted oceanic crust from the South Armorican Ocean.

The suture can be extended towards the Aquitaine Basin owing to an alignment of positive gravimetric and magnetic anomalies which prolong the basic intrusions modelled off S. Armorica. Further SE, modelling of the magnetic structures suggests that the suture becomes cryptic (de Poulpiquet, 1985) and that re-activation occurred at a later date. The 130–310°-trending Carboniferous graben detected beneath the Aquitaine basin (Fig. 32) appears to follow, more or less, the remnants of this fault-zone. In addition, it should be noted that further South of this region there are none of the volcanic, sedimentary or volcano-sedimentary formations so typical of the Lower-Mid Palaeozoic of Brittany; this supports the hypothesis of a major discontinuity in patterns of sedimentation and, hence, suggests the existence of an Ocean. The possibility of an eo-Devonian phase of deformation has not yet been confirmed by bore-hole studies in the Aquitaine Basin (Paris and Robardet, 1985), but major tectonic and metamorphic activity has been recorded at this time just to the North of the suture (Ters, 1979; Ters and Chantraine, 1980). All this evidence suggests that the inactive margin of the South Armorican Ocean lies to the SW of the geophysical suture, a suggestion that has already been made for regions further East (Autran, 1978).

To the West of Brittany, in the region of Shamrock Canyon and the Meriadzec Terrace, there is a strong, elongate magnetic anomaly which certainly represents the western extremity of the South Armorican suture (Lefort, 1983) (Fig. 36). Even though this anomaly is parallel with the contact between present-day oceanic and continental crust, it appears not to be due to magnetic edge effects since such phenomena are practically non-existent to the West of Brittany. This anomaly, situated 90 km North of the oceanic crust, appears to be interrupted by the northerly extension of the Porto – Badajoz – Cordoba transform fault which operated during the Palaeozoic (see below for further discussion). Magnetic modelling shows that this structure is about 4,000 m deep; it is thus clearly part of the intracrustal basement (de Poulpiquet, 1985) and cannot be correlated with certain Lower Cretaceous tholeiites dredged at 3,000 m water-depth in the Shamrock Canyon (Pastouret and Maury, 1982).

In this way, the geophysical anomaly which may represent the South Armorican suture can be followed over nearly 900 km; it is abruptly interrupted on the Meriadzec Terrace by three major N-S-trending fractures picked up by seismic reflection (Guennoc, 1978).

The difficulty of integrating the Baie d'Audierne basic-ultrabasic complex in this geodynamic model has already been discussed above; taking account of the 530 Ma ages given by Peucat (1983), it could equally well represent the initial stages of opening of the South Armorican Ocean as the remnants of back-arc spreading related to the end of subduction of the Cadomian Channel Ocean (Lefort, 1975; Auvray and Lefort, 1980). Nevertheless, since the submerged prolongation of the Baie d'Audierne "scar" (recognized from magnetism) seems to be cut off by the South Armorican suture (Figs. 19 and 22), there are perhaps some grounds for interpreting this complex as being linked to the late Proterozoic history of the English Channel. The Lower Palaeozoic sedimentary and volcano-sedimentary formations which lie above the basic-ultrabasic complex could, on the other hand, belong to the South Armorican Domain. In fact, there are strong analogies between the Palaeozoic offshore Baie d'Audierne succession and the St Georges-sur-Loire Group. The volcano-sedimentary formation off the Baie d'Audierne could, according to this criterion, represent an ancient zone of distension to the North of the subducted South Armorican Ocean plate.

5.2 The Porto–Badajoz–Cordoba Transform Fault Zone

At present, the residual relief of the Meriadzek and Trevelyan outliers is associated with basement horsts "floating" upon thinned crust (Montadert et al., 1979). These basement blocks are controlled by fractures which are generally E-W-oriented, probably representing extensions of the South Armorican Shear Zone towards the West, and N 40°-trending faults, detected by seismic reflection, in continuity with the northerly part of the Porto — Badajoz — Cordoba fault zone (Montadert et al., 1979; Lefort, 1983). The continuity between the Little Sole Bank (Guennoc, 1978) west of Brittany and the offshore extension of the Portuguese fault, off N.W. Spain (Fig. 33) (Lefort et al., 1981), is assured by a distensive fault zone that separates the Galicia Banks from the continent (Auxietre and Dunand, 1978).

It is known that the Porto — Badajoz — Cordoba Fault (Lefort and Ribeiro, 1980), which can be followed over a distance of 600 km, has played an important role in the Palaeozoic history of Iberia. At the surface, it is characterized by a 1–5 km-wide-zone of blastomylonites derived from the deformation of Pre-Cambrian basement and peralkaline intrusives of Upper Ordovician age. In addition, gravimetry has shown that this fault zone traverses the entire crust and separates two domains that were also faunistically quite distinct during the Lower Palaeozoic (Paris and Robardet, 1977). Relations between the proposed development of a South Armorican Ocean and the Porto — Badajoz Fault have been studied in great detail because of the geodynamic significance of such a fault (Lefort and Ribeiro, 1980).

According to the orientation of the "Sardinian" fold axes in adjoining areas, this fault should have produced a dextral displacement during the initial opening of the South Armorican Ocean. The Upper Ordovician peralkaline and alkaline volcanics, with their associated deformation, suggest there was a period of relaxation before the reversal in shear direction. During the Devonian, closure of this Ocean was contemporaneous with proven sinistral movements and appears to have led to crustal thickening in South Brittany and the obduction of ophiolitic klippes onto N.W. Spain. The ophiolitic klippes were not deformed until Carboniferous times. It was also after Devonian times that the original Iberian-Armorican Arc (the Easternmost Cadomian ridge n° 12, Fig. 18) began to tighten (Lefort, 1979). This final point is now confirmed by palaeomagnetic data (Perroud, 1980). The history of development of this fault is summarized in Fig. 35. In view of the stratigraphic similarities that exist between the Lower Ordovician of Crozon (West Brittany) and Bouçaco (North Portugal), it is important to stress that, according to the model in Fig. 35, opening of the South Armorican Ocean could only have occurred after deposition of the transgressive Grès Armoricain Formation. Cylindrical models have been proposed to explain the development of the Iberian-Armorican Arc during

Palaeozoic times — these will not be discussed here since none of them take into account the major structural discontinuities detected by offshore geology and geophysics as well as studies of the concealed basement beneath the Aquitaine Basin. It is possible that the Little Sole — Porto — Badajoz — Cordoba lineament was not the only fault involved in the development of the transform zone described above. Numerous parallel fractures, nowadays considered as typically Hercynian, were probably already operating as transcurrent faults during the Ordovician. Furthermore, it is probable that such transform faulting was accompanied by local distension and the generation of true oceanic crust in some places.

The southern extremity of the Porto — Badajoz — Cordoba fault terminates beneath Recent cover near the Beja and Aracena zones in SW Iberia. These zones will be discussed in terms of the "Collector anomaly" shown in Fig. 36.

5.3 The "Collector Anomaly" Suture and its Continuation Across the North Atlantic

A synthesis of smoothed magnetic data from a variety of sources has enabled the drawing up of a map showing the major magnetic discontinuities across the North Atlantic. This includes magnetic data from Portugal (G, Fig. 36) (Serviços Geologicos de Portugal, 1979), estimates of the vertical gradient reduced to the actual pole for the Channel and Bay of Biscay (Gérard, 1975; Horn et al., 1974) (A and C, Fig. 36), anomalies detected by aeromagnetic surveys (E, Fig. 36) (Sibuet, 1972) or towed magnetometer surveys in Europe (B, D and F, Fig. 36) (Segoufin, 1975; Guennoc, 1978; Groupe Galice, 1976) and compilations of data from Canada (H, Fig. 36) (Haworth and McIntyre, 1975) and the United States (I, Fig. 36) (Kane et al., 1972).

A comparison of Fig. 36 with the gravimetric compilation given by Lefort (1983) (see Fig. 37) shows that there are two generations of positive gravity anomalies which are nearly always superimposed on the magnetic highs. The gravimetric compilation is based on free air data (A, C, D and H, Fig. 37) (Sibuet, 1972; Groupe Galice, 1979; Roberts, 1970), Bouguer anomalies (F, G, I, J and K, Fig. 37) (Instituto geographico y cadastral, 1972; Instituto geographico e Cadastral, 1960; Haworth and McIntyre, 1975 and 1977; Kane et al., 1972), measurements corrected for edge effects (Sibuet, 1972) and non-published results from the Oil industry. The first generation of gravity highs is composed of

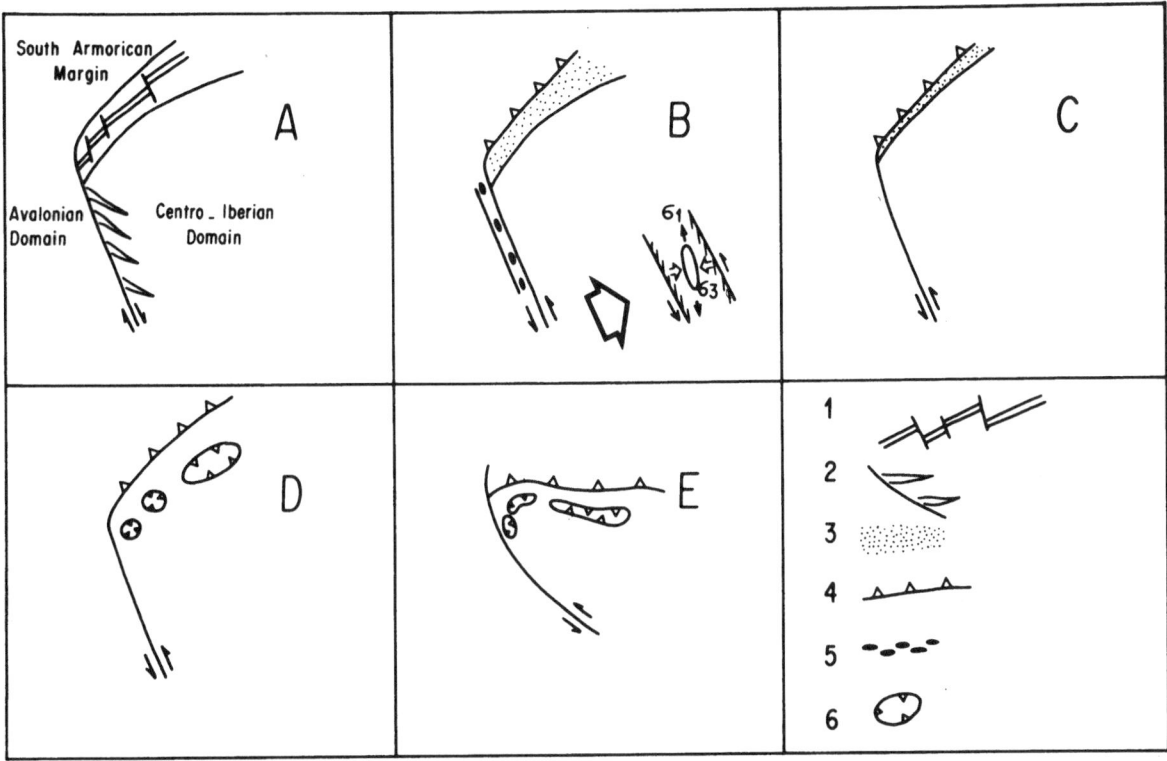

Fig. 35. Sketch diagrams showing development of South Armorican Ocean in relation to Porto-Badajoz-Cordoba Fault. A: Lower-Mid Ordovician, B: Upper Ordovician, C: Silurian, D: Devonian, E: Carboniferous. 1: rift-valley; 2: "Sardinian" phase fold-axes; 3: oceanic crust; 4: subduction; 5: hyper-alkaline igneous suite; 6: klippes. (After Lefort and Ribeiro 1980)

N-S or ENE-WSW trending anomalies which all correspond to Pre-Cambrian ridges on the Avalon Spur. The second generation forms two well-defined arcs — one is to the South of Brittany and the other is between Portugal and Massachussets. The first arc is particularly well picked out by magnetism and corresponds to the South Armorican Suture discussed in the previous section. The second corresponds to the "Collector Anomaly" in the sense defined by Haworth and McIntyre (1977). In addition, on land gravimetry reveals the presence of a strong horizontal gradient (stippled in Fig. 37) which is associated with the Porto — Badajoz — Cordoba fault in Spain and Portugal.

Finally, a third map shows the areas where magnetic and gravimetric anomalies are clearly superimposed (Lefort, 1983) (Fig. 38); some regional geological information is included in the inset maps. It is obvious from this map that the two arcuate belts (shown in black) are linked by the Little Sole — Porto — Badajoz — Cordoba fault and that these belts post-date the Pre-Cambrian ridges (shown shaded). To the South of Brittany, the concealed belt of basic

rocks should correspond to fragments of ocean-floor crust and not magmatic arc or back-arc material. This is because the material sampled on land and at sea, as discussed in Section 5.1, belongs to the active margin. In Southern Portugal and on the New-foundland Grand Banks, on the other hand, the "Collector Anomaly" may not necessarily be composed of the same ophiolitic material — especially since the rock-types observed on land, between the Ossa Morena and South Portuguese Zones, would suggest otherwise (cf. "ophiolitic" Beja Complex; Andrade, 1979).

The Beja Complex (op. cit.) (inset map B, Fig. 38) is composed of gabbros, serpentinites, dolerites and basalts which were emplaced before Frasnian times. This belt continues into the Aracena massif of S.W. Spain (inset map A, Fig. 38) which is made up of foliated amphibolites having the composition of abyssal tholeiites (Bard, 1977; Bard and Moine, 1977). Near Aracena, the basic rocks are interbedded with acid igneous rocks and basic tuffs of pre-Silurian age. These two zones are aligned and form a belt which has been interpreted as a geosuture

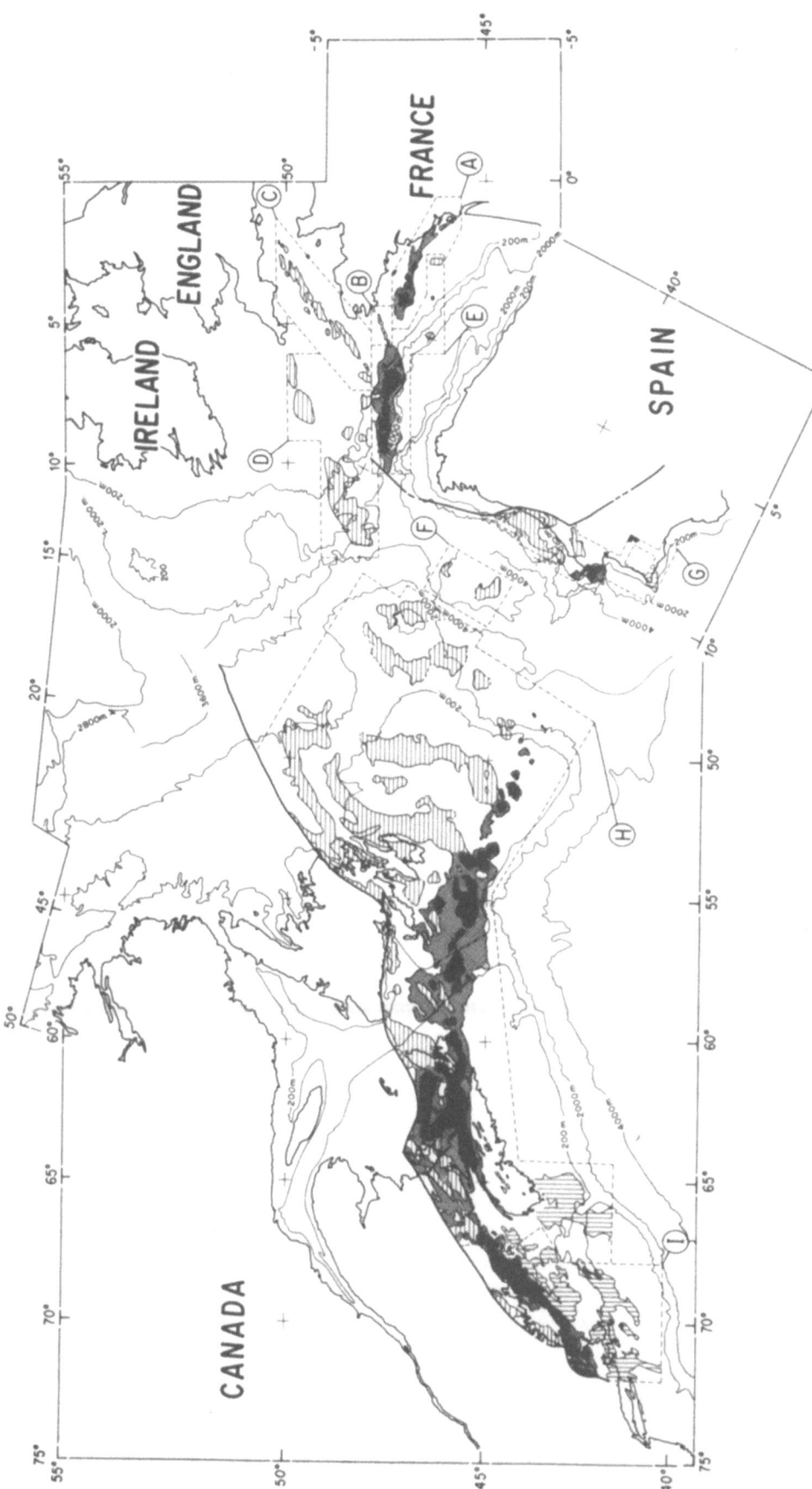

Fig. 36. Compilation of magnetic data from both sides of the northern North Atlantic. Circled letters refer to sources cited in text. High amplitudes attributed to the Ligerian

Acadian "suture" are shown in black. (After Lefort 1983; reproduced with kind permission of the Geological Society of America)

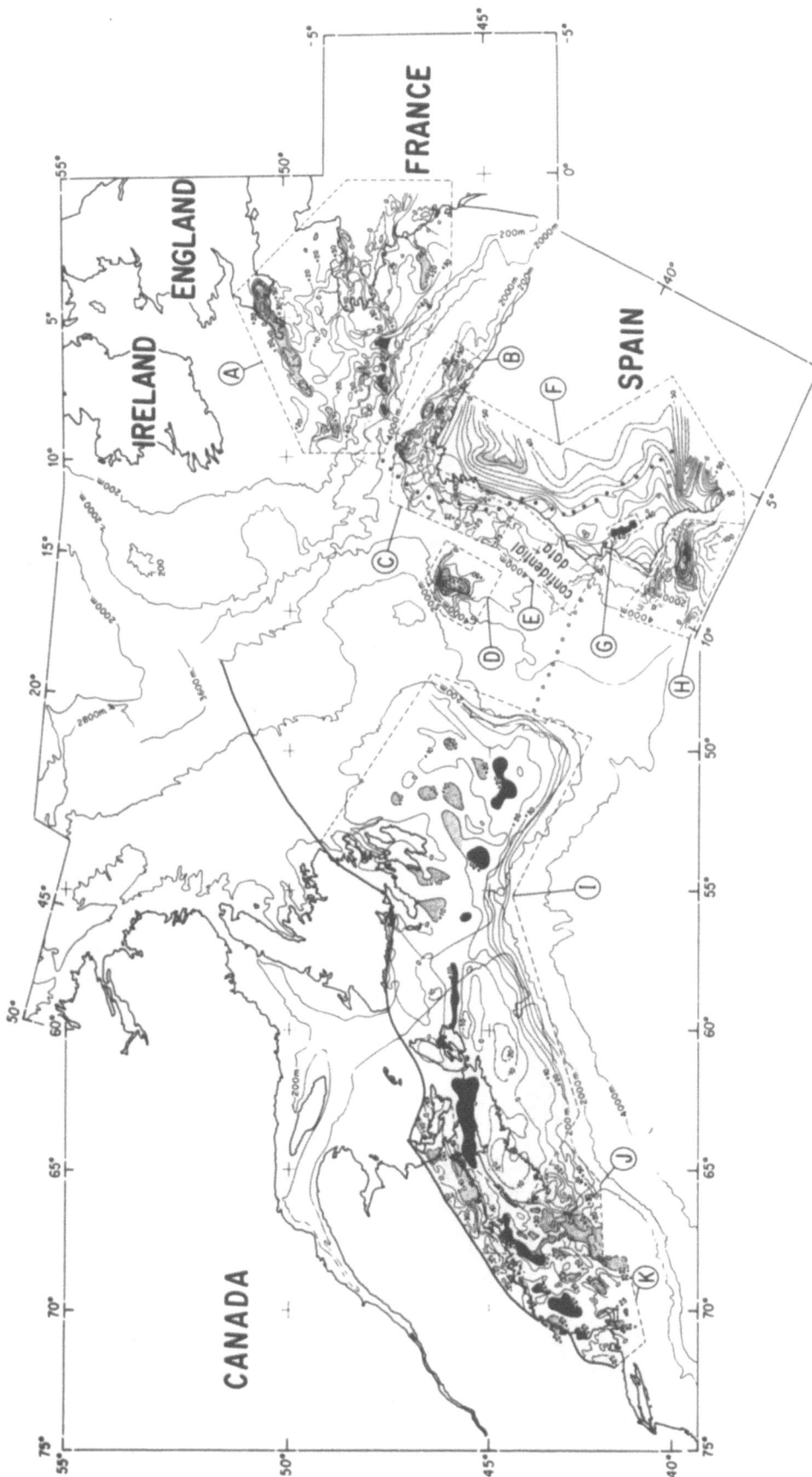

Fig. 37. Compilation of gravimetric data from both sides of the northern North Atlantic. Otherwise, same caption and reference as for Fig. 36

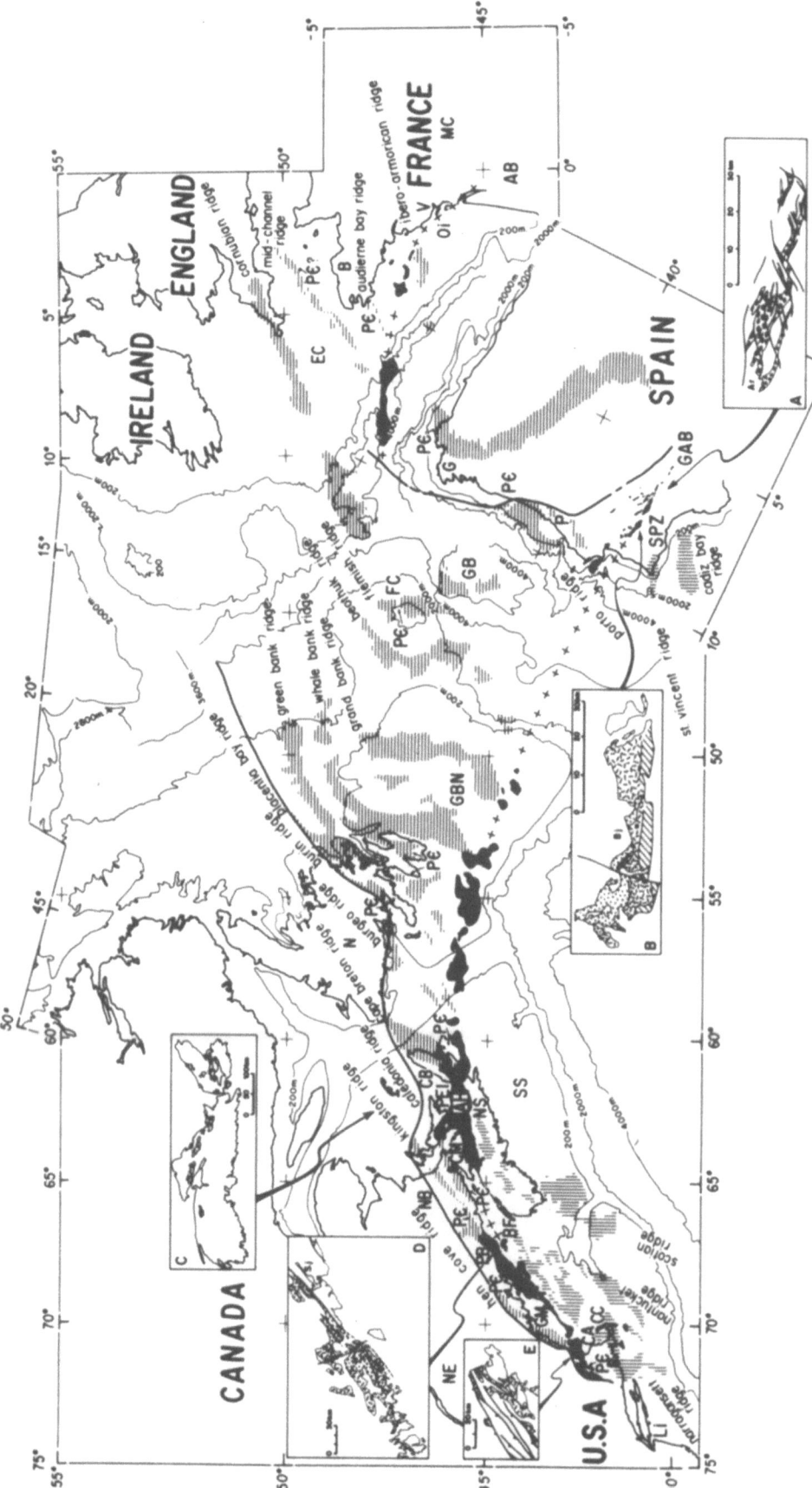

Fig. 38. Compilation of basic bodies making up or bordering the Ligerian-Acadian suture. (After Lefort 1983; reproduced with kind permission of the Geological Society of America). AB: Aquitaine Basin; AH: Antigonish Highlands; AR: Aracena; B: Britanny; BF: Bay of Fundy; BJ: Beja; CA: Cape Ann; CB: Cape Breton; CC: Cape Cod; CM: Cobequid Mountains; EC: English Channel; FC: Flemish Cap; G: Galicia; GAB: Guadalquivir Basin; GB: Galicia Banks; GBN: Grand Banks of Newfoundland; GM: Gulf of Maine; MC: Massif Central; N: Newfoundland; NB: New Brunswick; NS: Nova Scotia; OI: Oleron Island; P: Portugal; PB: Passamaquoddy Bay; PEI: Prince Edward Island; SJ: Nova Scotia margin; SPZ: South Portuguese Zone; V: Vendée. **Ornamentation,** horizontal: Cadomian basement ridges; vertical: basement ridges recognized in Gondwana. **Inset map A,** Aracena zone; open circles: basaltic tuffs; V: amphibolites. **Inset map B,** Beja zone; random dashes: gabbros; dotted: basalts; diagonal ruling: serpentinites. **Inset map C,** Nova Scotia zone; V: volcanic rocks. **Inset map D,** Passamaquoddy Bay zone; Random dashes: gabbros; dotted: volcanic rocks, **Inset map E,** Cape Ann zone; random dashes: gabbros. Black: Ligerian-Acadian intrusions; crosses extension of "Collector Anomaly"; PC: Pre-Cambrian outcrops

(Tamain, 1978). However, neither of these two massifs appears to be composed of ophiolites sensu stricto — they are more likely the result of back-arc spreading activity. Furthermore, the formations are almost autochtonous and lack any ultramafic tectonites at their base.

Gravimetric studies conducted on the regional scale suggest that the South Portuguese Zone corresponds to a zone of thinned crust (Gaibar-Puertas, 1976); even more interesting are the results obtained from seismic refraction (Mueller et al., 1973; Prodhel et al., 1976). According to this work, the South Portuguese Zone crust is characterized by the presence of low velocity channels, at least one of which may represent a layer of laminated rocks. The laminated layer appears to separate tectonic units characterized by different seismic velocities. Figure 39 shows a possible interpretation of the seismic log in terms of tangential tectonics. The existence of a 6.8 km/s layer above the Moho may be interpreted as revealing Pre-Cambrian crustal material that was, in part, intruded during a previous spreading event; velocities of 6 km/s could correspond to the interface between the Pre-Cambrian and the mylonitized Lower Palaeozoic. Terrains with 6.5 km/s could be correlated with metamorphosed and intruded Lower Palaeozoic formations. The 5.5 km/s layer may be correspond to a zone of "decollement" or slightly metamorphosed Lower Palaeozoic. As for the 6.3 km/s velocities near the surface, these may represent allochthonous cover units of Upper Palaeozoic age — i.e. thin-skin type tectonics (Ribeiro et al. 1983). Such high velocities suggest that basic material was involved in the thrust movements.

Southern Portugal has often been considered as the site of a possible plate collision (Bard, 1971; Carvalho, 1972). This has led to the development of various models (Vegas and Munoz, 1977; Bard et al., 1980; Lefort, 1983). However, these different authors are not in agreement concerning the precise location of the suture. Owing to seismic refraction studies, it is now possible to combine these different models into a unique solution (Caetano, 1983).

The South Portuguese Zone, according to this model, is covered with decollement nappes derived from oceanic or thin continental crust. Oceanic crust was consumed by subduction towards the N.E. underneath the continental crust of the Beja-Aracena region. This happened in such a way as to produce a cryptic suture with the preservation of volcanic-arc (or back-arc basin) fragments at the surface rather than remnants of ocean floor material (Munha, 1979). The Ossa Morena Zone would represent the northern continent, whereas the deep crust of the South Portuguese Zone corresponds to a southern continental block thinning towards the North. The magnetic map (Panchon Ruiz, 1978) clearly shows the boundary between the two plates, due to the presence of circular positive anomalies along the axis of the Beja-Aracena massifs. The magnetic anomalies show that there are many different types of basic material along this axis. If this is also the case on the Grand Banks of Newfoundland, the actual suture zone should lie rather to the South of the alignment of basic bodies. Since no major transform faulting is envisaged between the geophysical markers in Canada and Portugal, Ligerian-Acadian subduction should also have been towards the North beneath the Grand Banks. Otherwise, Pool (1976) had already proposed that the "Collector Anomaly", then incompletely

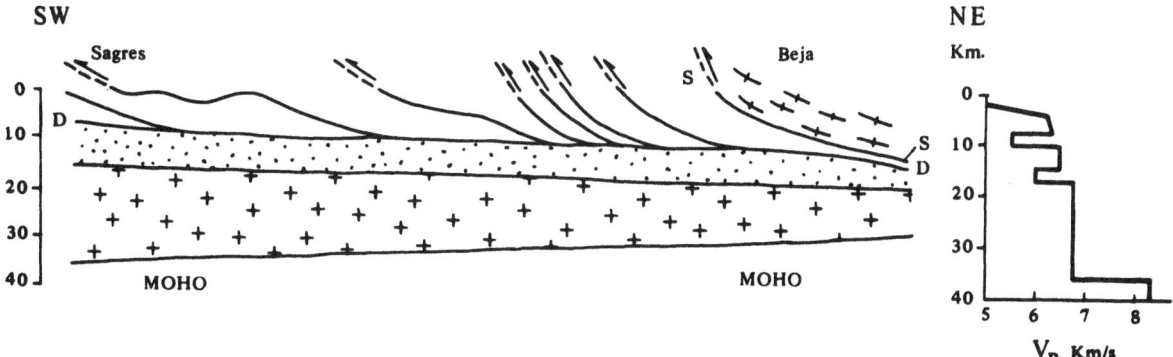

Fig. 39. Tectonic interpretation of a seismic velocity record (Prodhel et al. 1976) established in the southern South Portuguese Zone. No ornament: Upper Palaeozoic with basic intrusions; dotted: Lower Palaeozoic; vertical crosses: thinned Pre-Cambrian basement; diagonal crosses: Beja Massif. (After Ribeiro et al. 1983)

mapped, should correspond to a suture zone. It is important to remember that, to the North of this suture, the Acadian phase is apparent on the Grand Banks of Newfoundland and that, in the Ossa Morena and (southern) Central Iberian Zones, there is a Mid-Devonian sedimentary break (Julivert et al., 1983). This break is contemporaneous with low pressure regional metamorphism and Givetian deformation (Chacon et al., 1983) with a strong sinistral transcurrent component. The mainly basic volcanism of the Ordo-Silurian succession in the Almaden Synclinal of Spain reflects the great crustal mobility of this region (Saupé, 1973). Furthermore, the thick accumulation of tholeiitic basalts and dolerites during the Mid-Devonian in this same region demonstrates an important return of volcanic activity during the Ligerian-Acadian phase (Julivert et al. 1983).

The South Portuguese Zone is characterized by a series of curved thrust faults, concave towards the NW, which affect equally the low grade Upper Devonian and Carboniferous successions. As far as gravimetry is concerned, these thrusts are characterized by important horizontal gradients, regular transverse inflexions in the contours or a sudden re-orientation of isogams. The most important and continuous thrust is that which bounds the Ossa Morena Zone to the S.E. and which affects the Beja gabbro-dioritic complex. This thrust is well marked by a positive gravimetric anomaly that can be followed northward under Mio-Pliocene cover and also beneath the crystalline basement complex West of Evora. Other thrusts (Fig. 33) (Lefort et al., 1983) are easily detectable by gravimetry, but do not cut the cover formations of Upper Devonian age – this suggests that they are probably Acadian thrusts with no Carboniferous re-activation as in Canada.

The rocks which produce the "Collector Anomaly" do not crop out in Nova Scotia, unless one counts possible equivalents in the Cobequid Mountains and the Antigonish Highlands (Keppie, 1979; Keppie and Dostal, 1979) (inset map C, Fig. 38). Towards the West, this belt of anomalies disappears under the Bay of Fundy, probably because of the thick Triassic basin-fill (Belt, 1968). Keppie (1977) considers that this graben is the result of the reactivation of an ancient suture.

Basic and ultrabasic rocks re-appear near Passamaquoddy Bay in the United States, where the Triassic decreases in thickness. These rocks are responsible for a strong magnetic and gravimetric anomaly (Figs. 36 and 37) which extends offshore at a small distance from the coast before reaching Cape Ann near Boston (Fig. 39, map inset D) (Kane et al.,

1972). In the centre of Nova Scotia, basalts and rhyolites (ranging in age from Ordovician through Silurian to Devonian) are all characterized by features which suggest emplacement in a tensional tectonic setting (Keppie and Dostal, 1979). This is similar in part to what is known of geochemical trends in Spain and Portugal and, again, the suture would appear to be cryptic rather than manifest. Other analyses from the southern part of Nova Scotia (Sarkar, 1978), in the immediate vicinity of the Bay of Fundy, also suggest a tensional regime. In northern Maine, where the continuation of the "Collector Anomaly" runs along the coast, the Bay of Maine Igneous Complex crops out (Chapman, 1962). This complex is composed of andesites and rhyolites, ranging in age from the Lower Silurian to the Lower Devonian, as well as Devonian gabbros (Gates, 1969). Osberg (1978) considers this material as belonging to a typical island arc which is thought to have run parallel to the plate suture. At one locality, the study of basic and ultrabasic rocks has even led to the suggestion of an oceanic affinity (Gaudette, 1980).

A very similar igneous suite is responsible for the positive and continuous geophysical anomalies that extend NE onto the southern shore of New Brunswick in Canada. Here, the Silurian volcanic arc is composed of the Mascarene, Long Read and Jones Creek Formations (Ruitenberg et al., 1977) – again, the arc is thought to lie near a plate boundary. This volcanic group was intruded at a late stage by rare basic and ultrabasic bodies of Devonian age. Since this volcanic arc continues northward beyond the bend which joins the Bay of Maine Igneous Complex to the "Collector Anomaly", it is clear that the arc was offset later by dextral Hercynian shears that follow the seaboard of Canada and the U.S.A. (Ruitenberg and McCutcheon, 1982). These include the Belle-Isle, Kennebecasis and Clover Faults which lie in the New Brunswick Shear Zone (SBSZ, Fig. 40). Towards the South, these faults can be traced by seismic reflection under the Bay of Fundy and Bay of Maine (Hutchinson et al., 1985). The geophysical anomalies which serve as a marker (Figs. 36 and 37) reappear on land just North of Cape Ann near Boston, where they overlap with the Newburry volcanics – rocks known to be petrographically equivalent to the Bay of Maine Igneous Complex (Shride, 1976). The Newburry volcanics contain abundant andesites and have been dated also as Lower Silurian – Lower Devonian. Thus, there is perfect agresment between geophysical results and petrological considerations (Fig. 40).

In this manner, the main question in eastern New

England is whether, as in Portugal and Nova Scotia, the suture is concealed beneath its associated magmatic arc/marginal basin or whether it is expressed somewhere else. The seismic reflection profiles in the Bay of Maine and on the Long Island Shelf should enable an answer to this question (Hutchinson et al., 1985 and 1987). Furthermore, seismic interpretations in this area benefit from the detailed on-land geological data which are available. However, the evolution of SE and NE New England must be discussed separately.

The Nashoba Zone in SE New England (NFZ in Fig. 40) has always been considered as a limit of fundamental importance, even possibly as a plate boundary (Osberg, 1978; Bradley, 1982). This is partly because it is formed of Lower Palaeozoic andesites and basalts, but also because it is bounded by major thrusts such as the Clinton-Newburry, Bloody Bluff, Lake Char and Honey Hill Thrusts. For Robinson and Hall (1979), the tectonic transport is broadly towards the East during Mid-Upper Devonian times. Most other authors are in agreement with this, although the initial process leading to thrusting is still debated.

In NE New England, arguments involving subduction have been partially called into question since the geochemistry of the Siluro-Devonian batholith which intrudes this area is not typical of convergent plate margin igneous rocks (Wones, 1979). This corroborates the geophysical argument (Lefort, 1983) by which the greater part of the volcanic arc is situated offshore (Fig. 38); the arc disappears progressively on land beneath thrust units bordering the Nashoba Zone (Skehan, 1983).

Data concerning the Avalonian basement to the East of the Nashoba Zone certainly show a consistent pattern — stratigraphic, geochronological and petrographic comparisons made across the Zone demonstrate that the Avalon block could not have been part of the North American plate before the Lower Devonian (Barosh and Hermes, 1981). According to palaeomagnetic studies, this region was part of an isolated block which only joined Laurentia at about the time of the Acadian orogeny (Spariosu and Kent, 1983). Finally, Gaudette (1980) considers (from data collected in the state of Maine) that there could not have been a collision of Avalonian and Laurentian basement before 410 Ma ago. However, the repeated re-activation of faults bordering the Nashoba Zone has obscured the original process of accretion.

Offshore data are helpful in suggesting a unique solution for SE and NE New England together; this yields a good compatibility between geophysics and geological observations. The main suture was first discovered by seismic reflection (Hutchinson et al., 1985) off southern New England (see Fig. 41). In this area, the Block Island Fault dips West, in complete agreement with the direction of subduction established in Portugal but also compatible with the direction of thrusting in the Nashoba Zone. The Border Fault, situated immediately to the West of the Block Island Fault, is antithetic with respect to this suture and is probably of the same age.

An entirely analogous situation is to be found in the Gulf of Maine (Hutchinson et al., 1987), where the westward-dipping Nauset Fault is synthetic with respect to the deep East-facing thrusts running onshore (Unger et al., 1987). The associated Fundy Fault (Fig. 40) probably represents a contemporaneous antithetic thrust (see Chapter Eight, Fig. 65). This phenomenon is also particularly well seen in the SW of the Iberian Peninsula, where northward-directed thrusts bordering the Ossa Morena Zone are antithetic (and cogenetic) with respect to the Ligerian-Acadian suture (Chacon et al., 1983). The wedges of crust thrown upwards on each side of the "Collector Anomaly" are thus remarkably similar on both sides of the North Atlantic. In the Bay of Fundy, the form of the wedge has been mapped in plan view by seismic and magnetic methods; by contrast, off Portugal, W of Lisbon, the wedge can be mapped on partly morphological criteria (a basement spur projecting seaward opposite the Ossa Morena Zone) (Fig. 33).

The Block Island and Nauset Faults are directed eastward, in contrast to major units of the Southern Appalachian foldbelt which form westward-facing structures of Hercynian age. Since the Block Island and Nauset Faults dip westward, and because the South Portuguese suture also dips away from Gondwana, these thrusts must be considered as part of the Acadian suture. This is supported by their close association with a Siluro-Devonian mafic igneous complex. Their steep dip, associated antithetic faults and considerable extent in depth (as well as later reactivation) are all features that have been described several times for Ligerian-Acadian sutures in Europe. The thin skin tectonics as described onshore in the Carboniferous rocks of the U.S.A. are thus in complete contrast with the pre-Hercynian tectonic style.

It is interesting to note that there are two styles of deformation in the Ligerian-Acadian mobile belt so defined. Apart from the local problem concerning the displacement of the Ile de Groix blueschists, thrusts are always synthetic with respect to the presumed subduction zone off South Armorica. This is also the case in Spain and Portugal – except for the Ossa Morena Zone, which is an upfaulted tectonic wedge. This is in opposition to what is found in the U.S.A. and Canada (excepting the antithetic fault in the Bay of Maine), where there is a symmetry of overfold structures (facing direction) on the regional scale. For more clarity, these differences in facing direction are indicated on Fig. 40. In North America, it is in fact possible to distinguish those structures which are synthetic with respect to the "Collector Anomaly" subduction from those structures which are further away, showing antithetic relations. These westward-directed structures always occur in a zone on the northern and western sides of the Appalachian foldbelt. On the regional scale, this can be explained as if the synthetic thrusts were directly linked to Ligerian-Acadian subduction, whereas the westward-directed (antithetic) tectonics (Keppie et al., 1982) would correspond merely to the Acadian reactivation of previously existing Taconic structures. In fact, this only serves to emphasize the convergence of an Acadian belt having Caledonian affinities with an Acadian belt belonging to the Variscides – this occurs in the Gulf of St Lawrence area.

The recognition of a Ligerian-Acadian offshore suture does not conflict with the obvious discontinuity across the Nashoba Zone, but suggests that the latter is probably not the major suture. It is possible that the Nashoba Zone, with its bordering thrusts, has resulted from the closure of a back-arc basin which was partly superimposed onto the Bay of Maine volcanic arc during a compressive phase. Back-arc spreading in this zone could possibly be represented by the Merrimack Trough. It is also

possible to imagine that accretion of SE New England pre-dated the main collision with Gondwana – this is discussed in a later section.

The development of the Acadian orogeny culminates with major granite plutonism, especially on the western side of the Atlantic. In order to preserve clarity, only offshore evidence of plutonic activity will be presented. Amongst other rock-types, the pre-Devonian igneous suite contains rare riebeckite granites which are of particular interest due to the fact that they yield (apart from the Quincy Granite; Hermes and Zartman, 1985) Ordovician ages. It can be seen from Fig. 40 that riebeckite granites (Sorensen, 1974) on the southern part of the Avalon Spur are clearly aligned parallel to the Ligerian-Acadian suture. Since such granites are normally typical of the pre-rifting stage (Hermes et al., 1978), this suggests that here we have evidence of the initial tectonic regime leading to opening of the Theic Ocean. Thus, it is logical to find basic rocks fairly close which mark the suture. In this way, the basement outcrop in the middle of the Bay of Maine is not entirely made up Hercynian granites (Ballard and Uchupi, 1975), but also contains Ordo-Silurian peralkaline granites; these two types of intrusive rock thus reflect the genesis and destruction of the Theic Ocean at one and the same place.

5.4 The Meguma Zone Problem

Meguma-type terrains begin to crop out on the South side of the "Collector Anomaly" suture; they are widely exposed on Nova Scotia (Schenk, 1978). The succession is made up of quartzitic turbidites,

▶

Fig. 40. Schematic map showing relationships between Acadian and Ligerian (eo-Hercynian) foldbelts. 1a: zones with Acadian southerly tectonic transport; 1b: zones with probable southerly Acadian tectonic transport; 2: zones with Acadian northerly tectonic transport; 3a: Meguma zone facies; 3b: possible Meguma facies; 4: passive Iberian margin; 5: eastern limit of Acadian tectonic effects; 6: Moroccan Coastal Horst; 7: mylonitic zone; 8: Siluro-Devonian ultramafics and orthogneisses (Ordo-Silurian in Western Brittany); 9: gravity and magnetic highs; 10: thrust faults; 11: transcurrent faults; 12: island arc volcanics; 13: fault; 14: post-Upper Devonian faults; 15: back-arc or intra-plate volcanics; 16: drill-holes discussed in text; 17: blueschists; 18: riebeckite – granites (1: Quincy; 2: Cape Ann; 3: Bay of Maine; 4: Arronches; 5: Monforte); 19: drill-core sampling: granites: crosses; Meguma facies: M; Triassic volcanics: T (other crosses indicate geophysical evidence for granites); 20: geophysical trends (gravity and magnetics); 21: offshore limit of Moroccan Coastal Horst. **Large capitals,** CA: Cartwright Arch; G: Greenland; OK: Orphan Knoll; PB: Porcupine Bank; R: Rockall. **Small capitals,** A: Avalon; AA: Anti-Atlas; AB: Aquitaine Basin; AH: Antigonish Highlands; AL: Almaden Syncline; AU: Audierne Bay; BIF: Block Island Fault; BIM: Belle-Ile-en Mer; BMIC: Bay of Maine Igneous Complex; BOF: Border Fault; BR: Brittany; CH: Cobequid Mountains; CIZ: Central Iberian Zone; FB: Bay of Fundy; FF: Fundy Fault; GAB: Guadalquivir Basin; BG: Galicia Banks; GBNF: Grand Banks of Newfoundland; GR: Groix Island; L: Lisbon; MA: Massachusetts; MAI: Maine; ME: Merrimack Trough; MC: Moroccan Coastal Horst; MP: Mazagan Plateau; MZ: Meriadzec Terrace; MW: Murre well; NB: New Brunswick; NBV: Newburry Volcanics; NS: Nova Scotia; NSF: Nauset Fault; NFZ: Nashoba Zone; OM: Ossa Morena; PBCZ: Porto-Badajoz-Cordoba Shear zone; PEI: Prince Edward Island; PSB: Petite Sole Bank; RI: Rif; RT: Rabat-Tiflet Fracture Zone; SBA: South Armorican Shear Zone; SBSZ: New Brunswick Shear Zone; SGL: Saint Georges-sur-Loire; SH: Shamrock Canyon; SLG: Gulf of Saint Lawrence; SPZ: South Portuguese Zone; V: Vendée

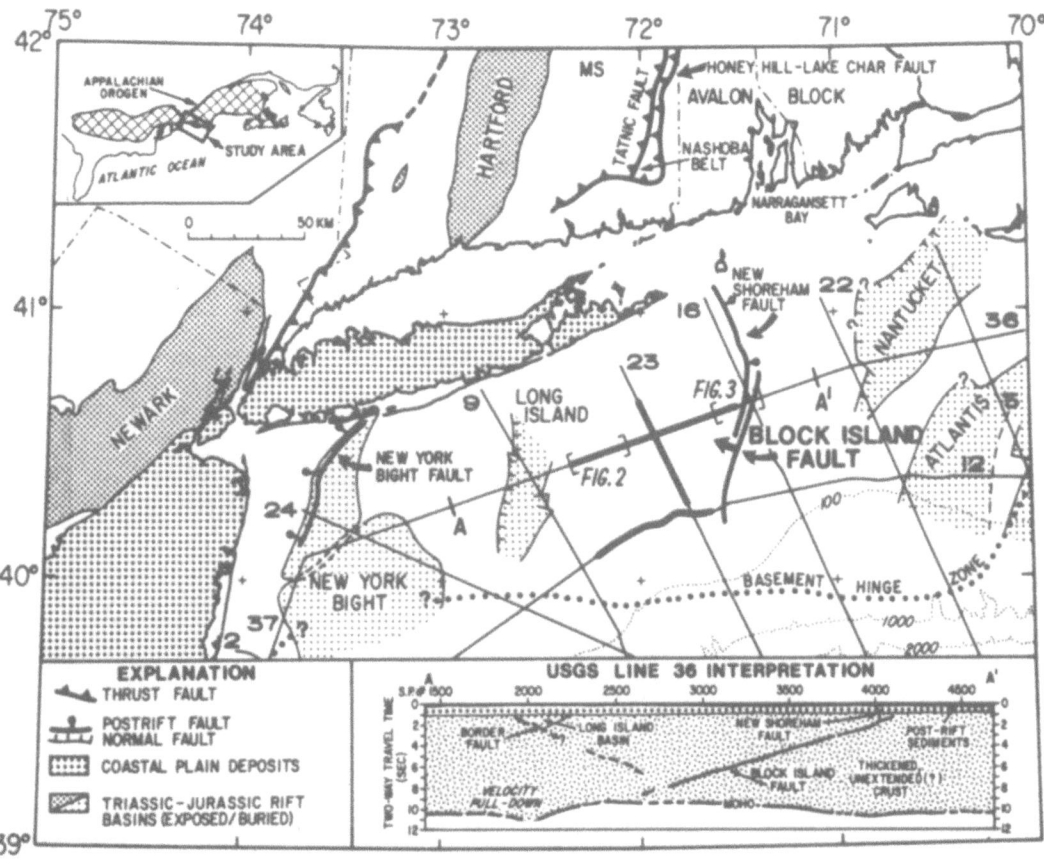

Fig. 41. The Acadian suture South of New England. Insets – Locality map, key and seismic profile. (After Hutchinson et al. 1985)

greywackes and shales of Cambrian age overlain by Tremadocian shales and thin turbidites which grade laterally from abyssal to coastal facies. In northern Nova Scotia, conglomerates and quartzose arenites are locally developed which contain volcanic members of Caradocian age. These terrains have probably not been affected by Taconic deformation, which suggests that they were not an integral part of the orogenic belt at that time. By contrast, Acadian deformation and granite intrusions are well developed in this zone. Schenk (1970) has interpreted these mainly Cambro-Ordovician sediments as deposits from the lower part of the continental margin having their source to the S.E.; the thickness has been estimated at 14 km, implying an important land mass as source-area. The succession is terminated by fine-grained Upper Ordovician facies that pass up into sandy facies of Siluro-Devonian age. It is generally recognized that the Meguma Zone deposits represent a prograding wedge on the passive, NW margin of Gondwana. These terrains are lacking in basic rocks, but, according to magneto-telluric measurements, it was thought that this

area could be underlain by crust with oceanic affinities (Cochrane and Wright, 1977). Recent seismic refraction studies from offshore areas have now limited the thickness of the Meguma Zone to 8 km (Keen and Haworth, 1986), and the existence of a partly oceanic crust beneath the margin is seriously put in question. At the present day, these terrains are considered to extend as far as the Tail of the Grand Banks of Newfoundland (Haworth and Keen, 1979) and perhaps even Portugal. This is because the 5.4 km/s velocity measured in Canada (Dainty et al., 1966) could be equivalent to the 5.5.–5.3 km/s material (Sousa Moreira et al., 1977) which corresponds to a layer intermediate between the thinned basal crust of the South Portuguese Zone and the Devono-Carboniferous.

The South Portuguese and Ossa Morena Zones plunge southward beneath the Guadalquivir basin (see Fig. 42). The concealed basement of this basin has a straight northern limit, which does not correspond to a fault (Perconig, 1960 and 1962), and displays a regular deepening of the roof beneath mainly Miocene cover towards the Betic Cordille-

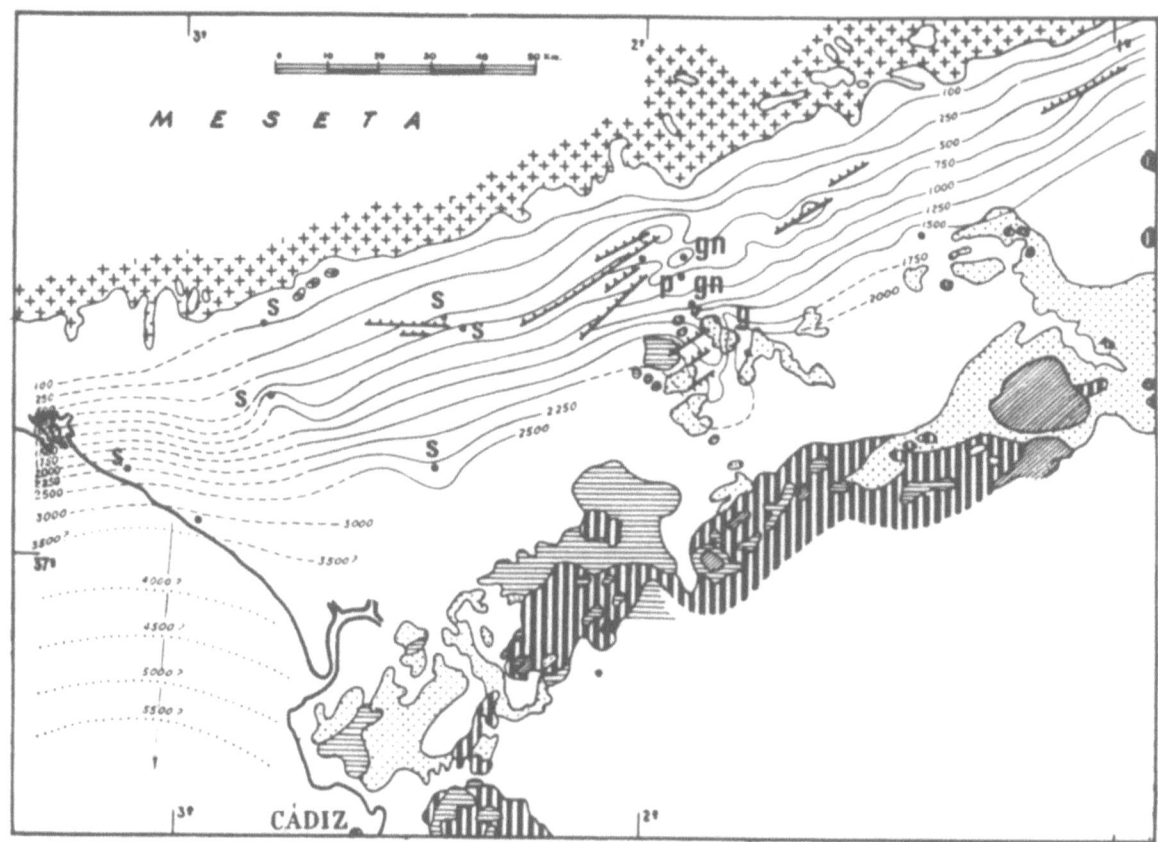

Fig. 42. Isobaths to roof of basement beneath the Guadalquivir Basin. Rock-types in drill-holes — s: different types of schist; gn: gneiss; g: granite; p: quartz porphyry. (Modified from Perconig 1960 and 1962; reproduced with permission of the Société Geologique de France)

ras. Drill-holes within the basin have reached formations similar to outcrops further North, thus providing simple correlations. There are mainly green and black sericitic schists and phyllites in the western part of the basin, which resemble certain facies of the South Portuguese Zone. These formations constitute the possible cover to the Meguma Zone. At 2°W, the metamorphic rocks and up-faulted granites are undoubtedly part of the Ossa Morena Zone. The tectonic discontinuity which separates the South Portuguese Zone from the Ossa Morena further North cannot be traced beneath the Guadalquivir basin.

To the East of Nova Scotia, drill-holes have reached Meguma Zone rocks showing Acadian metamorphism which have been intruded by Devonian granites; other granites have been located by gravimetry (Stephens et al., 1971; Kane et al., 1972) and magnetism (Kane et al., 1972) (Fig. 40). The cored granites are rich in biotite and resemble rock-types found on land. Metamorphic rocks are made up of either metaquartzites and sericitic schists with occasional garnet or phyllites (Jansa and

Wade, 1975). As a matter of comparison, it is interesting to note that seismic velocities recorded for the basement, according to a profile across the Nova Scotia shelf, are always comprised between 5.1–6.0 km/s (Sheridan and Drake, 1968). Other shot-points in the Bay of Maine (Ballard and Uchupi, 1975) demonstrate that terrains to the East of the suture show velocities always less than 6.2 km/s, which suggests that the Meguma Zone extends further South (Schultz and Grover, 1974; Poag, 1982). Velocities between 6.2 and 7.0 km/s, on the other hand, are restricted to the West of the suture; this proves the abundance of basic rocks in this area and probable disappearance of the Meguma Zone.

According to the plate sutures proposed in this study, seismic velocities greater than 6.2 km/s should be commonly observed South of Long Island; unfortunately, data from this area are rare. The 5.2–6.0 km/s velocities observed by Ewing et al. (1950) probably indicate the presence of metamorphosed Palaeozoic, or even Pre-Cambrian rocks which are known to occur in the vicinity (Skehan, 1983). Those low velocities do not constitute evidence for the existence of Meguma rocks to the West of the Block Island Fault.

Schenk (1978) has proposed that the Meguma Zone can be correlated with the Bou Regreg Formation of Morocco, where it displays a proximal turbidite facies. Palaeocurrent measurements from this area are, furthermore, compatible with data obtained from Nova Scotia. Such a result does not, however, imply that the Meguma Zone covered large areas of Morocco at some stage (Piqué, 1979) since it is known that the Bou Regreg Formation is a "suspect terrain". This part of Morocco has probably been tectonically displaced from the West due to dextral movements along the transcurrent fault zone joining Rabat with Tiflet (see Fig. 40).

Previous studies have sometimes claimed an important Acadian orogenic influence in Morocco, but it has now been established that such effects are mainly epi-orogenic and that clear evidence for a corresponding unconformity is very rare (Hollard et al., 1976). Acadian effects are mostly limited to sedimentary disturbances which are sometimes accompanied by slight volcanism (Hollard, 1967). This is generally thought to be the consequence of more intense orogenic activity situated further west. Zones weakly affected by Acadian events are illustrated on the schematic map of the mobile belt (Fig. 40) and include the Anti-Atlas, the western edge of the Meseta and the Rabat-Tiflet Zone in Morocco. Even though no major Acadian tectonic event is recognized in western Morocco, geophysical modelling (Ben Salmia and Lefort, work in progress) suggests a remarkable structural similarity between this region and Nova Scotia. The uplifted coastal block of Morocco corresponds to a horst where the overlying Palaeozoic shows little or no deformation (Guezou and Michard; 1976). This horst is well marked by a positive gravimetric anomaly (Van den Bosh, 1971) which is cut off abruptly along the Rabat-Tiflet Zone.

The Coastal Moroccan Horst shows a generally westward bend and is offset by E-W transverse shears which occurred at a late stage (Lefort and Haworth, 1981). Gravimetric trends within the horst also show the same curvature. Offshore, the Mazagan Plateau displays an homothetic structure (Ruellan, 1985) which can be extended under the Rif, using magnetic data (Saadi, 1975), if allowance is made for displacements across the Rabat-Tiflet Zone. This bend cannot be picked up by gravimetry in Canada since terrains in Nova Scotia (and on the margin) have been greatly influenced by Devonian granite intrusions and the formation of Mesozoic basins. Nevertheless, it has been possible to follow

Acadian fold axes by magnetic methods (Schwartz and McGrath; In: McGrath et al., 1973) owing to the occurrence of pyrrhotite in the Meguma Zone. These axes trace out an arcuate structure convex to the West which reflects the outline of the peninsula (Keppie et al., 1982). The perfect nesting of geophysical trend-lines in Nova Scotia with the Coastal Moroccan Horst clearly shows that there was a major bend in the Acadian foldbelt to the West of Morocco, even if near-surface tectonic effects are poorly-developed on the African side of the orogen. Isobaths on the continental margin between Lat. 34°–30° show a curvature in agreement with the observed bend in onshore markers. In this way, it is obvious that the terrains located west of the Moroccan Meseta, in common with Nova Scotia, have been subject to effects brought about by the closure of the Theic Ocean. This is not sufficient to support the hypothesis that Nova Scotia represents the continental margin of Morocco during Cambro-Ordovician times. However, it is evident that both sides of the Meguma Zone have been affected by the same deformations and that the southerly front of the Acadian orogeny in N. Africa should extend as far as the western limit of the coastal horst.

5.5 Conclusions Concerning the Ligerian (Eo-Hercynian)-Acadian Mobile Belt

On the eastern side of the Atlantic, development of the Ligerian orogenic belt occurred over a time-span of ca. 60 Ma between the early Silurian and the end of the Devonian. Pre-Silurian magmatic activity, although contemporaneous with Caledonian events in N. Europe, are in fact more closely related to Cambro-Ordovician crustal distension which occurred throughout the Variscan domain. It is often rather difficult to clearly separate crustal distensions due to the initial stages of Palaeozoic ocean spreading from those phenomena related to the development of Avalonian/Cadomian after-effects. It would appear that, South of Brittany and South of Portugal, Siluro-Devonian metamorphism and igneous activity corresponds to the onset of northward-directed subduction (more precisely SW towards NE) beneath the southern borders of two different Cadomian blocks. This subduction would have preceded a Mid-Late Devonian collision. However, it is still not known whether the South Armorican Ocean represents a branch of the Theic Ocean (Proto-Tethys) or whether it corresponds to the main ocean. A solution to this problem will arise from a better understanding of the basement under the Betic Cordilleras. This is because there is a convergence in this region between the South Iberian suture (which may extend further to the East) and the southern prolongation of the Porto — Badajoz — Cordoba Fault.

At present, there is still some disagreement concerning the direction of displacement for the Ile de Groix blueschists. According to structural studies, tectonic transport has occurred from South to North and is in opposition to the Ligerian overthrusts observed on land. Lefort (1979) has proposed that the blueschists were formed during flake tectonic-type processes (Oxburgh, 1972) — that is, the displacement is antithetic with respect to the South Armorican subduction direction. Another solution (Mattauer et al., 1980) is to ignore the theoretical possibility of such a mechanism and to refute the structural results obtained from the Ile de Groix. This solution leads inevitably to the proposition of a suture to the North of the island (Bard et al., 1980; Matte, 1986). In the absence of any convincing geological or geophysical evidence for such a suture onshore, it is preferable to base an interpretation on the offshore data cited in this book. It is clear that the South Armorican Shear Zone (on the mainland) has played an important role during plate convergence — notably as a transcurrent fault parallel to the strike of the subduction zone (Roeder, 1975). But this inland fault could not represent a plate boundary due to the absence of any directly associated basic rocks. In fact, the South Armorican collision zone is obviously not restricted to a single line but is developed over a wide belt which extends probably from the South Armorican Shear Zone as far as the geophysical marker recognized offshore. The few small outcrops of basic rock (showing evidence for oceanic affinities or high pressure metamorphism) that border the Ligerian belt on land are anyway insufficient to define the trace of a suture. They simply express limited spreading in areas of thinned crust followed by rapid closure during plate convergence.

The problems of deciding whether a within-plate scar is really a suture are rather similar in New England, insofar as the Nashoba Zone has also been proposed as a plate boundary (Skehan, 1973). Actually, it is necessary to clearly distinguish those boundaries which separate plates (and sometimes large microplates) from major shear zones which occur within plates — the former usually separate basements with contrasting structural histories, whereas the latter do not.

On the other hand, there is now general agreement on the mode of emplacement of the Galician klippes

and the transform character of the Little Sole — Porto — Badajoz — Cordoba Fault. As indicated on the schematic map (Fig. 40), the Ligerian-Acadian suture does not everywhere show the same degree of closure (perhaps because of a lack of detailed information on the Grand Banks).

It appears that the classic zonation from suture through volcanic arc to back-arc basin is still recognizable only in southern Brittany and south-eastern Canada. Volcanic rocks of the Chaleur Group (Ruitenberg et al., 1977) would appear to represent a preserved volcanic arc in northern New Brunswick. Elsewhere, it would appear that the suture disappears beneath its own volcanic arc or beneath various back-arc basins — this is probably the case in Spain and off southern Portugal. Alternatively, the suture can be concealed beneath back-arc basins as is probably the case in Nova Scotia. In New England, there is an intermediate case where the suture is apparent near a volcanic arc which is partly concealed beneath a major thrust. The existence of antithetic faults on both ends of the 'Collector Anomaly' is entirely remarkable and it should be stressed that this phenomenon appears to have developed where the suture shows a major bend. Finally, from the structural point of view, it can be seen that the mobile belt had a double origin on the western side of the Atlantic — the western belt was inherited from Caledonian (Taconic) events while the Eastern belt was related to subduction of the Theic Ocean. These two origins are so distinct that there is now practically a double foldbelt in the Maritime Provinces of Canada.

The outer limits of the Ligerian-Acadian mobile belt (i.e. fronts of deformation) are only discernible along Logan's Line and perhaps also in Morocco, since it is suspected that the Palaeozoic terrains located West of the coastal horst were slightly deformed during the Acadian orogeny. Elsewhere, these fronts have been strongly re-activated during the Hercynian and are unrecognizable.

It is reasonable to assume that the Meguma Zone deposits represent the passive margin of Gondwana — this is due to the sedimentary facies, the thickness of the succession and the presence of a possible oceanic suture on the side away from Gondwana. In addition, the Meguma Zone lies upon ancient and perhaps thinned crust on both sides of the Atlantic. The homothetic deformation which is apparent between Morocco and Nova Scotia suggests, in any case, that these two regions were opposite to each other during the Acadian Collision.

According to whether the Block Island Fault is correlated with the Clinton-Newburry-Honey

Hill-Lake Char Shear Zone (as proposed by Hutchinson et al., 1985) or the Nauset Fault (as suggested here), the S.E. corner of New England during the Lower Devonian was either a part of Gondwana or a fragment that was detached at an early stage from Gondwana – and only joined to Laurentia just before the main collision. Taking account of the structural arguments discussed above, and the probable absence of Meguma facies in this region, the second interpretation is favoured in the present work.

In any case, the intense Hercynian fracturing and abundant transcurrent displacements which have affected this belt have rendered the identification of pre-Acadian boundaries extremely difficult.

Chapter 6

The Concealed Pre-Hercynian Basement Between Africa and North America

The concealed pre-Hercynian basement between Africa and North America is far less extensive than the Cadomian, Ligerian and Acadian terrains described further North; furthermore, it crops out only rarely on the margins of these two continents. On land, the many drill-holes which reach the basement in the Coastal Plain area of the U.S.A. have certainly provided a great amount of data; however, this information is difficult to bring together in a synthesis (Thomas et al. 1988), being widely spread between many different companies and state surveys – the data were sometimes collected a long time ago and the descriptions may be incomplete. On the African side, the compilation of information is much easier since the rare documents in existence have practically all been published. Because of the lack of accurate data concerning the more ancient basement, all Pre-Cambrian formations will be treated together in this chapter, along with Palaeozoic terrains that pre-date the Hercynian orogeny.

6.1 Archaean and Pan-African Basement

Concealed Archaean basement (dated at ca. 2,700 Ma) should almost inevitably occur on the S.W.

border of the African continent in view of the fact that the Leo uplift is close to the margin (see LU marked on Fig. 43). However, the 6.1–6.5 km/s velocities recorded beneath the Palaeozoic off the Guinean Plateau (Sheridan et al., 1969) would appear to be more characteristic of the Pan-African Rokelide belt, which runs parallel to the ancient craton border, rather than the Archaen itself. These seismic data are the only presently available in the region.

According to oil prospection geophysics (Aymé, 1965), Pre-Cambrian basement exists at a depth of about 8 km beneath the Senegal basin but it is likely that this material is underlain by the same rocks as in the craton (Dillon and Sougy, 1974). Archaean rocks, in fact, come to the surface further to the North and may already be present beneath the northern part of the Senegal basin. Offshore from this part of the basin, velocities of 6.30–6.45 km/s have been recorded which certainly correspond to granodioritic material (Weigel et al., 1982). The Archaean basement can be said to come near the surface West of the Reguibat uplift where offshore gravimetric surveys (Fritsch et al., 1978) indicate that such ancient rocks form the substratum of the Arguin Bank as far as 19°W.

Fig. 43. Correlation of Pre-Cambrian, Lower Palaeozoic and Devonian structures across the southern part of the North Atlantic. 1a: Terrains generally older than 1000 Ma with unknown or thin Palaeozoic cover; 1b: pre-1000 Ma terrains; 1c: cores proving pre-1000 Ma basement; 2: "Grenvillian-type" terrains in Africa indicated by radiometric dating; 3: Grenvillian terrains in America indicated by radiometric dating; 4: Panafrican terrains showing local reworking of ancient basement; 5: Rokelides suture; 6: Hercynian fractures; 7: Archean gravity highs; 8: End-Precambrian and Cambrian basement composed of acid plutonic and volcanic rocks; 9: pre-Carboniferous basins slightly deformed during the Hercynian; 10: undifferentiated crystalline basement showing Appalachian deformations; 11: deep-seated basic intrusions; 12: zone of Taconic deformation; 13: zone of Acadian deformation; 14: various granites recognized within the zone of Hercynian deformation (SE Appalachians); 15: Senegal plate; 16: prong of African basement (see text); 17: limits of Mesozoic and Cenozoic basins; 18: crucial drill-holes; 19: Meguma zone; 20: Ligerian (Eo-Hercynian) — Acadian suture. **Large capitals,** GM: Gulf of Mexico; U.S.A.: United States of America. **Small capitals,** AA: Anti-Atlas; ARB: Arguin Bank; ATB: Aaïun-Tarfaya Basin; BA: Batafa; BB: Bové Basin; BJ: Bijagos Archipelago; BP: Blake Plateau; C: Charleston; CC: Cape Cod; CT: Carolinas Trough; DK: Doukkala Basin; EA: Emerald Arch; EJ: El Jadida; ES: Essaouira Basin; ESL: Eastern Slate Belt; F: Florida; GA: Gamon; GC: Guyanese craton; HA: Cape Hatteras; I: Ifni; K: Konakry; KV: Kidira-Velingara lineament; LU: Léo Uplift; MC: Coastal Uplift; MZP: Mazagan Plateau; NE: New England; PD: Piedmont; PG: Guinean Plateau; R: Rokelides; RU: Reguibat Uplift; SB: Senegal Basin; SUB: Suwannee Basin; TB: Tindouf Basin; YA: Yarmouth Arch

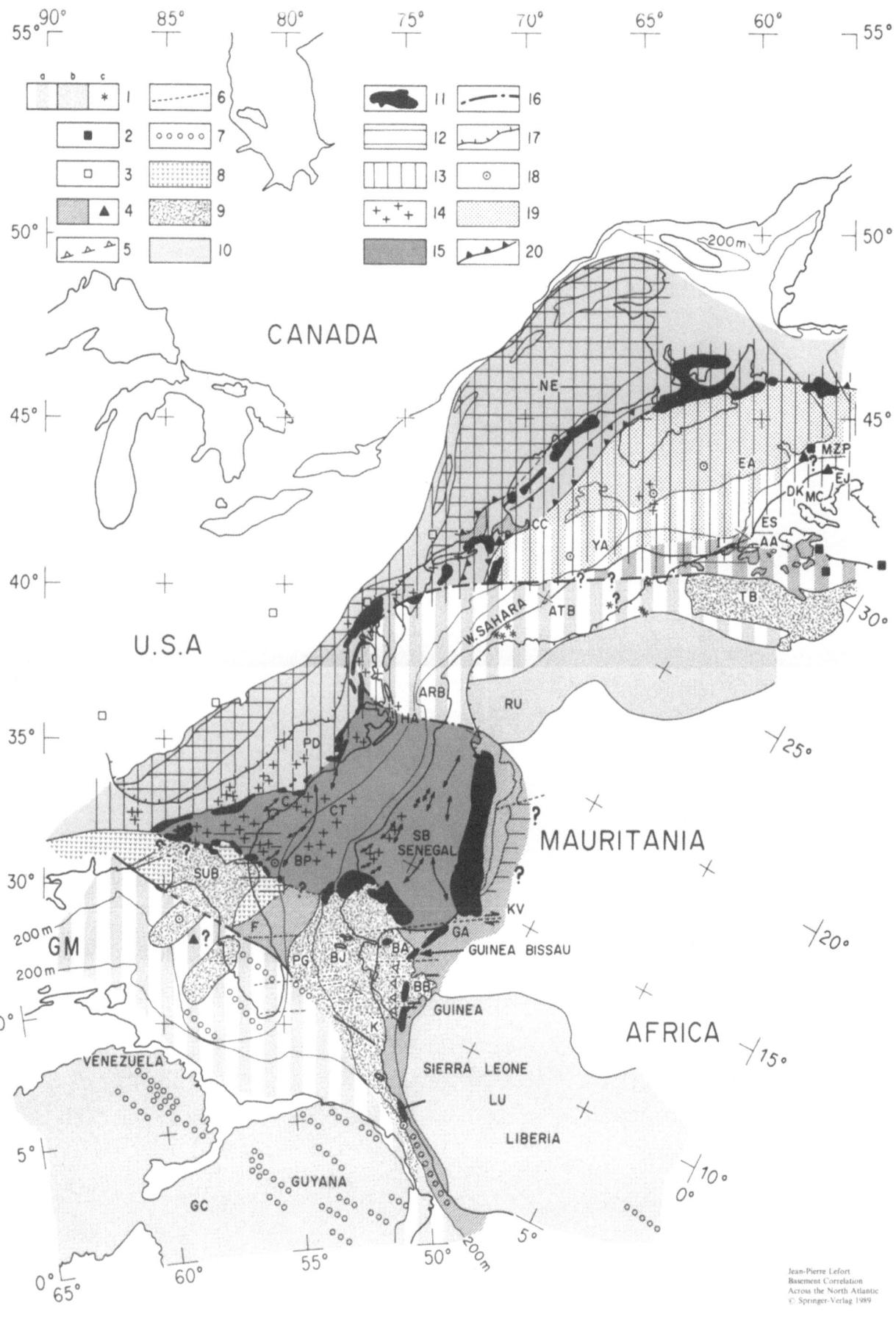

Jean-Pierre Lefort
Basement Correlation
Across the North Atlantic
© Springer-Verlag 1989

As a matter of fact, concealed basement older than 1,000 Ma has only really been sampled beneath the Aaiun-Tarfaya basin (Auxini, 1969; Ranke et al., 1982; Wissman, 1982).

The outcrops of charnockite on the Mazagan Plateau off Morocco are the only direct evidence for 980–950-Ma-old material on the African margin (Ruellan, 1985); it would be nevertheless incorrect to assign these age data to the Grenvillian sensu stricto, since identical results have been obtained in Southern Morocco and across Central Europe (Zwart and Dornsiepen, 1978).

The concealed Late Proterozoic and Lower Palaeozoic basement that has been sampled off W. Africa or North America always belongs to the Pan-African or its American equivalent, the Avalonian. The age spread of these terrains is generally between 650 and 550 Ma. In Africa, it has been shown that Pan-African terrains probably occur offshore from the Rokelide belt (Villeneuve et al., 1984), which is known to be related to an ancient plate suture. The Pan-African should also be found beneath the cover East of the Senegal basin and to the West of the Anti-Atlas, in the region of Ifni. However, it is surprising that the underlying material from the Coastal Moroccan Horst – composed partly of uppermost Proterozoic rocks (Michard, 1976) such as the El Jadida rhyolites – has not been dredged from the adjacent Mazagan Plateau. Of course, it is possible that the dates of 515 ±8 and 519 ±19 Ma, obtained on granodiorites which cut this spur (Wissmann et al., 1982), might not correspond to a true emplacement age. They could be interpreted as the partial re-setting (during the Hercynian, for example) of older Pan-African ages. Actually, it seems more reasonable to assume that the emplacement age was around 520 Ma. This would approach the age of a magmatic event between 540 and 520 Ma known in the Anti-Atlas (Charlot, 1978) and re-inforces the resemblance which is thought to exist between the Coastal Horst and regions in southern Morocco.

The Coastal Moroccan Horst, which behaves like a structural high (Guezou and Michard, 1976), occurs opposite the external margin of Nova Scotia and New England, where two areas of relatively uplifted basement are found – the Emerald Arch (Lefort and Haworth, 1981) and the Yarmouth Arch (Uchupi and Austin, 1979). These arches appear to be composed of crystalline basement (Klitgord and Hutchinson, 1985) partly capped with coral reef deposits (Poag, 1982). Drill-holes in this area have unfortunately not penetrated deeply enough to confirm the presence of rock-types typical of the Mazagan

Plateau basement.

Along the eastern seaboard of the U.S.A., the Avalonian basement re-appears just South of New England, where it is characterized by seismic velocities greater than 5.2 km/s (Mc Master et al., 1980); but this material cannot be traced with any certitude beneath the continental margin between Cape Cod and Florida. Neither the velocities (Sheridan et al., 1979) nor the densities (Grow et al., 1979) reveal any difference from metamorphosed Palaeozoic formations.

At sea, Avalonian terrains are not recognized again until the Gulf of Mexico (Dallmeyer, 1984), where gneisses and amphibolites have been found which are probably older than 500 Ma. In the present continental fit, this outcrop overlaps onto the Venezuelan coast – probably a consequence of stretching in the continental crust during the opening of the Gulf (Bufler et al., 1980) (Fig. 43).

On land, the most extensive and best-studied fragment of concealed Avalonian basement is found beneath Florida – here, drill-holes (Fig. 44) have revealed diorites, gneisses, granites and rhyolites (Gohn, 1983) very similar to those found in Western Africa (Smith, 1982; Thomas et al., 1988). The link between the "Avalonian Complex" of Florida (Banks, 1978) and the onland African outcrops (Dallmeyer et al., 1987) is not direct, but this does not present any difficulty for trans-Atlantic correlations since, to the West of Guinea, the Pan-African belt is hidden beneath Palaeozoic sediments (Dillon and Sougy, 1974) and is thought to more or less surround the West African Craton.

It has long been a matter of debate as to whether the basement rocks of S. Florida (mostly composed of acid volcanics) really belong to the pre-Mesozoic. According to Barnett (1975), the rhyolites of S. Florida are Triassic or Lower Jurassic in age, whereas other authors (Applin, 1951) have considered these rocks to be Upper Proterozoic or Lower Palaeozoic. Whatever the age of this material, certain geophysical correlations lead us to suppose that the deep basement of this part of Florida could be even older than the age proposed by Applin (1951). In any case, this basement complex is cut by transverse E-W trending faults (well seen on the magnetic map shown in Fig. 45) which are in perfect continuity with the ancient fractures (Simon et al., 1981) beneath the Bové Basin (Ponsard, 1984) and on the Guinea shelf.

The basement of southern Florida shows an average density of 3.0 g/cm³ derived from gravity modelling, which is much greater than the values of 2.73 g/cm³ for the Avalonian basement further North – the two

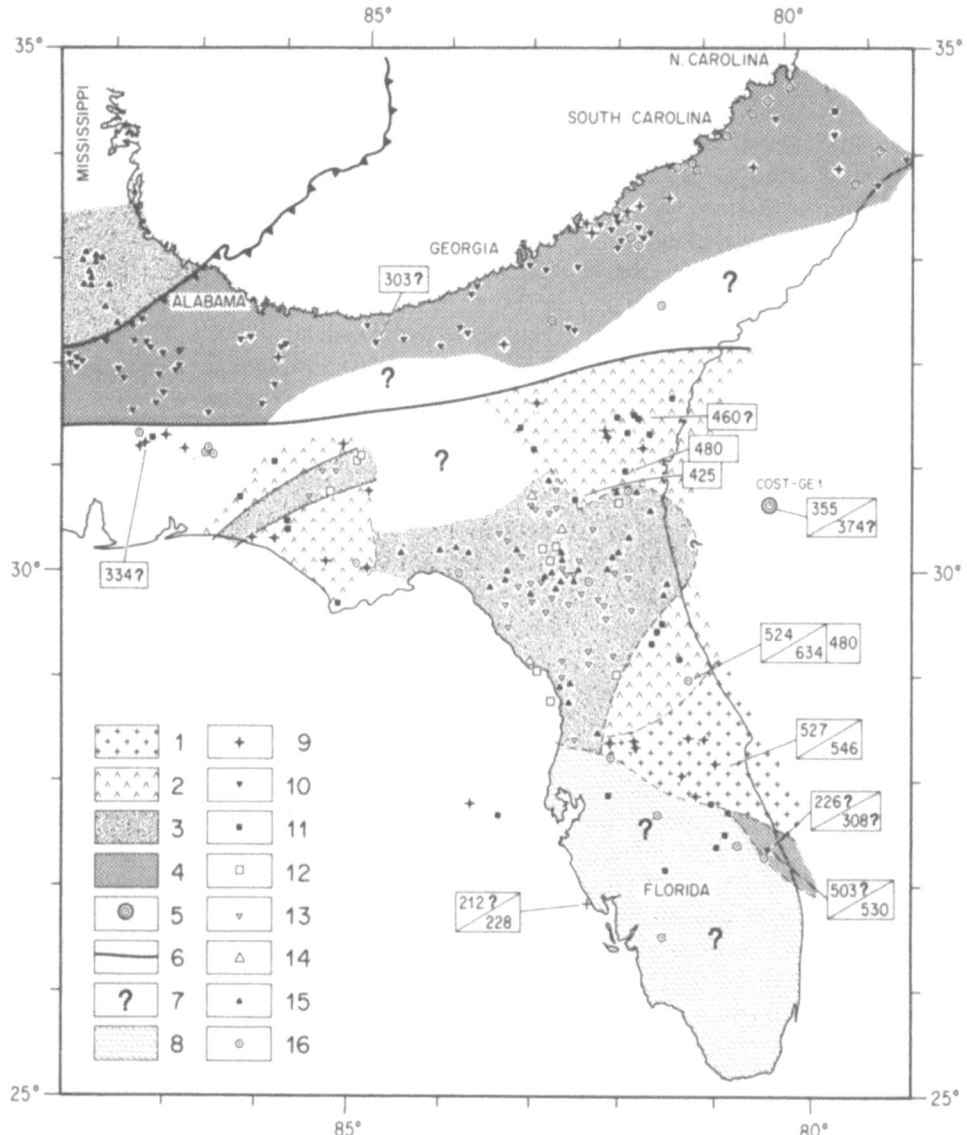

Fig. 44. Nature of the concealed basement beneath Florida. 1: granitic basement; 2: Proterozoic and Eo-Cambrian acid volcanics and granites; 3: undeformed Palaeozoic; 4: granitic and metamorphic Piedmont terrains; 5: drill-holes; 6: boundary between roof of unmetamorphosed and metamorphic terrains; 7: inferred graben; 8: possible Archaean at depth; 9: granites; 10: metamorphic rocks; 11: rhyolites; 12: Devonian; 13: Ordovician; 14: Silurian; 15: Palaeozoic; 16: undifferentiated basement. Figures in boxes refer to radiometric dates, question marks indicate doubtful data. (Modified from Gohn 1983 and Chowns and Williams 1983; reproduced with permission of the United States Government Printing Office)

domains are separated by a marked crustal discontinuity (Klitgord et al., 1984) and may be composed of terrains of contrasting type.

The positive gravimetric ridges trending N130°, observed in southern Florida, Brazil (Lesquer et al., 1984) and on the margins of Sierra Leone, Guinea (Jones and Mgbatogu, 1982) and Liberia (Behrendt et al., 1974), are parallel to structures which are older than the Pan-African in Guyana (Gibbs and Barron, 1983) or in Africa (Bessoles, 1977). This suggests that

the underlying basement of this part of the S.E. United States is made up of Archaean or Lower Proterozoic terrains. Such a correlation is backed up by the fact that densities calculated for the pre-Pan-African granulites off Liberia (2.87 g/cm^3; Behrendt et al., 1974) are relatively high and that velocities measured off southern Florida (6.7 km/s; Sheridan et al., 1981) are close to values usually obtained from granulites (6.4–6.7 km/s). The occurrence of granulites in Guyana (Gibbs and Barron, 1983) is also in

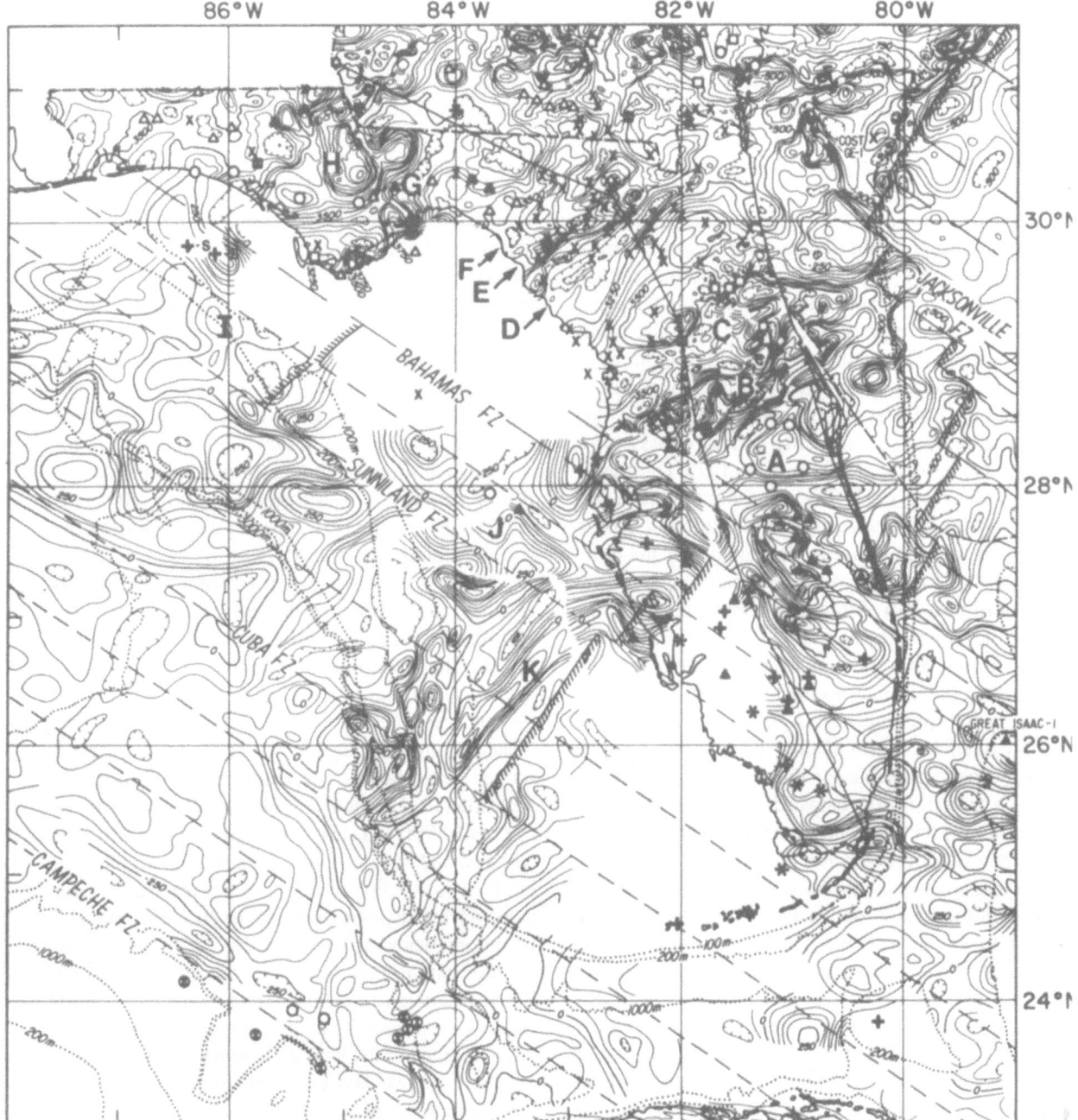

Fig. 45. Magnetic map of basement structures recognized in Florida and on the continental shelf. See text for explanation of letters A–I, symbols refer to drill-holes and rock-types given in Fig. 44. (After Klitgord et al. 1984)

favour of this argument since they occupy a position close to the Florida margin, as shown on a pre-Atlantic reconstruction (Fig. 43).

Obviously, this does not imply that the rhyolites cored at the top of the basement in southern Florida are necessarily very ancient – some of them, in fact, have been dated as Mesozoic (Barnett, 1975).

In Florida, the N 130°-trending structures have previously been considered as Jurassic transform faults (Klitgord et al., 1984), but the present author prefers to interpret them, in the light of the structures already mapped in the Guyana shield (Lesquer et al. 1984), as transform directions which controlled the location of transform faults during the opening of the North Atlantic. In Florida itself, Jurassic rejuvenation of these fractures must have been very weak or non-existent since there are no offsets in the E-W fault systems described above (Fig. 45).

Figure 44 shows the distribution of the main rock-types recognized in the Florida basement which are assigned to Pan-African or Lowermost Palaeozoic formations (Chowns and Williams, 1983). In general terms, there is a central granitic zone associated with acid volcanics that have produced the negative gravimetric anomaly and the low frequency magnetic anomaly shown as A on Fig. 45. This is bordered to the South by high grade metamorphic terrains (amphibolite facies). The central zone is thought to be a fragment of Pan-African basement (Chowns and Williams, 1983), even though it is not known which precise area in Africa is actually comparable. The southern Florida high-grade zone is considered as a prolongation of the Rokelide belt from Africa. A certain number of radiometric ages appear to confirm these attributions (Bass, 1969; Milton and Grasty, 1969; Barnett, 1975; Dallmeyer and Villeneuve, 1987). Taking account of the dating methods and the possibility of alteration in certain samples, most authors have retained only the older dates — this may be justified since there is no important post-Pan African metamorphism in the Florida basement (Bass, 1969). However, this approach has led to the omission of some local late-Hercynian rejuvenation effects which are, nevertheless, quite apparent in this American fragment of the Rokelides. North of the Central zone, granitic rocks are responsible for the strong positive ridge which extends north-eastwards onto the margin; the dominant rock-types here appear to be Pre-Cambrian — Cambrian diorites (shown as B on Fig. 45). The negative belt (shown as C on Fig. 45) corresponds principally with rhyolites, which also are seen to extend eastwards. The same formations, associated with the same magnetic effects, are found North of Lat. 30°30′ on the northern rim of the Palaeozoic Basin. On the western side of a Triassic basin located in Western Florida, there is a gravity low which is clearly related to a granitic batholith (H on Fig. 45). Taking all available data together, it appears that the metamorphic and intrusive rocks dated at between 634 and 524 Ma constitute a base-

ment for the acid volcanic suite. The existence of terrains in Florida which belong to the Rokelide belt is very likely, especially since dextral Hercynian shears are recognized and have probably displaced basement blocks westwards from the axis of the foldbelt in Africa.

At sea, Palaeozoic (or older) crystalline basement has been found offshore from the state boundary between Florida and Georgia — this material shows seismic velocities of 5.6–5.82 km/s (Dillon and McGinnis, 1983). Although the Blake Plateau was probably initially formed of Pan-African terrains, it has been so extensively intruded by basic dykes during the opening of the North Atlantic that velocities are now increased to 7.1 km/s (Sheridan et al., 1981) and are thus more typical of intermediate rather than continental crust. The same is true for the basement under the Carolinas Trough. In addition, similar types of intrusions could have contributed to the general increase in density for the southern margin of Florida.

Despite the paucity of petrographic descriptions (Maher, 1971), it is possible to recognize two Pre-Cambrian groups beneath the Coastal Plain of the U.S.A. The first group is composed of metavolcanic and metasedimentary rocks of Upper Proterozoic age, whereas the second is made up almost entirely of acid batholiths. Some of these latter have been dated, such as the Cape Hatteras batholith (Scott and Cole, 1975) which yields a typically Pan-African age of 610 Ma (Fig. 46); other ages ranging between 580 and 680 Ma have also been obtained (Thomas et al. 1988). A few rather rare granite massifs are well marked by negative gravimetric and magnetic anomalies and can thus be easily delimited. On the whole, however, there is little correlation between ancient plutonic massifs and the gravity field — perhaps due to the later development of tangential tectonics. The shallow roots of the southern part of the Hatteras belt are probably also related to this type of tectonics.

The heat flow map (Smith et al., 1981) could possibly provide some control on the location of concealed

▶

Fig. 46. Nature of basement beneath the US Coastal Plain and adjoining shelf. **Sampled rock-types,** 1: gabbro; 2: schists; 3: granites; 4: unknown; 5: Pre-Cambrian soap-stone facies. (After Maher 1971). **Basement terrains assigned to the Pre-Cambrian and Palaeozoic,** 6: Pre-Cambrian acid batholiths; 7: Pre-Cambrian metavolcanic and sedimentary rocks; 8: Palaeozoic acid batholiths. (After Bayley and Muehlberger 1968). **Basement detected by geophysics and proven by drilling,** 9: Eastern Slate Belt; 10: granites recognized from gravity. (After Klitgord et al. 1985). **Radiometric ages,** 11: dates given by various authors cited in Scott and Cole (1975). **Basement inferred from geophysics,** 12: coastal Plain limit; 13: gravimetric lineaments; 14: magnetic lineaments; 15: boundary between domains showing structural continuity with the Appalachians and domains showing no correlation with Appalachian structures; 16: magnetic intrusions showing trend correlated with Appalachian Structures; 18: ultrabasic intrusions marked by overlapping magnetic and gravity highs; 19: possible granite bodies recognized by gravimetry. ESL: Eastern Slate Belt; PG: Petersburg Granite

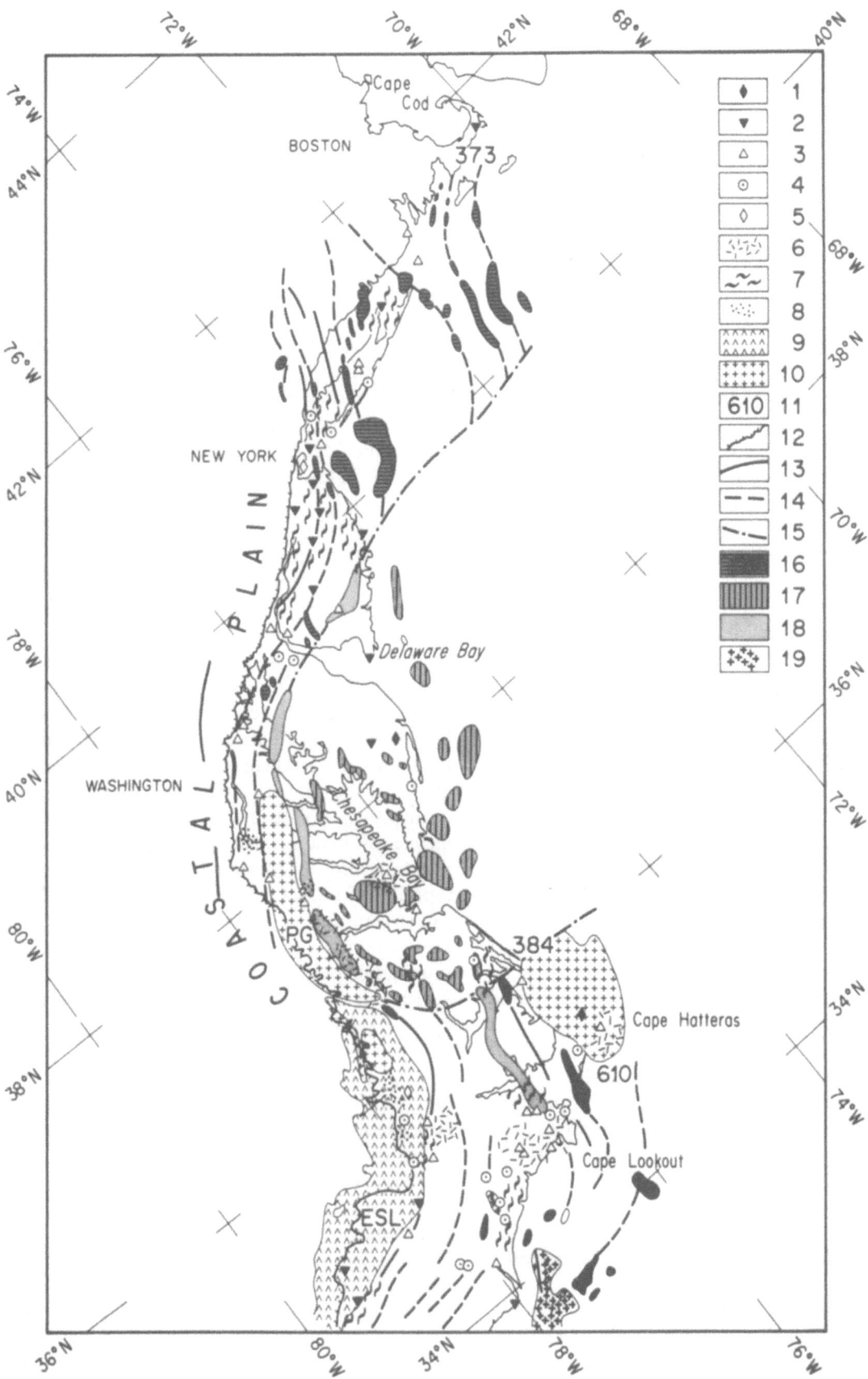

Avalonian terrains, since the rocks of this age which show little or not reworking during the Appalachian orogeny appear to give higher than normal heat flow. Unfortunately, this method is insufficiently precise to enable the mapping of geological boundaries beneath the sedimentary cover of this region.

The existence of blocks of Avalonian basement cropping out to the East of the Appalachians (Piedmont), or concealed beneath the Coastal Plain and shelf, leads to the same problem as posed for the Avalonian of the Maritime Provinces: are the similarities in age pattern and petrography between these blocks (which also reflect their Gondwanan affinities) a sufficient argument to affirm that they all joined Laurentia at the same time? The present author thinks it rather unlikely, especially since it is known that the Avalon Spur — representing a fragment which was detached from Gondwana around Ordovician times (Perroud et al., 1984) — was joined with Laurentia well before the rest of the continent from which it was derived. Nowadays, it is possible to imagine that there are blocks of African affinity beneath the margin and Coastal Plain which were amalgamated with the Grenvillian Craton well before the main collisional phase during the Permo-Carboniferous. These diachronous accretion events are probably responsible for the development of multiple micro-sutures beneath this region. This phenomenon was clearly predicted by O'Hara and Gromet (1985) for the accretion of the SE New England terrain, and has been similarly proposed for the Carolinas (Secor et al., 1983; Dallmeyer and Villeneuve, 1987) and Georgia (Dainty and Frazier, 1984). Even though the hypothesis of a single Avalonian plate has now been abandoned (Hatcher, 1972; Cook et al., 1981), this does not mean that a new microplate should be postulated each time a "suspect terrain" is detected East of the Appalachian axis. In certain cases, suspect terrains could arise from lateral displacements along transcurrent fault zones parallel to the Appalachian foldbelt; only detailed stratigraphic studies will enable a proper distinction to be made, as far as possible, between the various additions of Pan-African terrain along the eastern seaboard of the U.S.A. during the Palaeozoic.

6.2 Pre-Hercynian Palaeozoic Basement

These terrains can be subdivided into two groups according to whether they have suffered Carboniferous deformation or not. The following Palaeozoic terrains do not include those Cambrian formations which were affected by the latest Pan-African tectono-metamorphic events.

6.2.1 Pre-Carboniferous Terrains Showing Little or No Hercynian Deformation

The Florida Palaeozoic is mainly composed of quartzitic sandstones and black shales (Applin, 1951) — some of the sandstones were certainly derived from the erosion of granitic material. The base of the succession has been assigned to the Lower Ordovician on fossil evidence (Berden, 1964); the rest of the stratigraphic column is made up of Mid-Ordovician to Mid-Devonian strata, with a possible Lower Silurian disconformity. All these facies are different from those observed in the Appalachians, but compare very closely with successions in the Bové Basin, beneath the Senegal Basin, as well as Guinea, Siera Leone and Liberia (Aymé, 1965; Rodgers, 1970; Dillon and Sougy, 1974; Schlee, 1980) (Fig. 44). The Palaeozoic succession in Florida has been correlated, on palaeontological grounds, with certain formations in Brazil and Guinea (Cramer, 1971); the dating of detrital zircons from the Ordovician sandstones of Florida (yielding intercept ages of 685 Ma and 1800 Ma; Odom and Brown, 1976) indicates that these sediments are derived from nearby basement of Pan-African and Eburnean affinity. Furthermore, the faunistic assemblages (Cramer, 1971; Pojeta et al., 1976) show no ressemblance with the Appalachian province. Thus, a study of basement and cover has clearly demonstrated that the Florida basement is a fragment of Gondwana despite the fact that typically African carbonates or red beds are lacking. The structural correlations established above for the Pre-Cambrian and Palaeozoic suggest that the latitudinal displacement between Florida and West Africa was very small or non-existent before the opening of the North Atlantic; this justifies the hypothesis that these two basement blocks formed part of a single plate. The absence of Acadian orogenic effects in Florida or South of the Mauritanides complements the available information and shows that the western flank of Gondwana only came into contact with the Laurentian craton after Devonian times; this appears to be confirmed by palaeomagnetic data (Helsley; In: Smith, 1982).

On the other hand, the existence of a possible mild "Taconic" deformational phase (in fact, Upper Ordovician in age) in Guinea (Villeneuve, 1980) and to the South of Senegal (Lecorché and Sougy, 1978; Lecorché, 1983) poses a dilemma insofar as these movements cannot be related to deformation that was going on at the same time in the Appalachians – Gondwana and Laurentia were still separated during this period (Scotese et al., 1979). For the moment, it is therefore necessary to admit that, if this movement really took place – there is actually no reliable radiometric dating evidence – it was independent of the American Taconic event. Furthermore, the tectonic and metamorphic map of the Southern Appalachians, published by Williams and Hatcher (1983) and Glover et al. (1983), shows that there was no Taconic deformation to the East of the foldbelt in America. The "Taconic" disconformity in W. Africa – probably related to the drifting of a small continental block towards the West – is overlain by a thick Siluro-Devonian succession in the Bové Basin.

Taking account of the apparent unity of these regions throughout the Palaeozoic, it is interesting to compare a geological cross-section of the Floridan basement (Smith, 1982) with two seismic profiles in Africa; one is in the Senegal Basin (Aymé, 1965) and the other on the Sierra Leone margin (Sheridan et al., 1969) (Fig. 47).

There are, in fact, very few stratigraphic or structural markers which enable a continuous correlation between the Bové and Suwannee Basins. Towards the South, in Africa, approximately 1000 m of Lower Palaeozoic have been drilled above uppermost Proterozoic igneous rocks in the Liberian Coastal basins. At least one drill-hole, off Monrovia, has reached probable basal Devonian (palynological evidence; Cortesini and Minner, 1972).

The Palaeozoic of the Bové Basin probably extends as far as the Guinea Plateau, where velocities of 5.20 km/s have been recorded – these formations are also very likely to cover the Sierra Leone margin where 5.2–5.4 km/s material exists over a thickness of 3 km (Sheridan et al., 1969). These terrains lie above an Upper Proterozoic basement which is characterized in this region by velocities of between 6.1 and 6.5 km/s.

Finally, a bore-hole near the coast in Guinea Bissau has yielded black and grey shales of Wenlock age cut by dolerite dykes (Cramer and Cramer, 1972).

From the geophysical point of view, a compilation of magnetic data given by Jones and Mgbatogu (1982) reveals that anomalies situated off Conakry follow the prolongation of a NE-SW-trending belt of basic and ultrabasic rocks (Palaeozoic in age) which traverse the Bové Basin. After rotation of Africa to take account of the opening of the Atlantic, the trend of this belt becomes identical with the direction of the faults which cuts across Florida from East to West (Fig. 43) (see above for discussion).

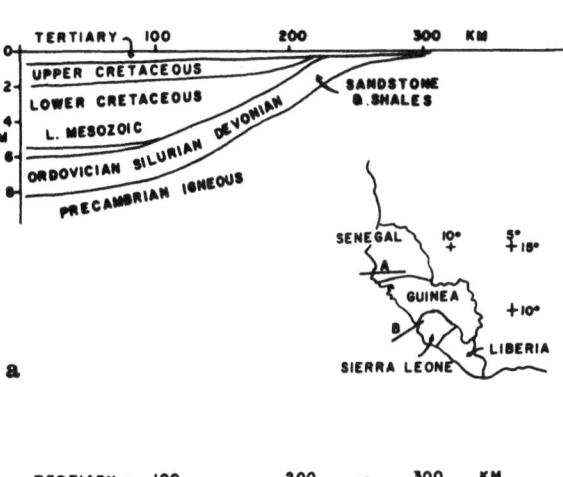

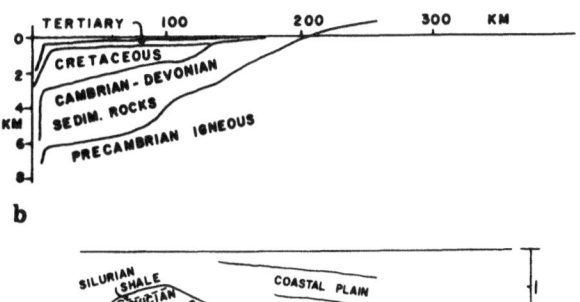

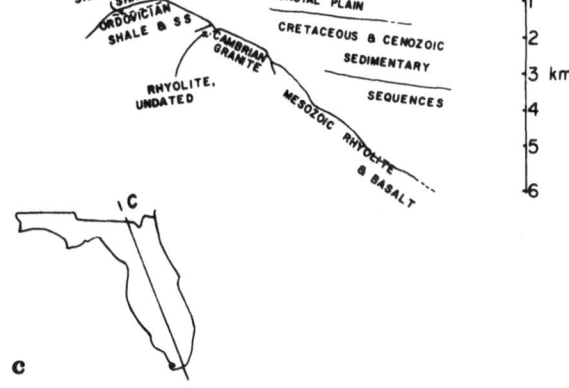

Fig. 47. Cross-sections through Palaeozoic basement. **a** Senegal basin, **b** Offshore Guinea, **c** Florida. (After Smith 1982)

In the past, certain authors (Sheridan et al., 1969) have suggested that the Mauritanides extended across the northern part of the Palaeozoic Bafata-Bové Basin as far as the SW edge of the Bijagos Archipelago. Ever since it was realized that the Mauritanides divide into two branches to the South of Senegal (Sougy, 1964; Bassot, 1969) it has become apparent that the western branch probably passes to the North of the Archipelago. This will be discussed further in a later section, since the concealed basic/ultrabasic belt which corresponds with this branch is also the northern limit of terrains showing little or no Hercynian effects.

Four other sedimentary belts of Palaeozoic age also show little or no Upper Palaeozoic deformation, although they are located in the Hercynian domain, but for different reasons. These are the Coastal Moroccan Horst, the southern part of the Anti-Atlas, the Tindouf Basin and southern Nova Scotia. The three African examples are undeformed because their underlying basement is made up of rigid Pre-Cambrian core-massifs (Guezou and Michard, 1976; Michard, 1976). The Canadian example corresponds to a region where suture formation on the northern margin of Gondwana occurred during the Acadian orogeny and where Hercynian movements were mostly transcurrent and located in a narrow shear belt (Lefort, 1983). Three of these sedimentary belts have recorded some weak Acadian effects, but decreasingly so away from Nova Scotia (Choubert, 1952; Michard, 1978). The Tindouf Basin, situated furthest to the East, remains totally undeformed.

6.2.2 Pre-Carboniferous Terrains Showing Hercynian Deformation

The terrains situated on the NW border of Gondwana have already been described in the chapter treating the submerged part of the Ligerian-Acadian mobile belt (Chapter Five). On the western side of Gondwana, however, the limit between pre-Carboniferous terrains that do or do not show Hercynian deformation is sometimes difficult to establish and merits further discussion. This limit locally overlaps with a pre-existing plate boundary. On land, it has already been established that the Hercynian Mauritanides are partly superimposed onto a strong positive gravimetric anomaly (letter A in Fig. 48) (Crenn and Rechenmann, 1965) that has been sometimes interpreted as a Taconic suture (Guetat, 1981) (Fig. 48). Since the Hercynian "orogeny" here has reworked Pan-African tectonites (Lecorché, 1983; Le Page, 1986), it has been suggested that it should be possible to trace the Pan-African suture, known further South (Villeneuve et al., 1984), into the Senegal suture. This is re-inforced by the existence of calc-alkaline and volcanic arc rocks of Late Proterozoic age along the gravity anomaly (Lecorché, work in progress). A "Taconic" event may have occurred in the region, but it is more likely related to a rejuvenation of the previous Proterozoic suture than to a true subduction. In any case, there is certainly a close relationship within the Mauritanides of E. Senegal between the Pan-African suture and the Hercynian belt.

To the South-West, in The Gambia, there is a gravimetric and magnetic anomaly which extends from Banjul to the eastern part of Casamance (Fig. 48) (Liger, 1979) – this anomaly shows many features in common with the Mauritanides anomaly. The Banjul anomaly, furthermore, separates the low grade Palaeozoic rocks of the northern Bové Basin from the micaschists, gneisses and diorites cored beneath the Senegal Basin. In addition, the western branch of the Mauritanides traverses this region and is seen to deform the northern part of the Bové-Batafa Basin (Lecorché, 1983). By analogy with what is known further East, the present author considers that the Banjul anomaly is also related to a suture. This conclusion is in agreement with Liger (1979), who proposes that the anomaly is linked to an intrusive body within the basement. Such a proposition does not totally contradict the results of Burke (1976) and Ponsard (1984) which favour, at the same place, the possibility of a decrease in the depth to Moho. In this way, it is quite possible that the Banjul anomaly has a double origin; the basic intrusion within the Gambian basement would be superimposed on an area of uplifted Moho, thus recalling the model put forward to explain the Mauritanides suture. Contrary to the suggestion of Burke (1976), the opening of the North Atlantic had probably little or no influence on the thinning of the crust in this area.

The Kidira-Velingara discontinuity is a major NE-SW-trending fault zone, well-marked by gravimetry (Fig. 48), which cuts the Mauritanides and Banjul anomalies at their southernmost extremities. Dextral shears on this fault zone are possibly Hercynian (Ponsard, 1984), but in any case post-date the intrusion of the basic bodies described above. South of this fault, the offset of the Gamon and Bataja intrusions with respect to the Banjul and Mauritanide sutures is about 50 km. Both the Gamon and Batafa intrusions are considered as the fragments of a suture zone (Ponsard, 1984) (Fig. 43). The Kidira-Velingara discontinuity represents a major transcurrent boundary which can also be recognized at Lat 28°N in Florida, where it appears to displace a N 130°-trending discontinuity marked on Fig. 43. This latter structure separates the Florida basement into two main zones whose Lower Proterozoic and Pan-African affinities have already been discussed (Fig. 44).

On the American side, the positive gravimetric and magnetic anomalies which link with the Gambian "suture" appear to separate different types of basement – in the Suwannee Basin (NW Florida), sediments are low grade or unmetamorphosed (Smith, 1982), whereas higher grade material from

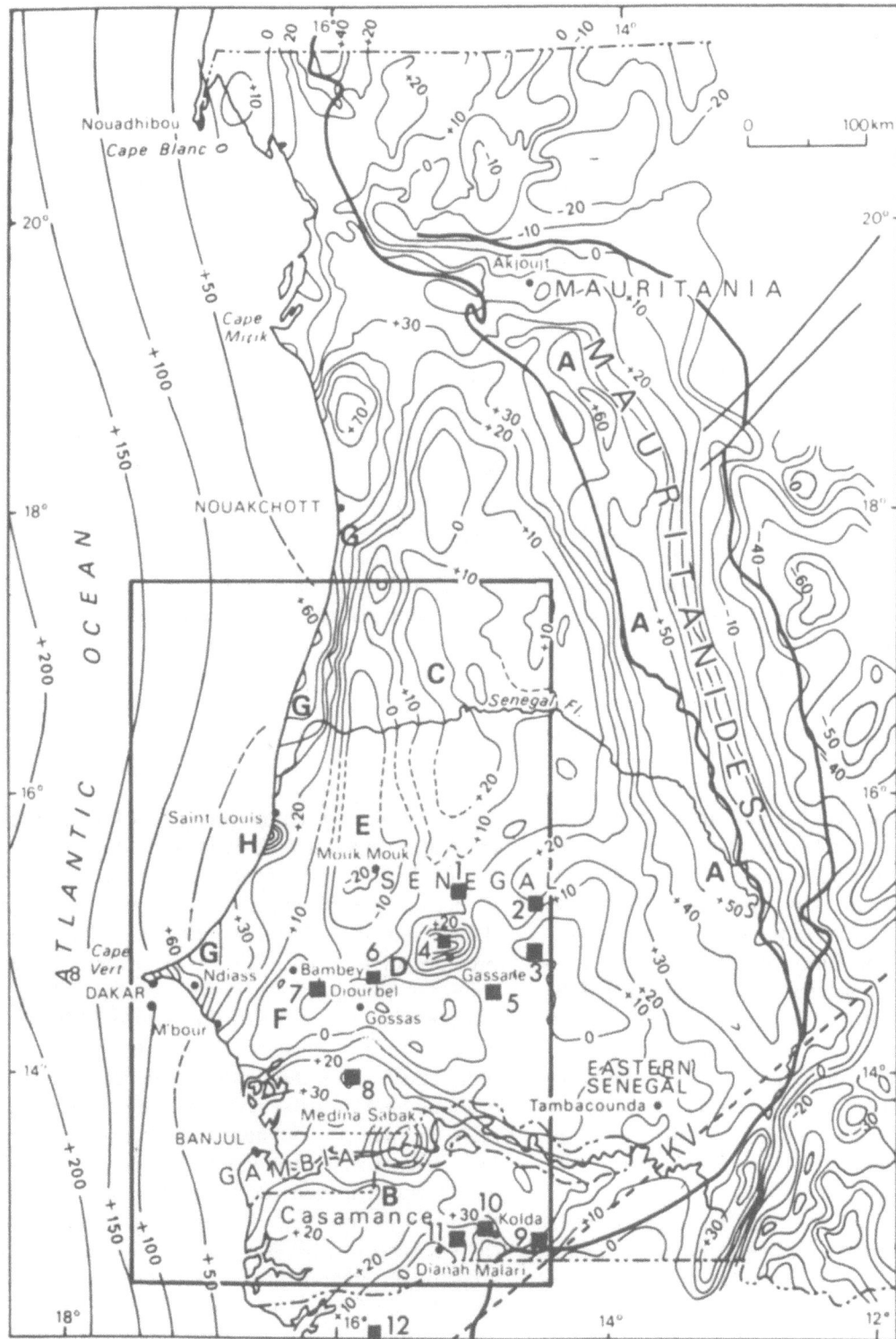

Fig. 48. Map of Bouguer anomalies surveyed in Mauritania and The Gambia. The limits of the Senegal Basin and Mauritanides orogenic belt are shown as bold lines. Offshore data are due to Uchupi et al. (1976). See text for explanation of letters A-H. Outline box shows area covered by magnetic survey. KV: Kidira-Velingara discontinuity; 1: micaschists (? Cambrian); 2: gneiss (Cambrian); 3: gneiss (?Cambrian); 4: gneiss; 5: micaschists; 6: biotite hornfels; 7: diorite; 8: basalt; 9: ?Palaeozoic; 10: ?Silurian; 11: ?Ordovician; 12: Silurian. Sample localities from Liger (1979). (Modified from Crenn and Rechenmann 1965; reproduced with permission of the Institut Français de Recherche Scientifique pour le Developpement en Cooperation − "ORSTOM")

the COST-GE1 well off the NE Florida coast (Fig. 44) is Palaeozoic, showing a metamorphic age of ca. 355 Ma (Scholle, 1979). This age of metamorphism corresponds with the K-Ar date of 360 Ma recorded in the Mauritanides (Lecorché, 1983). It is proposed here that the string of heavy and magnetic intrusions which extend from Senegal to Florida not only broadly represents the southern front of Hercynian metamorphism and deformation, but also traces the outline of an ancient microplate (Lefort and Van der Voo, 1981). This microplate could have drifted slightly away from Gondwana – alkaline through tholeitic basalts and rift sequences of Late Proterozoic age are known East of the Senegal suture (Lecorché, work in progress) – then was rapidly joined by the rest of Gondwana to produce the Pan-African deformation known in Mauritania.

Rhyolites in NE Florida (Fig. 44), close to the postulated microplate boundary, are associated with tuffs containing fragments of andesite and dacite. This material has yielded ages of between 480 and 425 Ma (Milton and Grasty, 1969; Chowns and Williams, 1983), thus supporting the hypothesis that the boundary of the Senegal microplate, which extended westwards into NE Florida and Georgia, had been remobilized by Mid-Ordovician/Mid-Silurian times. The geometrical form of this microplate explains why the "Taconic" activity controlled by Pan-African structures has no connection with the main Taconic events known in the S.W. Appalachians. Such an interpretation implies that the Senegal microplate was once again joined to Gondwana at the time of the Hercynian collision.

The excellent continuity of transverse structures recognized by gravimetry between the Senegal Basin basement and the West African Craton (Liger, 1979) is only interrupted by the Mauritanide belt – this shows that the drift of the microplate was obviously not important with respect to the rest of Gondwana since deep markers are still facing one another and are practically unaffected by the plate movements (Lecorché et al., 1983).

The identification of pre-Hercynian markers between northern Florida and the S.E. Appalachians is much more complex. This is due mainly to the difficulty in separating, from the geophysical point of view, those basic intrusions related to Hercynian collision from more ancient bodies which characterize the interior and borders of the Senegal microplate. The interpretation proposed by Klitgord et al. (1983), based mostly on magnetic data, addresses the problem of the western boundary of this microplate in the S.E. United States.

South of the magnetic highs which cut off the northern part of the East Suwannee Basin, there is evidence that the basement belongs to a Gondwanan domain showing little or no Palaeozoic deformation (Fig. 43). By contrast, North of the linear magnetic gradient marked as AA' on Fig. 49 (sometimes associated with oblong basic intrusions; Klitgord and Popenoe, 1983) there is a domain of magnetic high frequencies typical of the Piedmont terrain (Popenoe and Zietz, 1977; Daniels and Zietz, 1978; Daniels et al., 1983). The major tectonic boundary represented by the line AA' can be extended further westward (Horton et al., 1984) and is seen to clearly cut all the structures in the Southern Appalachians. As a consequence, it must correspond to a Hercynian or more recent feature; this is discussed further below.

Between the Piedmont and the Suwannee Basin, there is a terrain with high amplitude positive magnetic anomalies of broadly circular outline associated with positive gravimetric anomalies corresponding to basic intrusions (Daniels et al., 1983). In the same terrain, there are also magnetic and gravity low gradient zones (marked LGZ on Fig. 42) (Long et al., 1972) which are considered as indicators of the existence of acid plutons (Klitgord et al., 1984). Some of these anomalies, furthermore, are superimposed on granite bodies proven from drill-holes (Fig. 44).

The belt of narrow, elongate magnetic lows of large amplitude (including the Brunswick Magnetic Anomaly; Fig. 49) corresponds to an alignment of grabens, some of which are filled with Triassic (Hutchinson et al., 1983, Daniels et al., 1983) – the presence of this basin fill has occasionally been confirmed by seismic reflection.

The boundary established to the North of Florida on the basis of drill-hole data (Gohn, 1983) (Fig. 44) represents a major discontinuity between rhyolitic or unmetamorphosed rocks and metamorphic terrains. However, this boundary is not considered here to represent a major tectonic feature because all the metamorphic terrains do not necessarily show a Piedmont affinity, and because Lower Palaeozoic rhyolites are not always related to the Pan-African orogeny. The intermediate zone, termed the "Brunswick Terrane" (Williams and Hatcher, 1983) or "Charleston Terrane" (Higgins and Zietz, 1983), located south of line AA' and North of the Brunswick magnetic high, displays closer ressemblance (as far as magnetic markers are concerned) to the Florida basement than the Piedmont terrains. This is an important conclusion since it indirectly implies that the western extremity of the Senegal microplate, as defined above, is really a fragment of

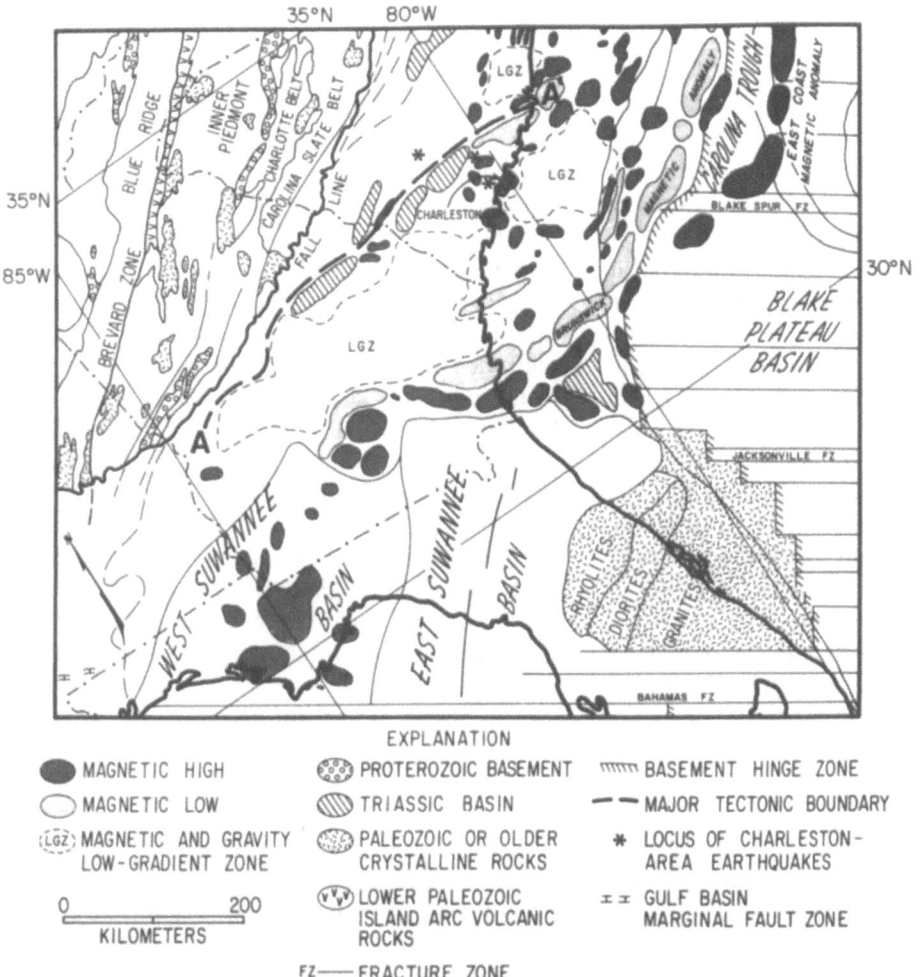

Fig. 49. Distribution of major geophysical markers to the NE of Florida. (Taken directly from Klitgord et al. 1983)

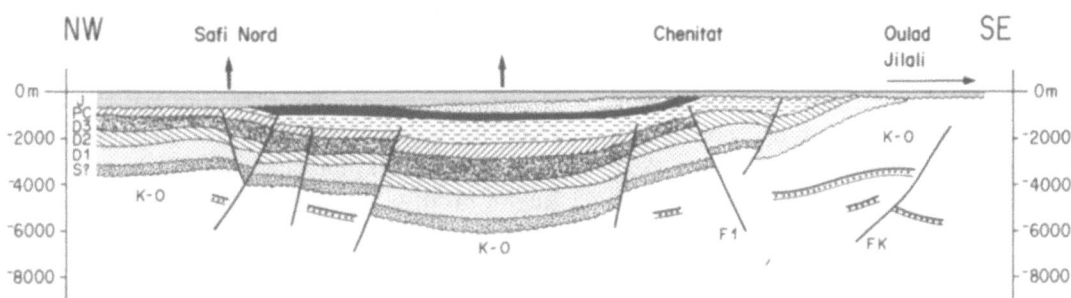

Fig. 50. Interpretative geological and geophysical section across the centre of the Doukkalas Basin (S. Coastal Moroccan Horst). KO: Cambro-Ordovician; S: Silurian; D_1: Lower Devonian; D_2: Middle Devonian; D_3: Upper Devonian; PC: Permo-Carboniferous; J: Jurassic. Vertical scale in metres. (After Barbu 1977)

Gondwana. The string of magnetic highs to the East of Charleston, following a more or less northeasterly trend (Fig. 49), is apparently homothetic with anomalies beneath the Senegal Basin (Fig. 61) and gives an idea of the arrangement of structures within the microplate. Whatever the age of deformation which has twisted these anomalies (Fig. 43),

it can be observed that they are clearly cut off at the microplate boundaries, thus supporting the hypothesis that such anomalies originate within the basement.

Some geophysical features, such as the East Coast Magnetic Anomaly and certain positive magnetic anomalies related to Mesozoic basic intrusions, owe their existence to the re-activation of zones of weakness within the Senegal microplate basement — this is probably also the case for the Brunswick Anomaly.

The "Charleston terrains", as described above, are bordered on their southern flank by gravimetric highs and a belt of positive magnetic anomalies especially well seen on the maps of Higgins and Zietz (1983). There are also some felsic rocks which occur to the North of the probable suture zone (Fig. 44). These undated rocks could be correlated with Cambrian igneous activity in central Florida and may correspond to the end of the Pan-African orogeny or, alternatively, the initial stages of separation of Senegalia from Gondwana. However, these rocks may also be comparable to the Ordovician formations known further East (Fig. 44). The rhyolitic activity around 425 Ma ago is considered here as resulting from the re-activation of the borders of Senegalia.

The magnetic and gravimetric boundary, taken here as a suture zone to the North of the Suwannee Basin, is superimposed on southward-directed thrusts seen in seismic reflection profiles — these thrusts are older than the Alleghanian suture, which cuts them at depth (Nelson et al. 1985), and dip off in the opposite direction (see sections 1 and 2 on Fig. 62). Furthermore, these southward-directed thrusts of inferred Pan-African age have a dip direction which is entirely compatible with the assymmetry of the Mauritanian suture as modelled beneath the Senegal Basin (Guetat, 1981) and The Gambia.

It is very likely that we are dealing here with the effects of an ancient subduction which enabled the "re-welding" of Senegalia to Gondwana. There is no evidence from available radiometric dating that the Senegal microplate suffered any rejuvenation during the Acadian orogeny — this is true not only for the Charleston region (Williams and Hatcher, 1983; Thomas et al., 1988), but also for Senegal (Lecorché, 1983). The only rejuvenation currently recognized has affected the Cape Hatteras region (Fig. 46) at the northernmost boundary of the microplate. This could be a possible consequence of the diachronic welding of Gondwanan fragments with North America.

The various bodies modelled from anomalies beneath the Senegal Basin are indicated as letters on Fig. 48. They include mafic rocks (B, C and D) and probably granites (E) at depth, but they do not help in placing any constraint on the age of metamorphism (i.e. Pan-African, "Taconic" or Hercynian).

The modelling of anomalies F, G and H (Fig. 48)

near the African coast comes up against an important problem of regional significance; that is, the probable existence of oceanic crust along the shore zone (Liger, 1979). Even though this particular problem is outside the general aims of this book, it is nevertheless useful to point out the existence of such a phenomenon since it explains why the proposed "fit" between Africa and America is so tight.

As regards the coastal Plain of the U.S.A., the main basic bodies are indicated on schematic maps (e.g. Figs.43 and 46) without any concern for their emplacement age, which may range from the Silurian or Devonian (as in the Gulf of Maine) to the Carboniferous (Fig.40). Figure 46 gives a more detailed picture of the possible distribution of concealed Pre-Cambrian terrains beneath the coastal Plain or margin, and has been supplemented by some geophysical information already brought together by Klitgord et al. (1985). Apart from granites of uncertain age, this eastern seaboard of the Appalachians can be subdivided into three types of structural domain.

The first type of structure corresponds to bodies which are obviously extensions of the magnetic lineaments recognized in southern New England and to the North of Florida and Georgia:

The second type traces out more or less concentric arcs which are sometimes incomplete — they are broadly centred on the Chesapeake Bay area.

Between these two domains, there is an arcuate belt of narrow anomalies which are undoutedly related to the existence of ultrabasic intrusions at depth.

Whereas most of the structures on the convex side of the Chesapeake Bay arc are probably pre-Carboniferous (because of their continuity with onland outcrops), the belt of ultrabasic bodies is younger since these structures cut all the pre-existing trends. Even though the complexity of these structural relationships has not been previously suspected, Taylor et al. (1968) have already observed some features of this tectonic pattern from magnetic survey results. Without going into a full description of the organization of these structures (given below), it is possible to affirm that the Chesapeake Bay Arc was created after formation of the Senegal microplate.

On the African side, Proterozoic or Archaean formations belonging to the Reguibat uplift are found further North beneath the Aaiun-Tarfaya basin (Auxini, 1969) (Fig. 43), thus suggesting that the northern extension of the Mauritanides, visible between basin and uplift, is a klippe (Marchand and Bronner; In: Lecorché, 1983).

Whether using geological or geophysical criteria, it

is difficult to separate structures on the concave side of the Chesapeake Bay Arc, which are related to Pre-Cambrian or Archaean terrains, from those which are Palaeozoic in origin.

The velocities of 6 km/s recorded off the Tarfaya basin do not really resolve the matter (Weigel et al., 1982) since Lower Cretaceous formations in this area yield velocities of 5.9 km/s (Maugis, 1955).

In fact, an important feature of the Tarfaya basin is that the basement has suffered no Acadian deformation — there is simply a temporary break in sedimentation during the Devonian.

The rather thin Palaeozoic cover on the western extremity of the Reguibat Spur appears not to have suffered the Taconic effects (Lecorché, 1983) which are nonetheless recorded further North in the Anti-Atlas (Michard, 1978). The lack of Taconic deformation may indicate that the Reguibat uplift behaved as a rigid block upon which the Palaeozoic cover remained flat-lying during possible rejuvenation of the boundaries of the Senegal microplate. However, weak "Taconic" events are also recorded further North in Morocco on the Meseta (Willefert, 1963; Piqué, 1979) and beneath the Essaouira and Doukkalas basins (Barbu, 1977) (Figs. 43 and 50) — this suggests that "Taconic" movements in this part of Africa were more than just local.

Since it is out of the question that the Mauritanide suture traverses or passes around the Reguibat Spur, it is necessary to find an explanation on the regional scale for movements in Morocco. Such an explanation is also needed because the "Taconic" movements in Morocco could not possibly represent a distant echo of tectonic events in the Appalachians during the Ordovician — the Meguma zone, which is interposed between the two regions, shows no Taconic deformation. The existence of ancient basement blocks, following different histories from Lower Palaeozoic times onwards (Piqué et al., 1985), lets us suppose that distinct cratonic segments initially developed in Morocco in a very similar way to that observed in Senegal. At a later stage, the reactivation of these blocks induced local epi-orogenic phenomena during the Ordovician. This hypothesis is supported by the fact that "Taconic" deformation developed is generally restricted to the boundary zones of rigid Pre-Cambrian blocks.

On the Mazagan Plateau (Ruellan and Auzende, 1985), the sampling of Hinz et al. (1982) has enabled a clearer picture to be drawn of the pre-Carboniferous history of this offshore region. Whole-rock ages obtained from dredged and cored samples are around 515 Ma (Wissmann et al., 1982), whereas fine-grained biotites record a re-setting event around 450 Ma which corresponds well with the age of certain on-land mylonites assigned to the Taconic "orogeny". By contrast, there is no record of rejuvenation during the Acadian orogeny — there was evidently no active mountain building in this area at that time. In the light of this situation, it is probable that the structures with large radius of curvature observed beneath the Coastal Moroccan Horst (see Chapter Five) were not formed during the Mid-Devonian but are much older — possibly Pre-Cambrian. This re-inforces the idea that the Meguma-type terrains are moulded around this ancient basement structure.

6.3 Conclusions Concerning the Concealed Pre-Hercynian Basement Between Africa and North America

The concealed pre-Hercynian basement between Africa and North America can be subdivided into six regions.

The first region includes most of West Africa (from Mauritania to Liberia) and Florida, corresponding to the western border of Gondwana. It is characterized by a basement generally older than 1,000 Ma and shows structural trends following a general N 130° direction. In places, this basement is re-worked by the Pan-African orogeny — notably in West Africa and the N.E. part of Florida. The Rockelides may be the result of a closure between West Africa and the Guyana Craton. The stratigraphy of the marine Palaeozoic deposits which overlie this basement shows that West Africa and the S.E. part of the United States were part of the same palaeogeographic province. These Palaeozoic basins show generally little or no Hercynian deformation.

The second region corresponds to the Senegal microplate, surrounded on all sides by deep basic intrusions. It became tectonically distinct after 700 Ma. The period of drift must have been short, since Senegalia rejoined Gondwana during the Late Proterozoic. This produced Pan-African deformation around its periphery. The possible African "Taconic" events previously discussed are perhaps related to the rejuvenation of this boundary and are different in origin from the Taconic tectonometamorphic effects recorded in the Appalachians.

The third region is represented by a protrusion of the West African craton towards the West (i.e. the Reguibat Spur) — basins within this cratonic area show no Lower Palaeozoic deformation because of the rigid nature of the underlying basement. This region displays many points in common with the rest of West Africa South of Senegal and is seen to interrupt or deviate some of the easternmost Appalachian structures on the American side. Thus, it is possible to establish that the westward impingement of the Spur must post-date the main fold phase in the American Appalachians. Otherwise, the Reguibat Spur also cuts the northern border of the Senegal microplate as well as the Acadian suture off southern New England.

The fourth region (described in Chapter Five) is the Meguma Zone, which is mainly composed of a turbidite-bearing Cambro Ordovician succession. It may be asked if these formations are also present on the NW border of the Senegal microplate? This question arises from the apparent lack of Meguma-type faciès on the micro-plate. One possible explanation is to consider the basement beneath the Meguma Zone as representing an ancient continental slope or transitional crust receiving sediment from Gondwana. If such sediments exist on the Senegal microplate, they should thus be restricted to the western border of the microplate, between northern Florida and Cape Hatteras. Their apparent absence could be explained by Alleghanian metamorphism in this region which may have rendered the original Gondwanan sediments hard to recognize.

The fifth region corresponds broadly to the Moroccan "Craton". Various periods of distension between basement blocks – similar in character to those described further South – did not, however, lead to separation and drift. But the later compression between Pre-Cambrian blocks that had been slightly disrupted produced epi-orogenic phenomena during the Ordovician similar in type to "Taconic" effects envisaged in Mauritania.

Finally, the sixth region is made up of several "suspect terrains" having a Pan-African basement which exist beneath the Coastal Plain and Piedmont of the U.S.A.

Taconic movements occurred on both sides of the North Atlantic, forming tectonic belts which are broadly parallel to the present coastlines. On the western side, the Taconic orogeny is continuous and was developed during an obduction of oceanic crust onto the Grenvillian craton (Williams and Hatcher, 1983). However, on the eastern side, the belt is more discontinuous and corresponds to rather limited crustal movements around disrupted basement blocks. This fragmentation of the basement was caused by distension or failed rifting – during the Palaeozoic, some of these fragments managed to drift away westwards from Gondwana, as witness the "suspect terrains" of the Carolinas (U.S.A.).

The apparent distribution of Acadian deformation is itself very informative since, contrary to Taconic deformation which developed independently on both sides of the North Atlantic, the Acadian orogeny is chiefly restricted to the present-day N. American plate. Whereas Acadian deformation on the northern flank of Gondwana is related to the closure of the Theic Ocean (see Chapter Five), tectonic effects of this age in the Southern Appalachians are of unknown origin, but may reflect collisions with detached fragments of Gondwana.

The most remarkable feature which can be seen on Fig. 43 is, however, the sharp interruption of the Mauritanide and Ligerian-Acadian sutures by the westward protrusion of the Reguibat uplift from the West African craton.

Chapter 7

The Submerged Part of the Hercynian-Alleghanian Foldbelt on the North Atlantic Margins

In contrast to the Caledonian-Appalachian and Ligerian (eo-Hercynian)-Acadian mobile belts, which occur offshore as fragments of Cordilleran-type orogenic belts, the concealed segments of the Hercynian foldbelt present none of the linear features typical of subduction-related orogenies. The Hercynian foldbelt, on the other hand, extends in a rather diffuse manner over almost the entire area of interest in this study. Furthermore, the intensity of metamorphism and deformation varies abruptly from place to place. In common with all other Palaeozoic orogenies, the Hercynian has resulted from the northward drift of Gondwana, or one of its detached fragments.

For greater clarity of presentation, deformation related to the Hercynian orogeny is discussed under two headings: the first concerns those terrains situated North of Gondwana whereas the other is restricted to basement involved in the collision between African and America.

7.1 The Hercynian Orogeny in the Northern Part of the North Atlantic Region

In this part of the Hercynian domain, as also further South, it is necessary to distinguish tectonic effects which result from the re-activation of pre-existing structures and those which developed for the first time during the Carboniferous.

A major pre-occupation of certain European geologists concerns the importance which should be attached to the "Bretonic phase" occurring around the Devonian-Carboniferous boundary. Events of this age are recorded in the Armorican Massif (Peucat et al., 1979), beneath the Aquitaine Basin (Paris, pers. comm. 1987), in Spain (Gil-Ibarguchi, 1983) and in Morocco (Piqué, 1983) — this phase is generally considered as the beginning of the Hercynian orogeny sensu stricto. The very rare events of this age recognized in England (Powell, 1983) are scattered over a wide area and have been interpreted as "late Acadian" in affinity — the same interpretation has been applied in Canada (Keppie et al., 1983). In the northern U.S.A. (Osberg, 1983), opinions concerning late Acadian effects are often divided — these disagreements are entirely justified, as we shall see below.

Offshore prospection and the investigation of concealed on land basins has not provided a resolution of this problem since "Bretonic" structures have always been reworked during the Hercynian; thus, only clearly established Carboniferous phenomena will be discussed in this study.

7.1.1 Re-Activation of Pre-Carboniferous Structures

Towards the North, late Caledonian faults (in the sense of Max, 1978) such as the Great Glen and Leannan Faults, show only slight evidence for Hercynian re-activation. In Scotland, eo-Hercynian and Hercynian sinistral shears would appear to have been demonstrated (Anderson and Owen, 1968; Haszeldine, 1984) but dextral displacements of this age are now proposed (Sanderson, 1984) for the Highland Boundary and Southern Uplands Fault as well as for the Iapetus Suture Zone. These re-activations are certainly developed in Ireland (Sevastopulo, 1981), but they are mainly vertical faults of small displacement developed during Dinantian times. For technical reasons, deep seismic profiles are not able to contribute to our understanding of this phenomenon.

Shallow seismic reflection methods, however, reveal the existence of Carboniferous deformation quite far to the North of the Variscan Front. There is an elongate, N60° trending Carboniferous basin between Ireland and Scotland (enlarged to the South between the Isle of Man and Anglesey) which show fold axes (N60°/N70°) that have resulted from the reworking of pre-existing Caledonian structures (Fig. 17) (Wright et al., 1971). Oil Company seismic surveys have, furthermore, shown that the westernmost part of this basin corresponds to a half-graben which was controlled by N-S-trending faults of probable Namurian age. As a matter of fact, there is a very wide concealed Carboniferous basin to the East of the Anglesey – Isle of Man ridge which has been affected by deformation equivalent in time to the earth movements recorded in the Pennines (Bennison and Wright, 1970).

In the Southern Irish Sea (Fig. 16), the general trend of the Upper Carboniferous basin is controlled by N 30°-striking faults, but the associated fold axes are oriented N 60° – this suggests the existence of an "en échelon" structure linked to transcurrent shearing. However, there is no certainty that these movements were really Carboniferous, especially since shearing in this case should have been sinistral and the southerly extension of the lineament shows mainly normal faulting at that time (Gardiner and Mac-Carthy, 1981).

Two main successions have been recognized in this basin; the lower group shows facies similar to the folded Carboniferous of Ireland whereas the upper group (occurring mainly in the NE) may correspond to the Upper Carboniferous. This basin may thus partly constitute a link between the Visean-Tournaisian successions in Ireland and Wales (Dobson et al., 1973).

The Caernarvon Basin (Fig. 16), forming a synclinal interdigitation to the SE of the "Central Basin", is itself probably also composed of folded Carboniferous strata. Finally, the Kish Bank Basin further west shows a thickness of 3-4 km with concealed Stephanian lying above folded Westphalian C and D (Naylor and Shannon, 1982). In a general way, Hercynian deformation appears to become more intense towards the Variscan Front (Fig. 51).

Knowing that the overthrust units of the Lizard are probably ophiolitic in nature (Kirby, 1984) and that the protoliths are possibly Devonian (Fitch et al., 1984), one can consider the thrust movements as Hercynian. This is despite the uncertainties in age data and mode of emplacement (Rattey and Sanderson, 1984) – the thrusting probably developed out of an initial zone of distension which, however, does not imply that we should consider this structure as an interplate suture (Le Gall, 1984). It should also be noted that, further East, the Lizard Thrust takes up a typical E-W Hercynian trend (Lefort, 1975). By contrast, to the West, the thrust has undoubtedly re-activated Cadomian or Caledonian crustal weaknesses since it is evidently parallel to structures of pre-Hercynian age.

In the southern part of the Western Channel Approaches, continuous and symmetrical thrusts are observed which are of the same age as the Lizard Thrust (Lefort, in press 1988b). These structures can be followed by seismic reflection and some of them actually reach the coast (CL and CH on Fig. 23). The axis of symmetry between these southward directed thrusts and the northward directed Lizard Thrust does not co-incide with the Mid-Channel Cadomian suture (Fig. 22) even though it is clearly parallel and continuous over a considerable distance. Amongst the northward dipping faults, some are certainly Hercynian due to the fact that they cut the mid-Channel suture. But there are others which re-utilize ancient structures that were antithetic with respect to Cadomian subduction (Balé and Brun, 1983; Lefort and Bardy, 1987; Lefort, in press 1988b). In the English Channel, the general Hercynian style of thrusting suggests a process of upward tectonic ejection which occurred at the same time as expulsion of a wedge towards the West. This is supported by the sinistral transcurrent component observed in

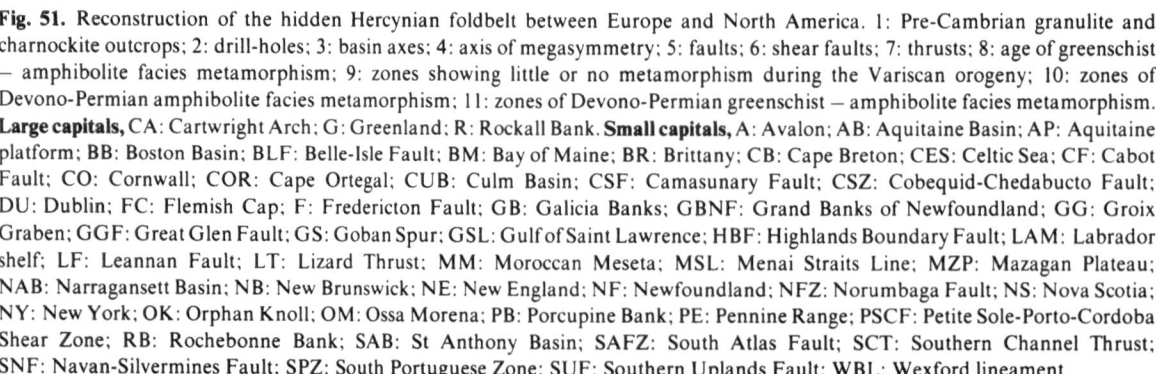

Fig. 51. Reconstruction of the hidden Hercynian foldbelt between Europe and North America. 1: Pre-Cambrian granulite and charnockite outcrops; 2: drill-holes; 3: basin axes; 4: axis of megasymmetry; 5: faults; 6: shear faults; 7: thrusts; 8: age of greenschist – amphibolite facies metamorphism; 9: zones showing little or no metamorphism during the Variscan orogeny; 10: zones of Devono-Permian amphibolite facies metamorphism; 11: zones of Devono-Permian greenschist – amphibolite facies metamorphism. **Large capitals,** CA: Cartwright Arch; G: Greenland; R: Rockall Bank. **Small capitals,** A: Avalon; AB: Aquitaine Basin; AP: Aquitaine platform; BB: Boston Basin; BLF: Belle-Isle Fault; BM: Bay of Maine; BR: Brittany; CB: Cape Breton; CES: Celtic Sea; CF: Cabot Fault; CO: Cornwall; COR: Cape Ortegal; CSF: Camasunary Fault; CSZ: Cobequid-Chedabucto Fault; DU: Dublin; FC: Flemish Cap; F: Fredericton Fault; GB: Galicia Banks; GBNF: Grand Banks of Newfoundland; GG: Groix Graben; GGF: Great Glen Fault; GS: Goban Spur; GSL: Gulf of Saint Lawrence; HBF: Highlands Boundary Fault; LAM: Labrador shelf; LF: Leannan Fault; LT: Lizard Thrust; MM: Moroccan Meseta; MSL: Menai Straits Line; MZP: Mazagan Plateau; NAB: Narragansett Basin; NB: New Brunswick; NE: New England; NF: Newfoundland; NFZ: Norumbaga Fault; NS: Nova Scotia; NY: New York; OK: Orphan Knoll; OM: Ossa Morena; PB: Porcupine Bank; PE: Pennine Range; PSCF: Petite Sole-Porto-Cordoba Shear Zone; RB: Rochebonne Bank; SAB: St Anthony Basin; SAFZ: South Atlas Fault; SCT: Southern Channel Thrust; SNF: Navan-Silvermines Fault; SPZ: South Portuguese Zone; SUF: Southern Uplands Fault; WBL: Wexford lineament

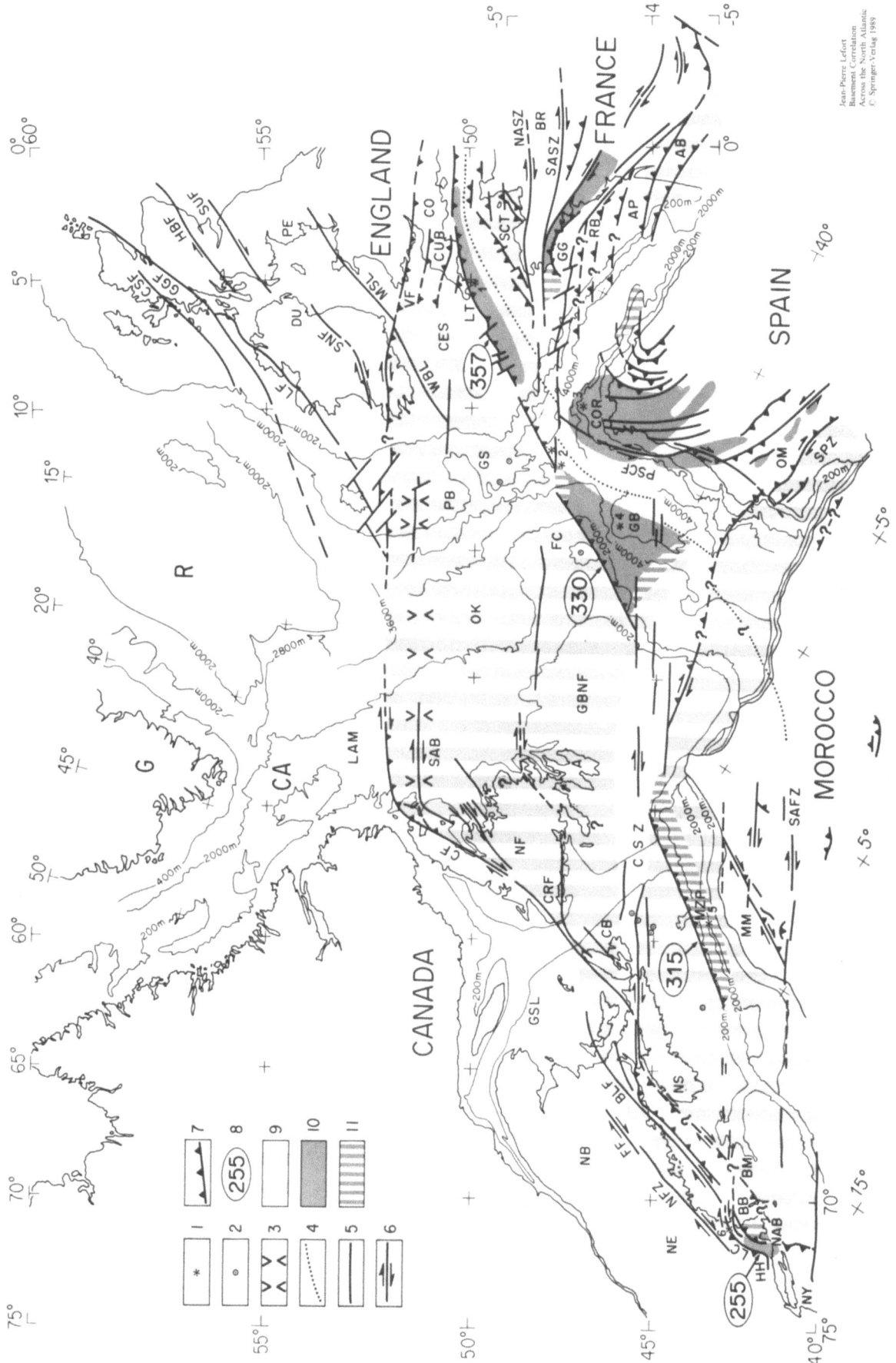

Jean-Pierre Lefort
Basement Correlation
Across the North Atlantic
© Springer-Verlag 1989

Northern Brittany and the very small dextral components recorded in England in several places (Sanderson, 1984).

In the Armorican Massif, it is known that the South Armorican Shear Zone was initially at the origin of sinistral displacements of Devonian age (Audren, 1986) before becoming a dextral shear zone during the Hercynian orogeny. The westward extension of this shear zone is based on the sampling of leucogranites (Lefort and Peucat, 1974) (Fig. 29) and magnetism (Lefort, 1975). Contact metamorphism associated with these shear zone granites has produced spotted slates and magnetite-rich hornfelses.

Following on from a tightening-up of the original Cadomian arc (Fig. 19), the formation of the Iberian Armorican Arc occurred during the Hercynian orogeny (Julivert, 1971). Offshore data has only enabled mapping of the northern branch of this arc, but the information is crucial because it finally reveals (and for the first time) the actual shape of the structure. This is made possible by gravimetry and seismic reflection results which indicate a solution profoundly different from the theoretical models of numerous authors (e.g. Bard et al., 1971). The deep seismic profile recorded on the Aquitaine Shelf (research group ECORS-GASCOGNE and Lefort, in preparation) enables the recognition of three major thrusts (AP on Fig. 51), showing a northerly dip, which can be correlated with tectonic discontinuities observed at depth beneath the Aquitaine Basin (Fig. 32) (Le Pochat, 1984). The northernmost thrust (not represented in Fig. 32) traverses the Gironde estuary and separates garnet micaschists (Well R, Fig. 32) in the North from very low grade metamorphic terrains (Well ART, Fig. 32) in the South. The two other thrusts further South follow the major stratigraphic boundaries deduced from well-log data. Taking account of the gravimetric correlations between Iberia and South Brittany (Fig. 34), as well as the structural organization of the Iberian part of the Arc, it is possible to assume that the tectonic zonation of the South Armorican margin reflects the original sedimentary domains (Fig. 24). On this basis, the thrusts situated between the South Armorican suture and the northern part of the Aquitaine plateau have been inferred. One of these thrusts, however, has been proven by the sampling of mylonites and the presence of a strong magnetic discontinuity (Lefort, 1975) (RB on Figs. 24 and 51).

By analogy with the South Armorican Shear Zone, with which it is parallel, the Ligerian suture should have been re-activated as a dextral shear during the Hercynian — this has not yet been demonstrated, but there is evidence that the Ligerian suture has controlled the Carboniferous graben (Fig. 32) which traverses the Aquitaine Basin and which is made up of terrains showing a strong post-Namurian schistosity. Neither is there any decisive argument in favour of a "North Pyrenean Fault" passing between the Aquitaine Basin and the Iberian Massif at this time (Mattauer and Séguret, 1971).

The sinistral re-activations of the Little Sole — Porto — Badajoz — Cordoba Fault during the Carboniferous have already been discussed, but it is important to point out that these movements were developed at the same time as the dextral displacements on the South Armorican Shear Zone. The tectonic convergence produced by such shearing has undoubtedly greatly contributed to buckling of the Iberian-Armorican Arc (Lefort et al., 1982; Brun and Burg, 1982), thus adding to the effects of collision which were developed between Gondwana and Laurussia (Chapter Eight).

Finally, in Southern Europe, the South Portuguese Zone is characterized by a series of thrusts turning towards the SW which affect both low grade Upper Devonian terrains and the Carboniferous. These thrusts are easily recognizable from gravimetry (Lefort et al., 1981); one of the most important and continuous is the thrust which bounds the Ossa Morena Zone to the SE and which affects the gabbro-diorite complex of Beja. This complex is well marked by a positive gravity anomaly which can be followed northward under the Mio-Pliocene cover (Fig. 33). In this region, as a matter of fact, Carboniferous deformation merely prolongs (from Siluro-Devonian times) the processes of Ligerian-Acadian suture formation described in Chapter Five (Ribeiro et al., 1983). Furthermore, there is a clear pattern in the migration of sedimentation and deformation between Upper Devonian and Upper Namurian times in the Ossa Morena and South Portuguese Zones.

Otherwise, the Hercynian crustal shortening recognized in Southern Portugal developed in a dextral (Andrews, 1983) transpressive shear regime whose effects can be detected on the Grand Banks of Newfoundland. Here, the southernmost tips of the N-S trending Pre-Cambrian ridges were actually deflected towards the West (Figs. 19 and 36) before late Hercynian shearing (Lefort and Haworth, 1978). This suggests that the clamping together of the Avalon Spur and Gondwana was oblique, continuing through into Carboniferous times the processes of Acadian plate collision (Lefort, 1983). The dextral Hercynian re-activations in Canada

(Howie, 1984), New England (Bradley, 1983) and Morocco (Michard and Piqué, 1980; Cailleux et al., 1983; Lagarde, 1985) generally follow discontinuities which show Caledonian trends. Some of these can even be traced offshore to the North of New-foundland (Hyde, 1984), in the Gulf of St Lawrence (Howie, 1984) and in the Gulf of Maine (Hutchinson et al., 1985). However, it is rather difficult in this region to tell whether the Acadian suture was re-worked as a shear zone, even though strong evidence is provided by the dextral transpressive régime observed along the eastern border of the Appalachians (Gates et al., 1986).

One of the most remarkable features of the North Atlantic is found in the southwestward extension of the Lizard Thrust. This structure, which controls the morphology of the margin South of Flemish Cap and to the East of Nova Scotia, is cut by three East-West transcurrent belts of late Hercynian age. Here, the structure is identified as a major fault which systematically places high grade Pre-Cambrian terrains in contact with Palaeozoic formations showing little or no Carboniferous metamorphism. The terrains situated to the South of the thrust have, furthermore, suffered Hercynian effects which are sometimes intense. In this way, one can follow two series of parallel sites (from Cornwall to Nova Scotia) which show contrasting geological histories. Thus, there are quartz-bearing granulites immediately South of the Lizard Thrust, assigned to the Pre-Cambrian (Edmonds et al., 1975) (Site 1, Fig. 51), which are associated with amphibolite facies gneisses metamorphosed during the Devo-no-Carboniferous (Styles and Rundle, 1984). To the North of the Thrust, on the other hand, Devonian sediments show only a very weak degree of metamorphism (Le Gall, 1983).

Off western Brittany (Site 2, Fig. 51), the outcrops of granulites and charnockites of probable Pre-Cambrian age, which are sampled along with Hercynian granodiorites (Guennoc, 1978), are seen to occur opposite Devonian rocks in drill-holes on the Goban Spur. Here, there is no trace of either Caledonian or Hercynian metamorphism (Lefort et al., 1984).

To the North of Cape Ortegal, and on the western Galicia Banks, there are further outcrops of Pre-Cambrian granulites and charnockite (Sites 3 and 4, Fig. 51) which form a basement to formations that have suffered amphibolite facies metamorphism during the Hercynian. The high grade rocks on the European side can be fitted near granodiorites on Flemish Cap that have suffered no Carboniferous rejuvenation (King et al., 1985).

Charnockites dated at 980 Ma (Ruellan, 1985) have been found on the Mazagan Plateau (located off Morocco) and are associated with granodioritic gneisses of Upper Proterozoic – Cambrian age (Site 5, Fig. 51). These rocks have undergone upper greenschist facies metamorphism during the Hercynian and occur opposite the pre-drift position of two drill-holes into younger and apparently less metamorphosed formations on the Canadian side. One of these wells has reached Cambro-Ordovician sediments showing Meguma-type facies (Jansa and Wade, 1975) and displays evidence for reheating around 300 ± 14 Ma ago. The other well has sampled Devonian granites (Cormier and Smith, 1973) which appear to have suffered resetting of their biotites around 329 ± 14 Ma (Given, 1977). The jump in metamorphic grade from one side to the other of this major fault is fairly large in the NE, but appears to diminish in magnitude towards New England. Any doubt about the sampling on Mazagan Spur should now be cleared up since granulitic zelics have been found in Hezcynian granodiorites of the Moroccan Meseta (Chemsseddoha, in preparation).

Finally, the granulites in the Nashoba Formation of southern New England (Skehan, 1983), of probable Upper Proterozoic age, are very close to the Narragansett basin where rocks show a late Hercynian amphibolite facies metamorphism (Dallmeyer, 1982). The systematic occurrence of high-grade ancient basement terrains along the southern border of the Lizard Thrust and its westward extension does not appear to have resulted from the erosion of the peri-Atlantic margins. This is because the belt of uplifted basement is not symmetrically disposed with respect to the present-day Atlantic Ocean basin, but rather seems to follow the trace of a major fault zone that was active before the Late Hercynian. It is quite possible that, like the Lizard Thrust, this fault zone is Hercynian. Since this fault is strictly parallel to the trace of late Caledonian faults, it could have been controlled by one of them. On the European side, this structure corresponds to a metamorphic front of greater importance than the Variscan Front. Southward, it cannot be traced beyond the south-eastern edge of the Avalon Spur (Chapter Four) because there was extensive remobilization along its length during the Upper Carboniferous. The East-dipping thrusts recognized in seismic reflection Southwest of Flemish Cap (de Voogd and Keen, 1987) could be associated with this main feature.

7.1.2 Development of Late-Stage Hercynian Deformation

The northernmost major late-Hercynian structure is represented by the Variscan Front, but this structure is also the least well marked in terms of gravimetry and magnetism. In Western England, Wales and Eire, this fault zone corresponds to a discontinuous thrust, directed towards the North, which is rooted in the Celtic Sea area (Fig. 52) (Cheadle et al., 1986). In contrast to the Lizard Thrust, the Variscan Front was associated with "thin skin" tectonics similar to the structural style of the thrusts concealed beneath the Northern part of the Paris Basin (Cazes et al., 1985) — this difference in style constitutes some evidence for the pre-late Hercynian structural development of the Lizard Thrust. As a matter of fact, the very weak geophysical signature of the Variscan Front can be better understood if one considers that it represents only an erosional boundary. The true thrust Front was previously developed much further North, as witness the klippes in Eire almost as far as the Dublin area (Cooper, pers. comm., 1987).

The Variscan Front thrust also shows the local development of a dextral transcurrent component, notably in western Eire (Max and Lefort, 1984) and in England (Sanderson, 1984). Indirect evidence of this dextral offset can be found in the Celtic Sea by comparing the negative gravimetric anomalies recognized on each side of the Front (Fig. 26) (Blundell, 1979). Even though these anomalies are really linked to the presence of Mesozoic and Tertiary basins, their displacement is significant insofar as these basins were controlled by equivalent basement faults (Fig. 51). Because of its very weak influence on potential field, the Variscan Front can only with difficulty be followed onto the Porcupine Bank, thus giving rise to much speculation. The interpretation proposed by Cherkis et al. (1975) is based solely on magnetic data from the oceanic crust off the Bank, and cannot be directly tested. Riddihough and Max (1976) have, for their part, proposed two possible solutions which are founded on rather sparse marine magnetic data. At that time, the role of late-stage offsetting was not appreciated. A more recent interpretation (Lefort and Max, 1984) is based not only on a new magnetic survey (Max et al., 1982) but also the recognition of N 130°-trending Permo-Triassic fractures which offset the Front towards the North (Fig. 53). The East-West flexure at 53°N, mapped by Bailey (1975) using seismic reflection, follows part of the proposed trace of the Front in the South. On the Bank itself, the interpretation of the fracture pattern is complex since several generations of fault are seen to intersect each other.

The first fault set, oriented N 60°, is mainly found North of Lat. 53°N where it prolongs a fault zone mapped in Northern Ireland — these are Caledonian structures which are sometimes truncated against E-W-trending structures of probable Hercynian affinity. However, the best-represented structural trends follow a N 130° direction which, as in the Celtic Sea and on the Armorican margin (where age control exists), should have been developed during the Permo-Triassic (Leutwein et al., 1972). This fault set is seen to displace all previous structures in the continental crust in a dextral sense, and has even acted as a transform direction during the opening of the Porcupine Sea Bight (Lefort and Max, 1984). In this way, the N 130° fault system offsets magnetic anomalies (partly of oceanic origin) that are detected in the Bight as well as the trace of the Variscan Front, various Caledonian faults and the northern edge of the Bank.

Finally, the N 130° set is cut by N-S-trending fractures which probably originated from the reactivation of ancient structures during opening of

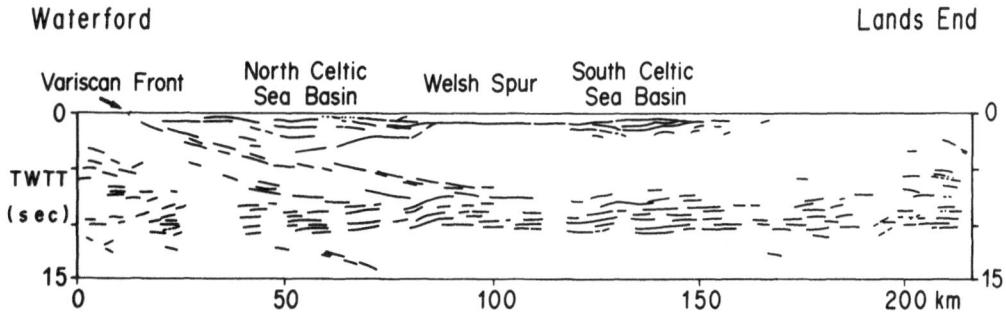

Fig. 52. Seismic reflection profile (SWAT) showing cross-section of Variscan Front in the Celtic Sea area. (After Cheadle et al. 1986)

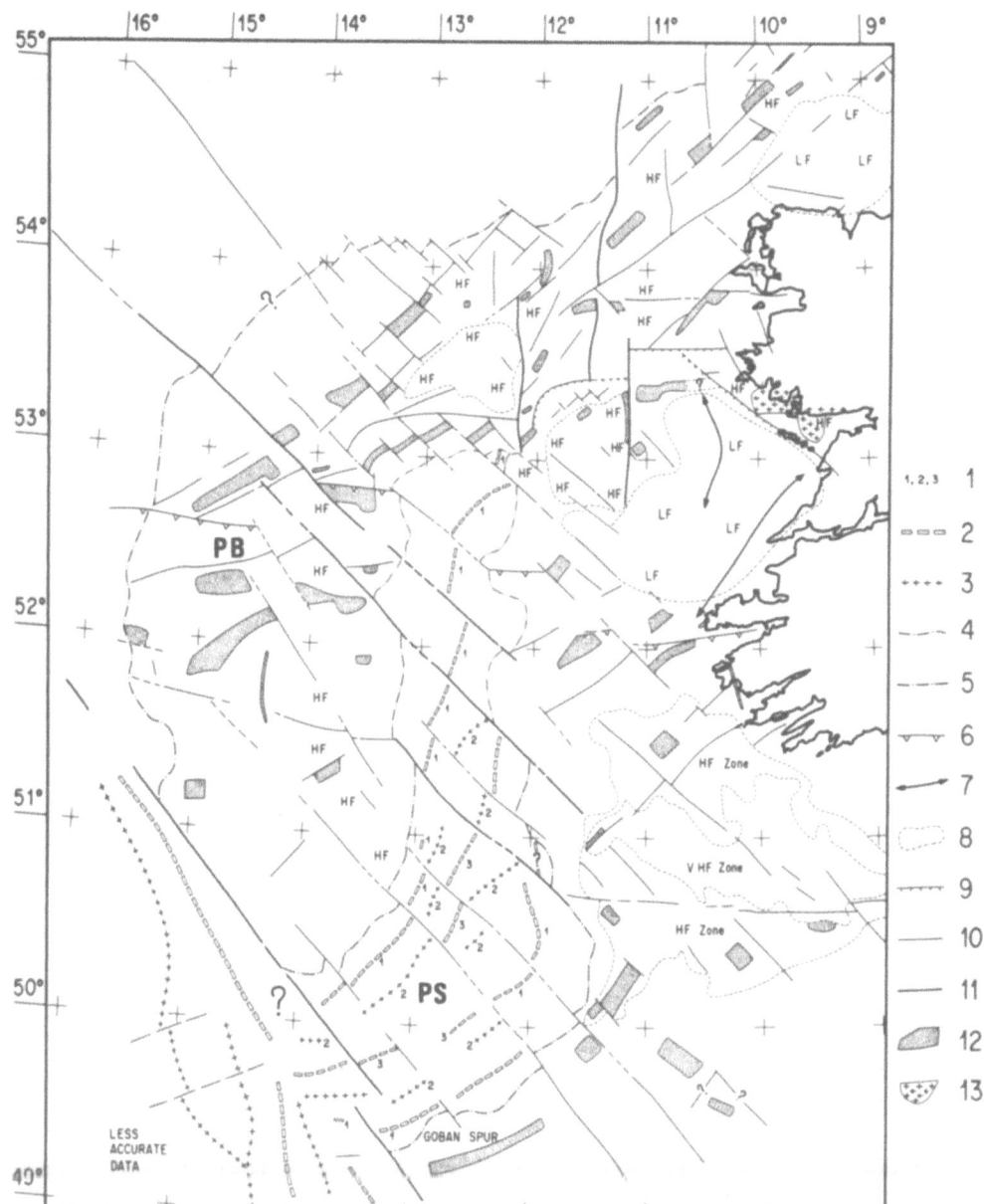

Fig. 53. Sketch map of basement West of Ireland, based on magnetic data. 1: magnetic lineations; 2: magnetic lows; 3: magnetic highs; 4: limit between normal and thinned continental crust; 5: limit between oceanic and continental crust; 6: probable trace of Variscan Front; 7: structures in the low frequency zones; 8: limits of zones, HF (high frequency), VHF (very high frequency) and LF (low frequency); 9: flexures or deep faults; 10: second-order structures; 11: major structures; 12: magnetic bodies; 13: granites; PB: Porcupine Bank; PS: Porcupine Seabight. (After Lefort and Max 1984)

the Northern part of the North Atlantic. These fractures controlled the general trend of the Porcupine Bight and form the walls of the bayonette-shaped graben which continues the axis of the Bight towards the North. Furthermore, such orientations have been observed on land (Russel, 1968).

The Variscan Front is bordered on the South by a sub-parallel structure which traverses a zone of

magnetic high frequencies produced by recent intrusions (Caston et al., 1981). Because of the straightness and orientation of this structure, it is probably a transcurrent Hercynian shear. South of the Porcupine Bank, the basement may be crystalline in nature or deeply eroded, covered only by a thin silver of Palaeozoic (Auzende, pers. comm., 1987), since there is a lack of seismic reflectors beneath the Mesozoic cover (Bailey, 1975). To the South of the Variscan Front, the presence of reflectors, by contrast, suggests a possible westerly extension of the Irish Devono-Carboniferous succession.

Opposite this zone on the other side of the Atlantic, there is a Mississippian-Pennsylvanian succession discordant upon the Ordo-Silurian off the North Coast of Newfoundland. This succession is affected by a N-S-trending fold phase which, on the one hand, does not follow pre-existing Appalachian structures (Haworth et al., 1976) but, on the other, is cut by broadly E-W-oriented structures (Lefort and Haworth, 1984). The southernmost E-W structure

(Fig. 54) corresponds to a dextral shear which follow the axis of the "Carboniferous" St Anthony Basin (Fig. 51) – as described in Chapter Four, this basin can be traced eastwards as far as the northern part of Orphan Knoll (Fig. 24).

The thrust showing a dextral shear component (recognized toward the North, in the southern Labrador Sea – Fig. 54) could be correlated with the Variscan Front on both structural and geometrical grounds, and is seen to curve round into the Taylors Brook Fault (a northerly equivalent of the Cabot Fault). This is the first of many examples showing that the rejuvenated ancient fault systems which follow the Appalachians may link up with the

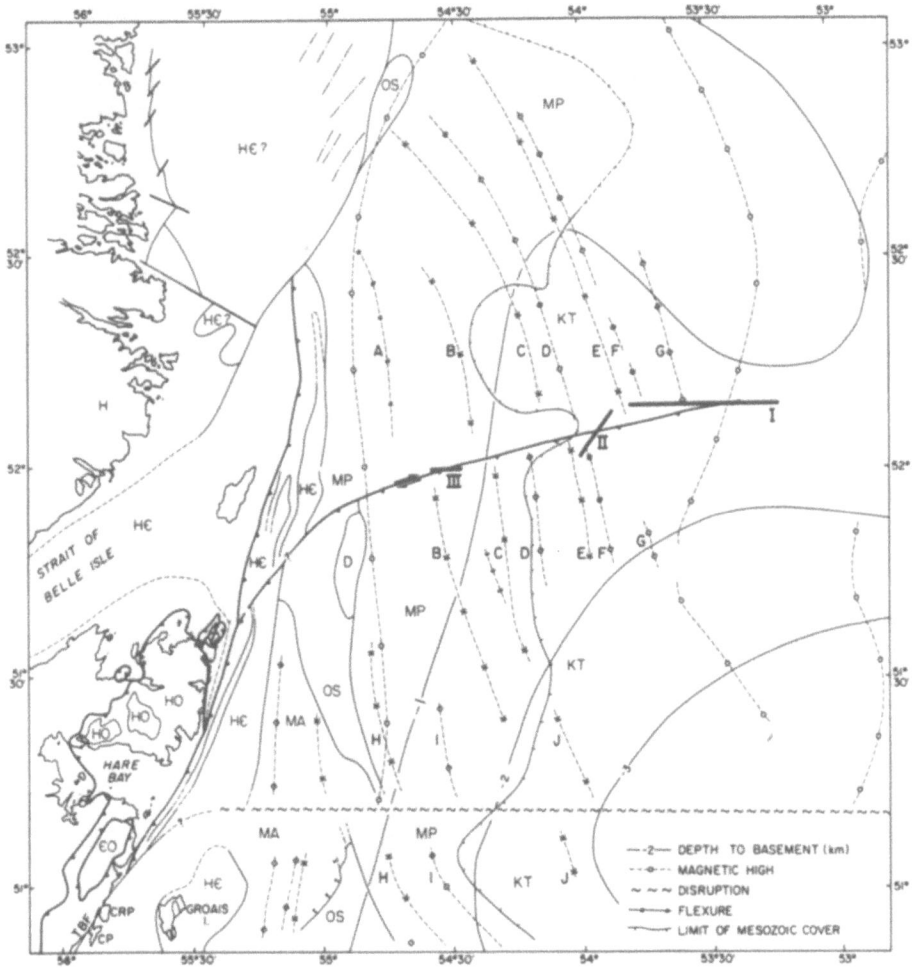

Fig. 54. Hercynian deformation North of Newfoundland. Letters A-J refer to fold axes correlated across transverse discontinuities. Seismic profiles which intersect thrust faults are given Roman numerals. Dotted lines with crosses: synclinal axes; dotted lines with open diamonds: anticline axes; TBF: Taylors Brook Fault; CP: Conche Peninsula; CRP: Cape Rouge Peninsula; H: Helikian or older terrains; HE: Hadrynian and Cambrian; HO: Hadrynian-Mid Ordovician; EO: Upper Cambrian-Mid Ordovician; OS: Ordovician or Silurian; D: Devonian; MA: Mississippian; MP: Mississippian and Pennsylvanian; KT: Cretaceous-Tertiary. (After Lefort and Haworth 1984; reproduced with permission of the Geological Society of London)

dextral Hercynian shears known on the European side (Lefort and Haworth, 1984). Such a correlation based on plan view observations does not, however, imply that the Variscan Front necessarily developed the same tectonic style on both sides of the Atlantic. The Culm basin in North Devon and Cornwall (SW England) is well known for its southward- and northward- divergent structures and the occurrence of symmetrical thrust faults of Hercynian age (Coward and Smallwood, 1984). This basin is superimposed on a strong positive magnetic anomaly which extends toward the SE Celtic Sea, but the material in the basin cannot explain the observed anomaly (Fig. 51). Magnetic modelling associated with the interpretation of seismic data has led to a resolution of this apparent contradiction. In actual fact, the Culm basin (Fig. 55) is seen to overlap a slice of lower crust, of Hercynian affinity, which can be modelled as a basic slab 7,500 m deep with a magnetic susceptibility of 2×10^{-3} e.m.u. (c.g.s. units). This is the first described example of an offshore Upper Palaeozoic basin showing tectonic control by contemporaneous deep crustal thrusting. There appears to be a correlation between the on-land Culm basin and offshore Carboniferous basins on the Canadian side. This would suggest, furthermore, that the "pop-up structure" known in Europe

(i.e. the Culm Basin) represent the remnants of an elongate E-W Carboniferous basin which was subject locally to upward tectonic ejection.

On the other hand, it is not known whether the pre-existence of a Devono-Carboniferous basin controlled the development of the Variscan Thrust Front, or whether the existence of early E-W crustal trends controlled both the formation of the Carboniferous basins and the location of the Front. The available seismic refraction and reflection data between England and Canada simply suggests that there is a more or less continuous basin, characterized by seismically slow basement, occurring to the South of the Variscan Front (Fig. 56).

Other dextral transcurrent belts can be followed across the Atlantic, such as the shear zone which traverses Brittany (Gapais and Le Corre, 1980) and joins up with the Newfoundland Grand Banks (Lefort and Haworth, 1978). The geophysical signature of the Canadian segment is characterized by a strong E-W-trending magnetic anomaly probably produced by a recent intrusion. Recent magnetic data (Lefort and Lapointe, work in progress) suggests that this fault zone may come onshore North of the Avalon Peninsula due to a southward bend.

The Cobequid-Chedabucto shear zone, which cuts Nova Scotia in two, has been shown to juxtapose the

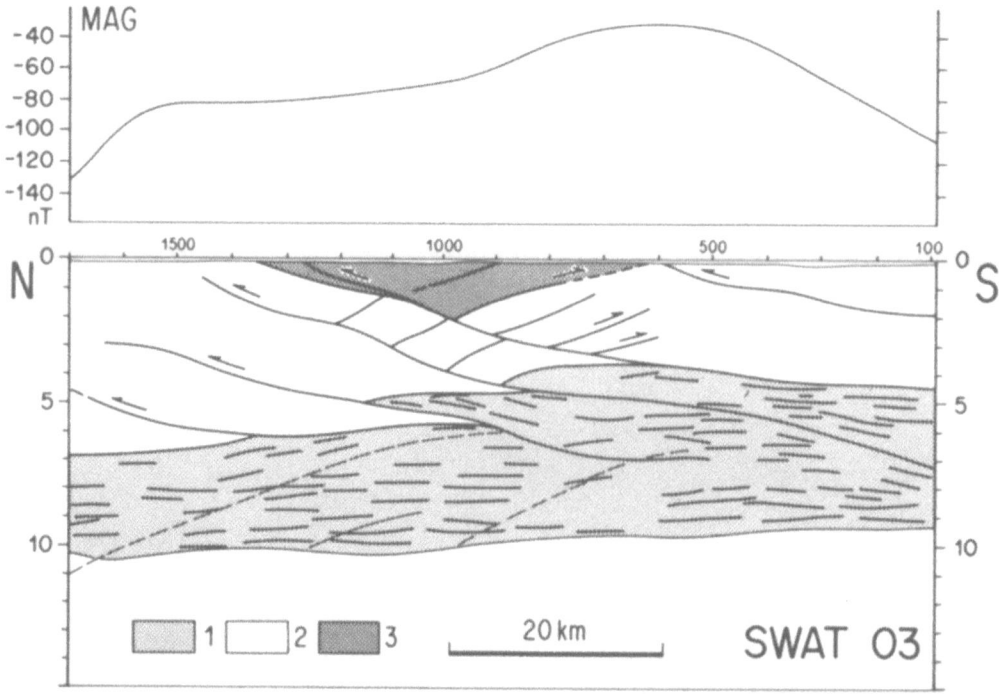

Fig. 55. Seismic reflection profile obtained across offshore extension of the Culm basin, showing example of a "pop-up structure". The superimposed magnetic profile shows a positive anomaly produced by thrusting of lower crustal slices. 1: wedges of lower crust; 2: upper crust; 3: Culm basin

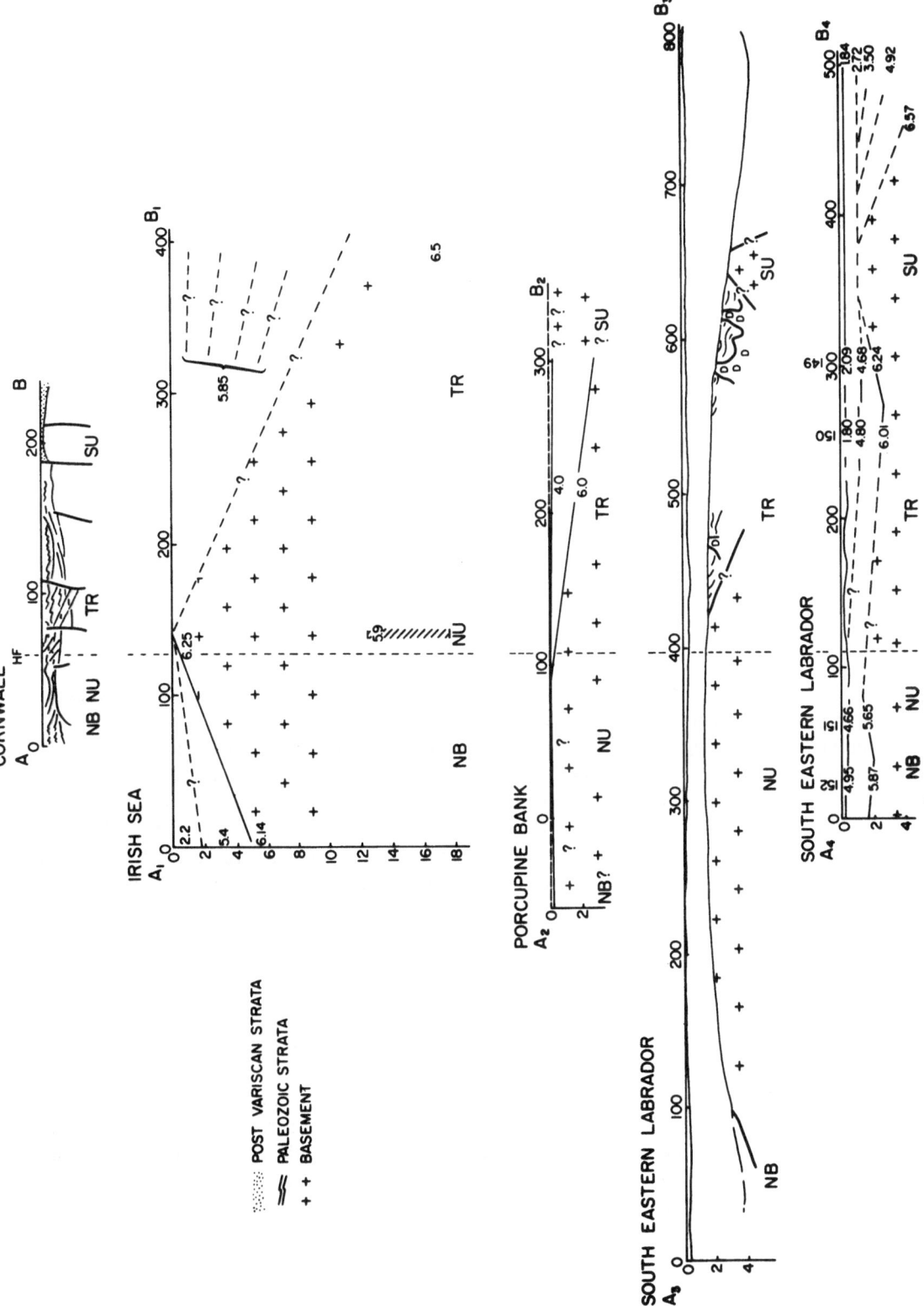

Fig. 56. Seismic refraction and seismic reflection profiles perpendicular to the Variscan Front. A–B – Geological cross-section across Cornwall, A1–B1 – Seismic refraction profile across the Irish Sea and Celtic Sea, A2–B2 – Seismic refraction profile along the Porcupine Ridge, A3–B3 – Seismic reflection profile parallel to the Labrador margin, A4–B4 – Seismic refraction profile to the SE of Labrador. Key – NB: Northern basin; NU: Northern Ridge; HF: Variscan Front; TR: thrust plane or ramp; D: diapirs. Horizontal and vertical scales indicated in km and velocities in km/s. (After Lefort and Haworth 1984; reproduced with permission of the Geological Society of London)

Avalonian terrains of Cape Breton Island with Meguma zone rocks further South (Eisbacher, 1969). Movements on the shear zone were dextral in sense and occurred during the Late Carboniferous. This has been confirmed by palaeomagnetic studies (Spariosu et al., 1984). Towards the East, this fault zone has been shown, by the analysis of magnetic and gravimetric data (Lefort and Haworth, 1978), to extend at least as far as the Galicia Banks. The displacements produced by this shear belt are apparent just as well in the Pre-Cambrian ridges of the Grand Banks as in the Ligerian-Acadian suture (Fig. 57), and show that movement continued after the coming together of Gondwana and Laurussia. Again, the connection with Appalachian discontinuities in the West is brought about through a curved thrust front (Rast et al., 1979).

A third Hercynian belt links structures in Morocco with structures in the Bay of Maine. The late Hercynian dextral shears which traverse northern Morocco and the coastal horst are well-defined (Lagarde, 1985), in contrast with the diffuse magnetic structures in the Bay of Maine (Lefort and Haworth, 1981; Simpson et al., 1979). Furthermore, this belt appears to join up with the curved fault traces which extend between the Boston Basin and the Clinton-Newbury Fault (Billing, 1976) — it provoked a re-activation of ancient structures either as thrusts or transcurrent shears. Figure 58 shows an interpretative sketch of the different geophysical markers recognized in the Gulf of Maine (Lefort and Haworth, 1978); it stresses the importance of E-W fracturing in the control of Triassic basins (Ballard and Uchupi, 1975) and the shaping of the continental slope.

The correlation between the South Atlas Fault (bounding northern Morocco in the South) and the 40°N Fault has long been proposed (Drake and Nafe, 1968; Drake et al., 1963; Drake and Woodward, 1963); although the South Atlas Fault was clearly active at a late stage of the Hercynian orogeny (Mattauer et al., 1972; Michard, 1976; Proust et al., 1977), the fault trace on the American side is still open to interpretation (Root and Hoskins, 1977). The 40°N Fault continues westward into the Appalachian foldbelt, linking with the curved fault which controls the Gettysburg Basin in Pennsylvania (Lefort and Lapointe, work in progress); the offshore part of this fault is less well defined, probably corresponding with a kink in the margin (Fig. 41) and an associated magnetic anomaly which follows it (Klitgord and Behrendt, 1977). On the African side, the South Atlas Fault can be extended offshore according to the trace proposed by Lehner

and de Ruiter (1977) — this trace has been chosen since it has here been defined with the greatest precision. However, it remains very difficult to link up the South Atlas Fault with the 40°N Fault — this problem will be discussed further in a later section. Nevertheless, apart from this problem, it is generally observed that the continuation of late-stage Hercynian transcurrent belts from Europe and Africa onto the North American plate nearly always produces curved zones of thrusting and shearing, sometimes spectacularly well developed, which link with re-activated pre-existing Appalachian discontinuities. Such structures are normally associated with the development of dextral re-activation in the hinterland along faults which were in many cases originally Taconic or Acadian. In the United States, this re-activation was first noted by Webb (1969) and has been recently taken up again by Gates et al. (1986).

Offshore, there are some Carboniferous transcurrent structures resulting from reactivation of N 60°-oriented faults in the prolongation of the Belle Isle (Rast et al., 1979), Fredericton (Ruitenberg and McCutcheon, 1982) and Cape Ray Faults (Wilton, 1983) (Fig. 51). In Morocco, parallel shears are developed at this time along the Coastal Horst (Michard, 1976; Barbu, 1977) and on the Mazagan Plateau (Ruellan, 1985).

Thus, the faults which border the northern part of the North Atlantic domain may equally well be re-activated as shears or thrusts when the original discontinuity is pre-Hercynian in age. By contrast, late-stage structures are normally organized as parallel shear belts which broadly truncate the previous structural grain. Despite differences in style and age, these deformations were all developed in the context of an anti-clockwise rotation on the regional scale. This rotation has to be studied on an even larger scale, however, in order to identify the geodynamic process involved (Chapter 8).

Another type of Hercynian structure has been recognized off South Brittany; it is distinct from the previous cases in that it results neither from re-activation of ancient Faults nor from late transcurrent shearing. In fact, there are offshore and on-land outcrops of Hercynian leucogranite in this area dated at 300 Ma which occur on either side of the Groix Graben — these granites define an assymmetric arc concave towards the Bay of Biscay. Studies on the mainland granite massifs (Bouchez et al., 1981; Audren and Jegouzo, pers. comm., 1987) have shown that, in the East, magma was emplaced into a series of flat-lying shear zones which are directed towards the South. This shearing was as-

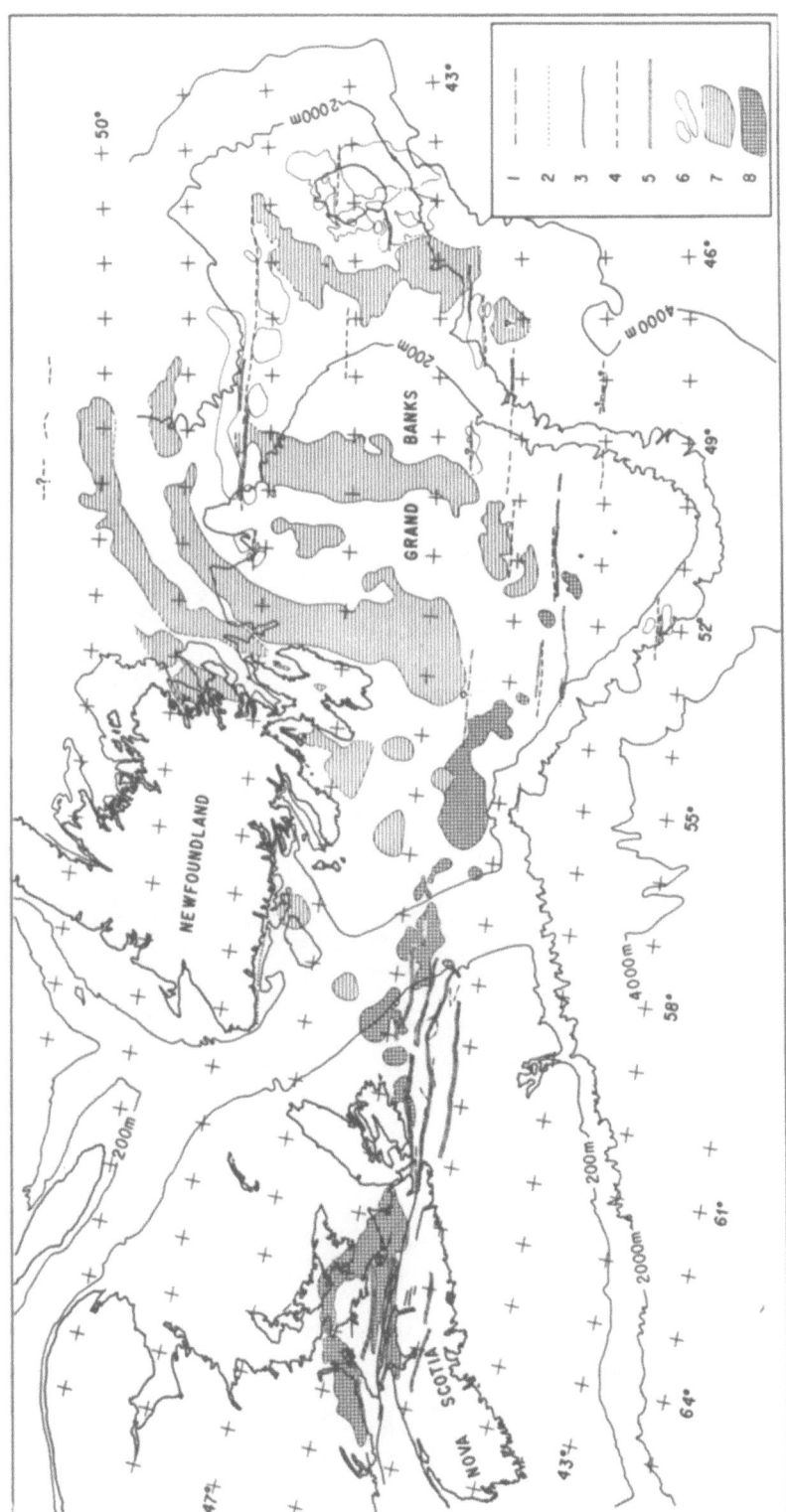

Fig. 57. Sketch map illustrating intersections between Pre-Cambrian ridges, the Acadian suture and late Hercynian transcurrent shear belts to the South of the Newfoundland Grand Banks. 1: faults recognized by seismic relfection; 2: faults based on morphology; 3: on-land faults; 4: faults recognized from magnetism; 5: faults recognized from gravimetry; 6: magnetic anomalies of various ages; 7: anomalies linked to Pre-Cambrian ridges; 8: anomalies belonging to the "Collector Anomaly". (After Lefort and Haworth 1978)

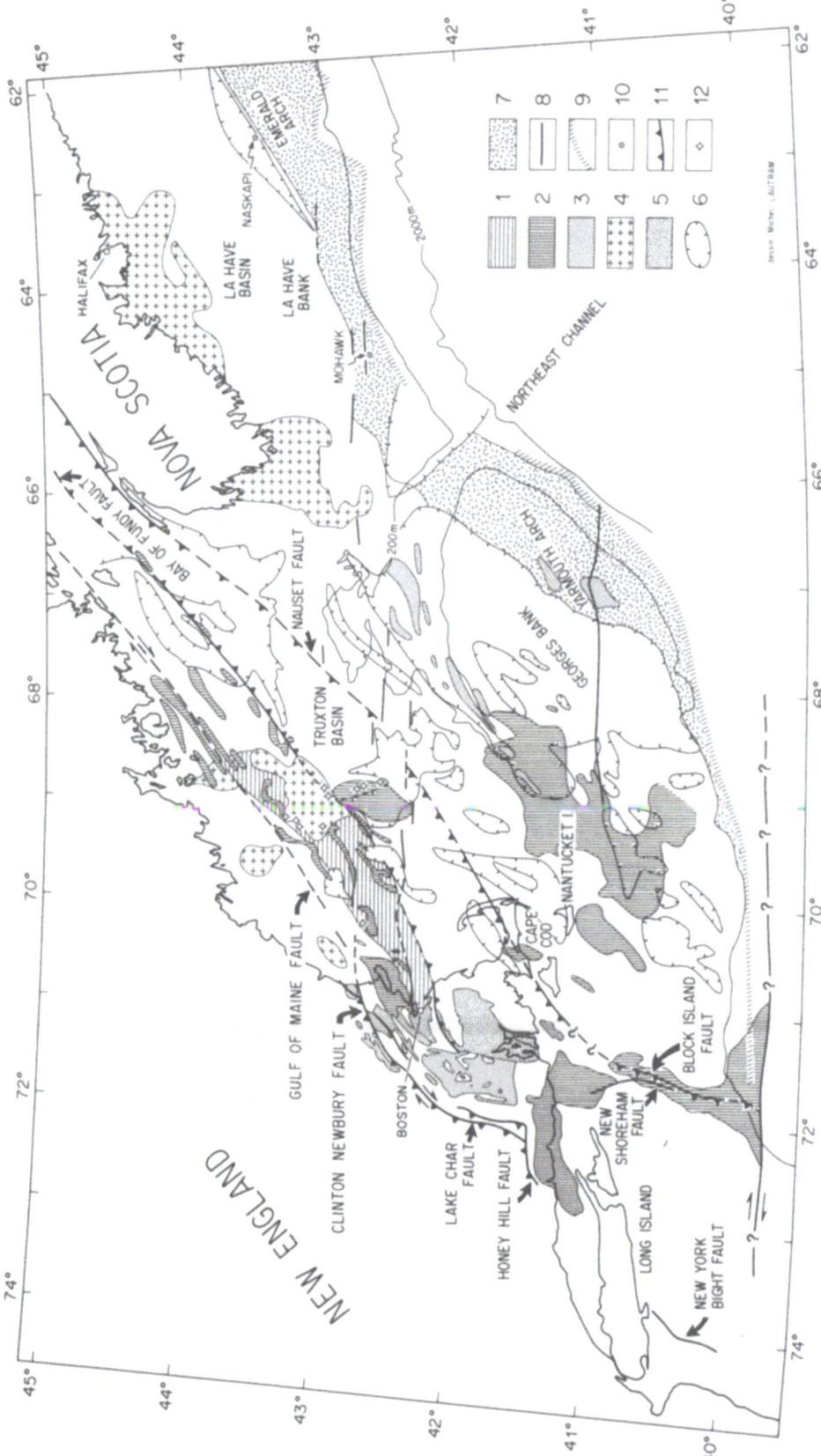

Fig. 58. Interpretative sketch map of geophysical data collected in the Gulf of Maine. 1: Permo-Carboniferous basin; 2: magnetic intrusive bodies; 3: intrusions mapped by seismic methods; 4: granites recognized from gravimetry; 5: Mississippian-Pennsylvanian; 6: Triassic basins recognized by seismic methods; 7: shelf-edge arch; 8: faults recognized from magnetism; 9: continental slope; 10: drill-holes; 11: thrusts; 12: cored wells. (Modified from – Ballard and Uchupi 1975; Kane et al. 1972; Uchupi and Austin 1979; Lefort and Haworth 1981; Hutchinson et al. 1985)

sociated with a dextral transcurrent component. In the West, the same shear regime is apparent but the transcurrent compcount is sinistral. The arcuate form of granite sheets in South Brittany (Vigneresse, 1978), as well as their mode of emplacement, is certainly due to a divergence in the tectonic transport directions (Fig. 59).

Diver observations on offshore micaschist formations betweeen the Ile de Groix and Belle-Ile (Lefort et al., 1982) show that the fold axes of the youngest Hercynian deformation in this area follow a shaped trace which reflects the eastward bend in the granite outcrops (see fold axis orientations on Fig. 59). Using geophysical information, it is possible to put forward a general model to explain why late deformation of the metamorphic terrains occured at the same time as the emplacement of the adjoining granites. It is apparent that the Z-shaped bend in late

fold-axis orientation is not only homothetic with the granitic arc but also with a pronounced indentation in the magnetic body which nowadays marks the South Armorican suture of eo-Hercynian (Acadian) age (see high apparent susceptibility body on Fig. 59). Since the late Hercynian folding and granite emplacement was contemporaneous, one is forced to admit that late-stage tightening of an irregularity (see Chapter Five) in the South Armorican suture was responsible for the curved fold trends and arcuate structures described above. This kind of deformation could be termed posthumous impingement (or identation) because late Carboniferous deformation was the result of intraplate stresses that continued after the main phase of collision between Iberia and Armorica during the Devonian.

In conclusion, it is appropriate to give some idea of the metamorphic grade attained during the Her-

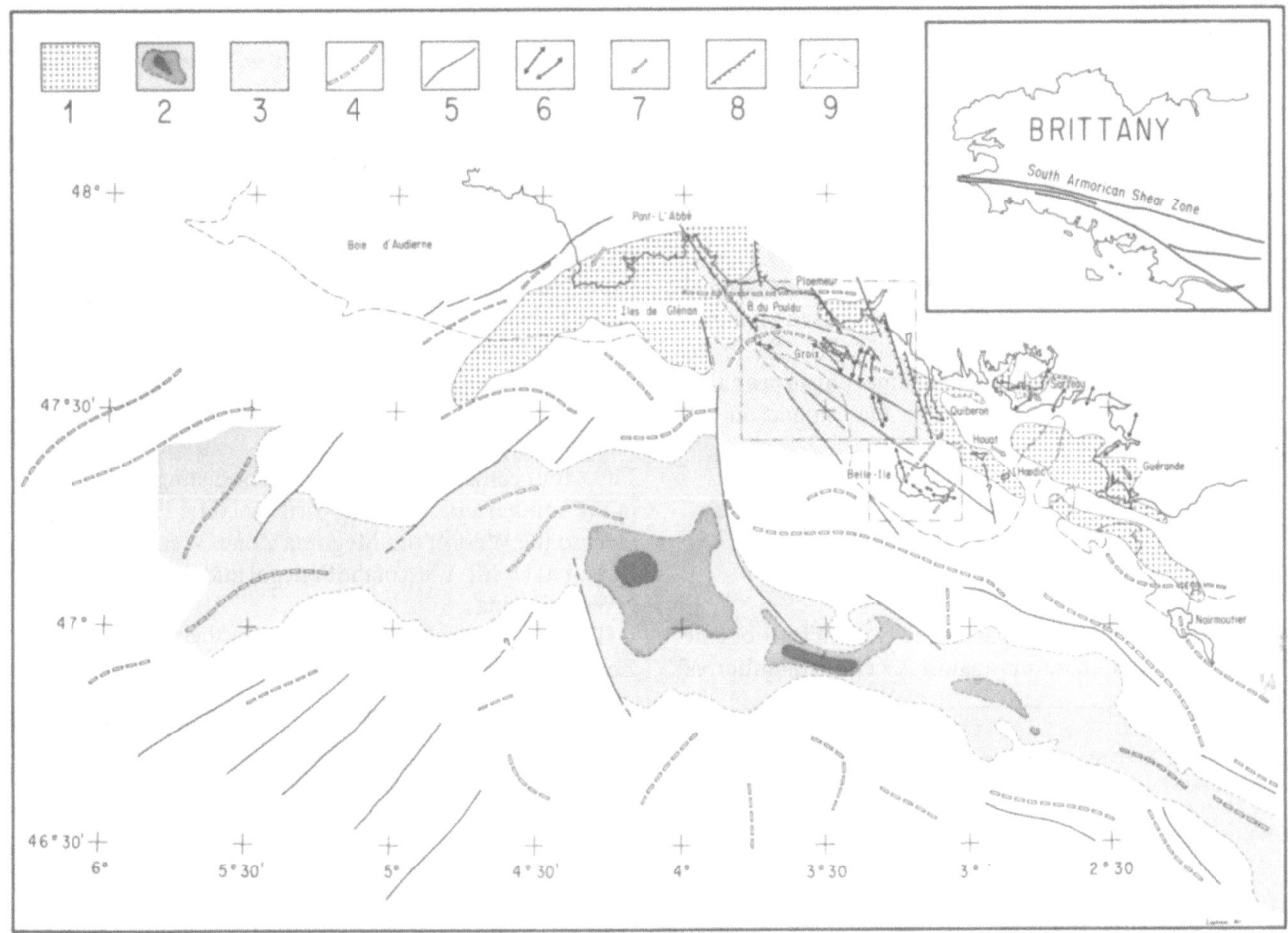

Fig. 59. Sketch map showing juxtaposition of major units off the southern part of the Armorican Massif; the southerly shaded zone corresponds to the suture. 1: 300-Ma-old granites; 2: magnetic bodies with high apparent susceptibilities (darker shading indicates higher susceptibility); 3: Ile de Groix graben; 4: axis of magnetic bodies calculated from vertical gradients reduced to the pole; 5: faults; 6: fold-axes; 7: shearing in the granites; 8: normal faults; 9: boundary between basement and cover. (After Lefort et al. 1982)

cynian orogeny in basement terrains around the northern part of the North Atlantic domain. Because of the geophysical methods employed in offshore areas and the frequent lack of precision in drill-hole descriptions, it is difficult to establish metamorphic facies and age relationships. In the absence of better information, it is nevertheless possible to distinguish zones showing little or no metamorphism from zones where metamorphic grade reaches the limit between the greenschist and lower amphibolite facies. A third class is represented by terrains showing rather higher grade amphibolite facies assemblages. However, these grade comparisons are meaningless unless it is realized that amphibolite facies metamorphism is of Late Carboniferous age on the western side of the Atlantic and generally pre-Carboniferous or Lower Carboniferous in the East. Towards the South, off Morocco, the regional metamorphism (Ruellan, 1985) is Late Carboniferous (Kreuzer et al., 1984) whereas the Hercynian metamorphic event in New England (Dallmeyer, 1982) occurs at the Permo-Carboniferous boundary (Wintsch and Lefort, 1984).

Thus, it can be seen from Fig. 51 that the Lizard Thrust (Edmonds et al., 1975), and its offshore prolongation, upfaults progressively younger Variscan metamorphic units as one proceeds SW across the Atlantic. (The estimated average ages of metamorphism for the Galicia Banks and Cape Ortegal are deduced from an extrapolation based on a diagram of Gil-Ibarguchi et al., 1983). This diachronism in the thermal peak of metamorphism probably goes hand in hand with the migration of deformation.

7.2 The Hercynian-Alleghanian Orogeny in the Southern Part of the North Atlantic Region

A description of this part of the Hercynian orogenic belt would come up against a certain number of problems if presented in the same manner as the preceding section. Taking account of the intensity of thrusting in certain zones, it is often difficult to say, for instance, whether Hercynian structures are "new" or re-activated. Furthermore, even though it is fairly clear that tectono-metamorphic events started with a "Bretonic phase" dated at between 360 and 340 Ma (Dallmeyer et al., 1986) and continued through from 330 to 230 Ma in the Southern Appalachians (Glover et al., 1983), such is not the case for Africa. The existence of 360 Ma dates in the Mauritanian orogenic belt of W. Africa may not necessarily indicate a Bretonic phase since the K-Ar age data (Lecorché, 1983) for these rocks may be in question. Furthermore, the ages of last movement are cut off at about 280 Ma (Dallmeyer and Villeneuve, 1986) and there are probably no Permian events. In Morocco, the time bracket for Hercynian deformation is well constrained, commencing with "Bretonic" movements and terminating with events around 300 Ma (Piqué, 1983); there, there is no clear record of the metamorphic climax around 255 Ma recognized further West in New England (Dallmeyer, 1982).

In this way, it would appear that the two flanks of this particular orogenic segment did not develop simultaneously and over the same time span. This asymmetry is further complicated on each side of the Atlantic by a certain diachronism along the foldbelt from North to South. Such is the reason for discussing the contact between Gondwana and Laurentia segment by segment in order to take better account of the progressive process by which these two plates became welded together to form Pangaea.

Thus, four collisional segments can be distinguished in the southern part of the North Atlantic Hercynides: to the West of the Meguma Zone, West of the Reguibat Uplift, West of the Senegal microplate and NW of Florida.

— The suture located to the West of the Meguma Zone has been described in the previous chapter. It

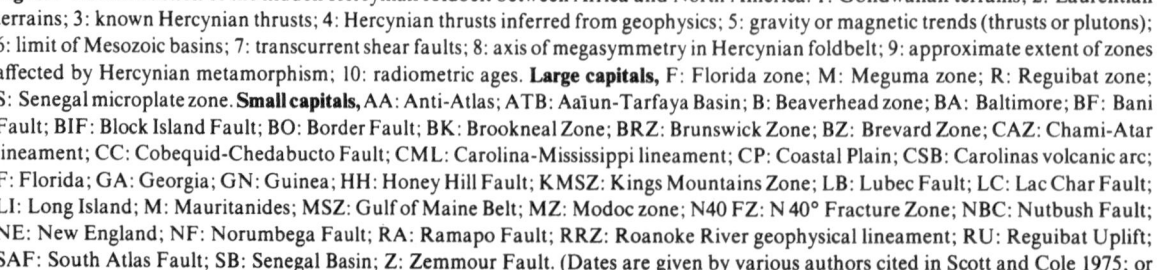

Fig. 60. Reconstruction of the hidden Hercynian foldbelt between Africa and North America. 1: Gondwanan terrains; 2: Laurentian terrains; 3: known Hercynian thrusts; 4: Hercynian thrusts inferred from geophysics; 5: gravity or magnetic trends (thrusts or plutons); 6: limit of Mesozoic basins; 7: transcurrent shear faults; 8: axis of megasymmetry in Hercynian foldbelt; 9: approximate extent of zones affected by Hercynian metamorphism; 10: radiometric ages. **Large capitals,** F: Florida zone; M: Meguma zone; R: Reguibat zone; S: Senegal microplate zone. **Small capitals,** AA: Anti-Atlas; ATB: Aaïun-Tarfaya Basin; B: Beaverhead zone; BA: Baltimore; BF: Bani Fault; BIF: Block Island Fault; BO: Border Fault; BK: Brookneal Zone; BRZ: Brunswick Zone; BZ: Brevard Zone; CAZ: Chami-Atar lineament; CC: Cobequid-Chedabucto Fault; CML: Carolina-Mississippi lineament; CP: Coastal Plain; CSB: Carolinas volcanic arc; F: Florida; GA: Georgia; GN: Guinea; HH: Honey Hill Fault; KMSZ: Kings Mountains Zone; LB: Lubec Fault; LC: Lac Char Fault; LI: Long Island; M: Mauritanides; MSZ: Gulf of Maine Belt; MZ: Modoc zone; N40 FZ: N 40° Fracture Zone; NBC: Nutbush Fault; NE: New England; NF: Norumbega Fault; RA: Ramapo Fault; RRZ: Roanoke River geophysical lineament; RU: Reguibat Uplift; SAF: South Atlas Fault; SB: Senegal Basin; Z: Zemmour Fault. (Dates are given by various authors cited in Scott and Cole 1975; or Dallmeyer 1982 and Dallmeyer and Villeneuve 1987)

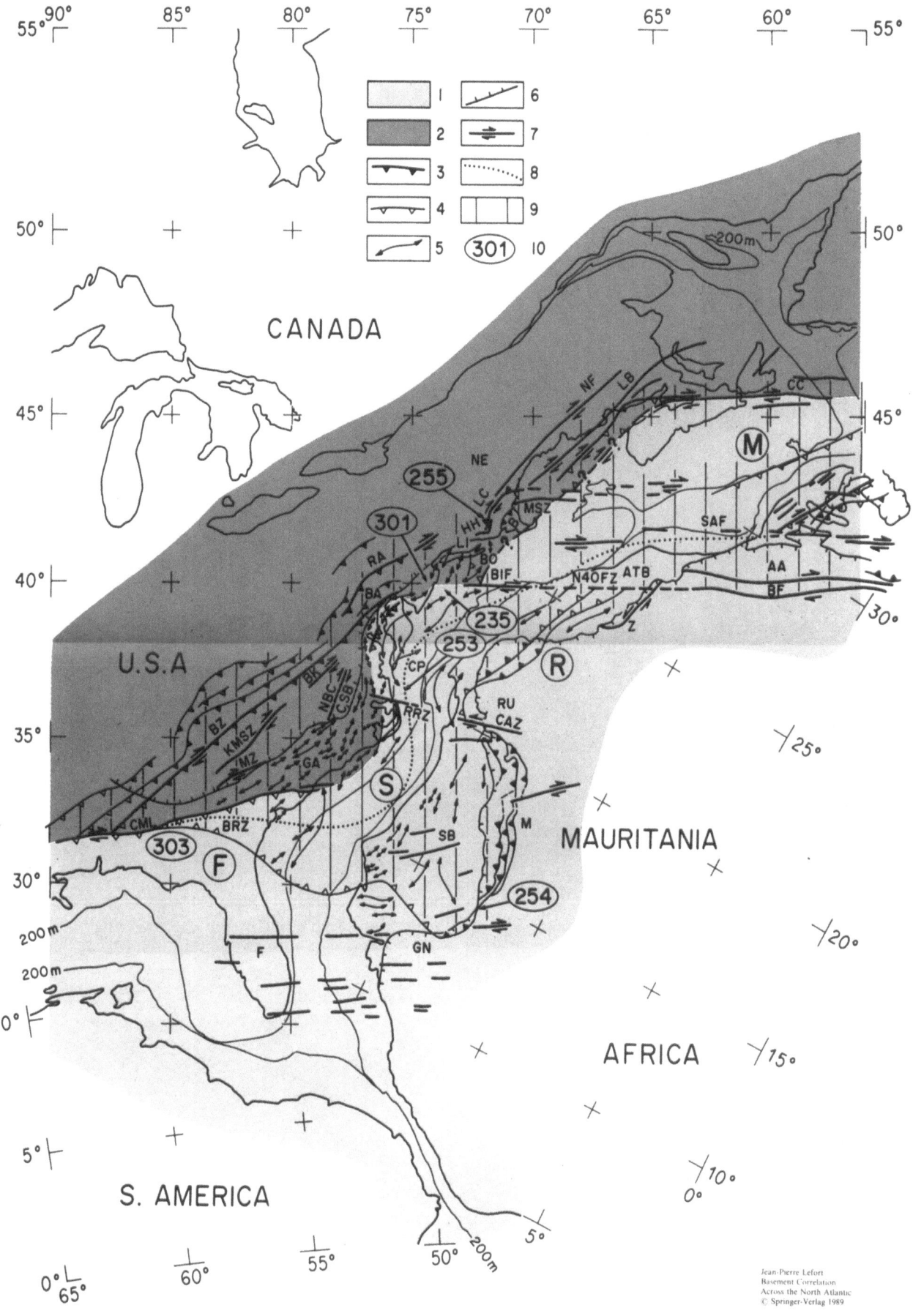

Jean-Pierre Lefort
Basement Correlation
Across the North Atlantic
© Springer-Verlag 1989

is an Acadian suture re-activated along its entire length during the Hercynian. It is known that the initial phase of convergence of Gondwana was, for a certain period of time, oblique. This obliquity continued during the Hercynian orogeny; all the trans-Atlantic fractures affecting the border of Gondwana show dextral displacements during the Carboniferous. The problem of a possible connection between these dextral Hercynian shears (or shear zone) in the south Atlas region (Proust et al., 1977) and the 40°N Fault has been only briefly discussed in the previous section, but there are still important difficulties to resolve. Despite the proposition of Lefort and Haworth (1981) previously discussed, and despite an attempted correlation due to Simpson et al. (1979), there appears to be no clear equivalent of the South Atlas Fault on the American side, unless one counts a minor discontinuity between Long Island and the mainland (Klitgord et al., 1985). Because it is out of question to modify all previously established correlations, one must admit that the effects of the South Atlas "Fault" are much less intense on the other side of the North Atlantic (Fig. 60).

According to the continental fit used in this study, the 40°N Fault is not situated opposite the South Atlas Fault Zone but rather adjacent to the Bani faults which cut the Anti-Atlas (Jeannette and Piqué, 1981). The Bani faults appear to have moved in a dextral sense during the Westphalian (Michard, 1976). This discontinuity is a major boundary since it probably corresponds to the southern limit of the Meguma Zone as indicated by drill-hole data (Fig. 43); it also controls the shape of the margin South of New England (Fig. 46) and cuts the Ligerian-Acadian suture to the South of the Block Island Fault (Fig. 43).

Taking account of the age of this suture, it is likely that the basic material marking the suture South of Cape Cod is pre-Carboniferous in age (Fig. 43). The Border Thrust associated with the suture is considered as an Acadian antithetic structure directed towards the West (conclusion also proposed by Phinney, 1986) – it should have been slightly re-activated at the Hercynian as a dextral shear thrust in order to accomodate the tectonic style prevailing further West and further South. This re-activated thrust seems to have been locally fairly important, even truncating the original Acadian suture at depth (Phinney, 1986).

– The suture located to the West of the Reguibat uplift has only been recognized from geophysical data (Lefort, 1984; Lefort, in press 1988a). Figure 61 shows, on a regional scale, the trace of magnetic and gravimetric lineations beneath the Coastal Plain of the U.S.A. and beneath the Senegal Basin. On the one hand, there is a westward curvature of the structures located North and South of this area (as already described as regards Fig. 46) but, on the other, there is also a bend in certain geophysical markers which are homothetic with known structures in W. Africa. Some lineations are even seen to closely follow the present-day outline of the Reguibat uplift and are superimposed on the Mauritanide trends concealed beneath sedimentary cover rocks (Dillon and Sougy, 1974). This is the main reason for considering the Reguibat Spur, along with its concealed extensions (Fig. 43), as an indenter structure which impinged onto the Appalachian foldbelt (see Chapter Eight).

The boundary between terrains having African or North American affinities is difficult to pick out beneath the Coastal Plain of the U.S.A. Using the criteria established for the construction of Fig. 46 (on a more detailed scale), we may conclude that this limit can be recognized by the presence of ultrabasic rocks at the roof of the basement. Such material can be identified in the westernmost arc as corresponding to overlapping gravimetric and magnetic anomalies of high amplitude (Fig. 46; key ornament 18). This choice of boundary is very close to that favoured by Klitgord and Popenoe (1983). The ultrabasic body marked by these anomalies may correspond to suture material transported westward from the Block Island Fault – in that case, the ultrabasic rocks would be Siluro-Devonian in age. If there had been some oceanic crust at this latitude during the Hercynian orogeny, the ultrabasic body could even be Carboniferous. It is entirely possible that there is an association between these two different types of ultrabasic formation.

According to the interpretation of Hatcher (1981), the westward-directed thrusts known in the Appalachians would have been rooted further West than the suture zone defined above (region of Kings Mountain Belt). If this is actually the case, then the basic suture corresponds also to the impingement front of the Reguibat Spur.

If, by contrast, the entire Piedmont terrain is considered allochthonous (Secor et al., 1986), then the suture at the roof of the basement has very likely been displaced from its original position at depth. This westward displacement implies that the apparent location of the suture is separate from the Reguibat Front. Recent seismic data (Pratt et al., 1987) recorded in central Virginia show west-facing thrusts to the East of the Alleghanian suture zone and very thick crust (50 km) near the coast which

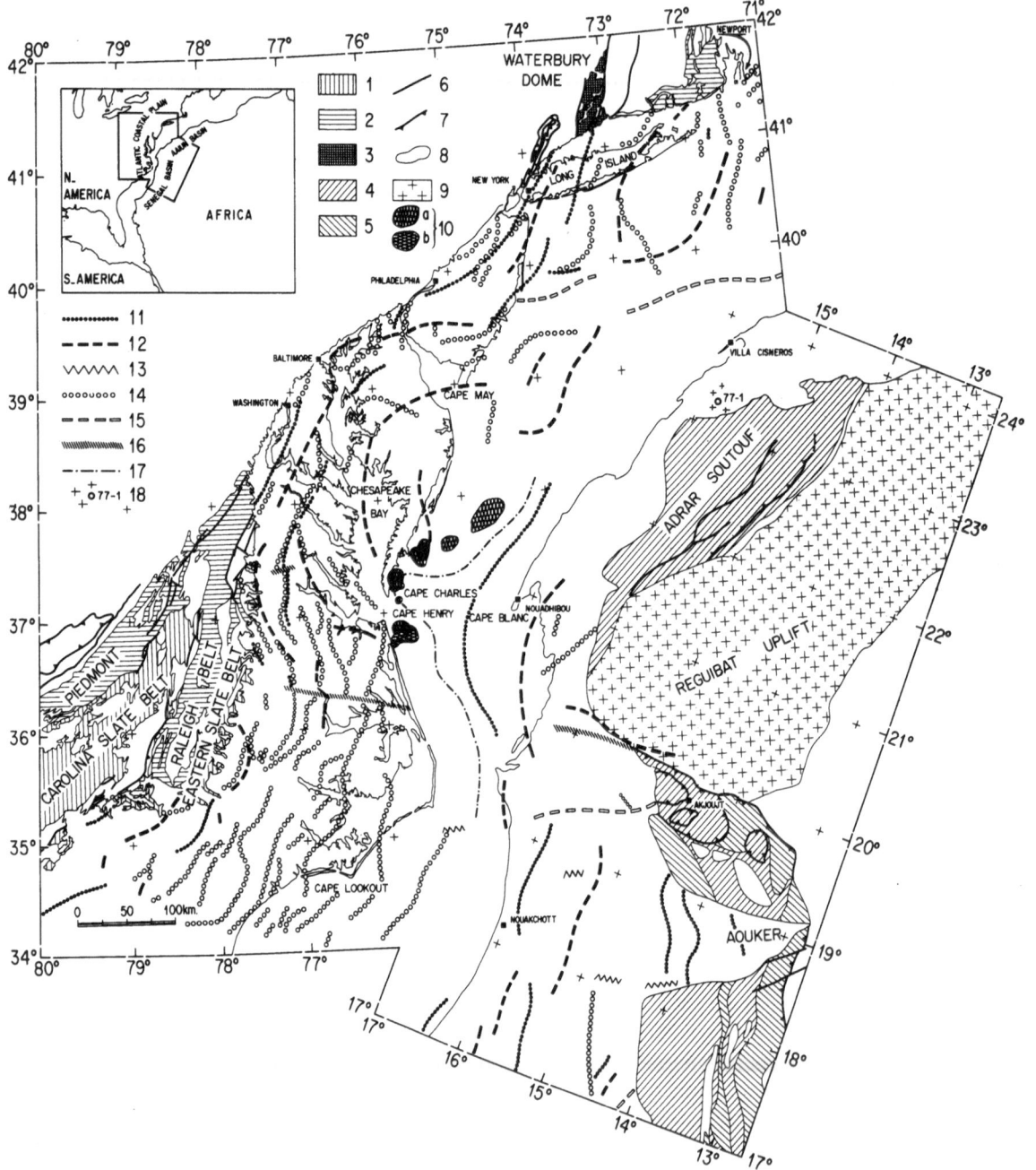

Fig. 61. Sketch map showing distribution of magnetic and gravimetric lineations in a re-assembled segment between the Reguibat Uplift and Chesapeake Bay. 1: volcanic arc and associated rocks; 2: amphibolite facies terrains; 3: tholeiitic or calc-alkaline volcanic suites; 4: internal zone of the Mauritanides; 5: external zone of the Mauritanides; 6: faults; 7: thrusts; 8: granites; 9: Reguibat Uplift; 10a: localized intrusions recognized from magnetics; 10b: localized intrusions recognized from gravimetry; 11: gravity highs; 12: gravity lows; 13: gravimetric discontinuities; 14: positive magnetic ridges; 15: negative magnetic troughs; 16: magnetic discontinuities; 17: probable limit between oceanic and continental crust; 18: basement reached by drill-hole. (After Lefort 1984)

might represent the tip of the Reguibat Uplift. These data favour an allochthonous character for the suture (Fig. 62.3).

In addition, it can be seen from Fig. 61 that, South of the Alleghanian suture, there is an alignment of fractures between the Roanoke River area (N. Ca-

rolina) (Klitgord and Behrendt, 1977) and the Chami-Atar region of Mauritania (Crenn and Rechenmann, 1965). This fracture zone is well marked by gravimetry and magnetics (Lefort, 1984) and cuts Appalachian and Mauritanian structural trends. The displacement across this fracture zone appears

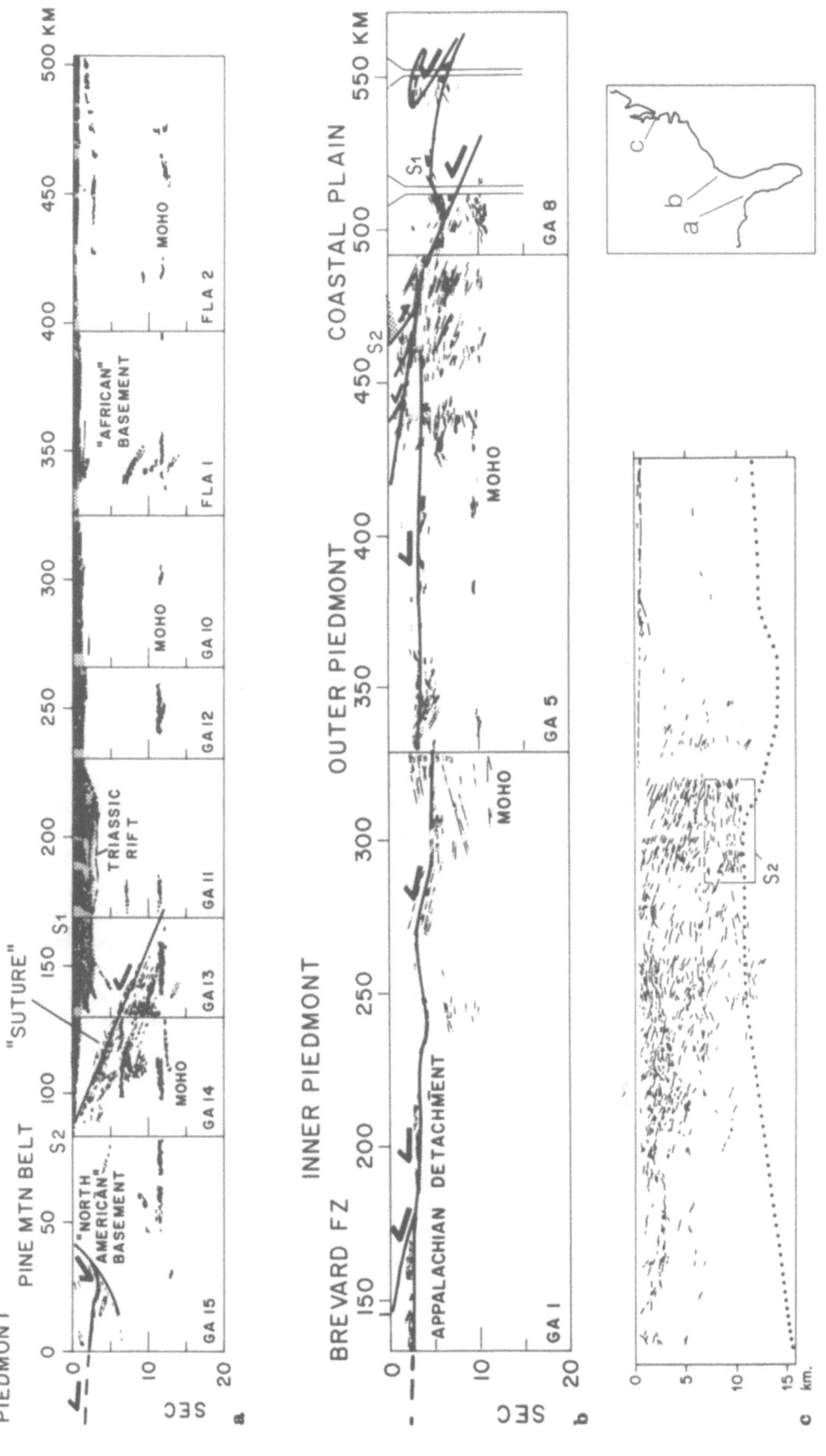

Fig. 62. Seismic profiles across Alleghanian and Pan-African sutures in the Southern Appalachians. S_1: inferred Pan-African suture; S_2: Alleghanian suture. **a** Section across the Alleghanian and inferred Pan-African sutures in Western Georgia. (After Brown et al. 1984; reproduced by permission of the American Geophysical Union); **b** Section across the Alleghanian and inferred Pan-African sutures in Eastern Georgia. (After Brown et al. 1984; reproduced with permission of the American Geophysical Union); **c** Section across the inferred Alleghanian suture in Western Virginia. (After Pratt et al. 1987)

to be sinistral (Figs. 46 and 60) – the apparently dextral shear associated with the 40°N Fault Zone is compatible with a symmetrical arrangement of faults on each side of the Reguibat Spur. All deformation structures between the two fracture zones are probably Hercynian, but it is not known whether the terrains beneath Chesapeake Bay (Fig. 46) represent an extension of the Mauritanides onto the North American plate or whether they show pre-existing Appalachian structures affected by Hercynian reworking.

– The Hercynian suture situated to the West of the Senegal microplate is not always easily mappable. Its location was originally established using the same criteria applied West of the Reguibat zone. This position was subsequently confirmed North of the Brunswick terrains, in the Southern Appalachians, by seismic reflection (Ando et al., 1983). A comparison between the cross-sections represented on Fig. 62 (sections 1 and 2) shows some interesting features. The section across the Piedmont of Georgia and Florida (Fig. 62.1) reveals a northward directed major thrust associated (at the 150 km-mark) with a reflector that is considered as a southward-directed thrust (Nelson et al., 1985). This latter thrust, as described in Chapter Six, overlaps with the southern suture of the Senegal microplate (western extremity). The section across Eastern Georgia (Fig. 62.2) shows a northward-directed Hercynian thrust (ca 500 km mark) similar to the northward-directed thrust in the previous section and an incipient divergence at 510 km which is considered by Cook (1986) as representing a northward dipping suture. In this way, it is possible to find the same structural arrangement twice – that is, an intersection of a probable Senegal Pan-African suture by the Hercynian suture. The different types of reworking and local rejuvenation of this Pan-African suture are now well characterized North of Florida (Nelson et al., 1987). Accordingly, the "pop-up structure" known North of Florida is different from that discussed in the context of the Ligerian – Acadian suture, since it could partly result from the intersection of two sutures (Nelson et al., 1985; Cook, 1986) rather than an antithetic thrust associated with a suture. The comparisons made by Cook (1986) between the Block Island "pop-up" (Phinney, 1986) and the crustal "pop-up" shown on Fig. 62 (sections 1 and 2) are only applicable in terms of tectonic mechanism – the original ages of the divergent faults concerned being entirely different in each case.

The Hercynian deformation superimposed on the Pan-African suture of Mauritania is evidently a result, or an indirect consequence, of the collision which occurred in the Southern Appalachian domain (Fig. 63) (Lecorché et al., 1983). All the terrains within this triangular zone, furthermore, would appear to have been deformed during the Hercynian orogeny – this is true for the "Brunswick" zone (Williams and Hatcher, 1983) as well as to the South of the Senegal basin, where Devonian rocks have been recognized amongst the metamorphic series (Maugis, 1955) (this proves the existence of post-Devonian metamorphism).

– The Alleghanian suture crossing Alabama, Georgia and South Carolina cuts straight across the "Older" Appalachians (Horton et al., 1984) and forms part of a extensive lineament known as the "Carolina-Mississippi Fault". The linear trace of this structure suggests that it first acted as a transform zone which accommodated the shearing movements between Gondawana and the Southern Appalachians, probably at a late stage of the Carboniferous closure (Higgins and Zietz, 1983).

The E-W fractures recognized between Florida and Guinea (Fig. 60), some of which are dextral (Le Page and Campredon, 1981), were active after deformation in the Mauritanides and are probably associated with this transform faulting.

In order to make a comparison of the Alleghanian suture on the American side with the Mauritanian suture in Africa, geological and geophysical information have been collected from a variety of sources – this has led to the compilation of schematic cross-sections (Fig. 63). The main interest of these sections is to show the symmetry of Hercynian thrusting directions on both sides of the Atlantic. These sections also give some idea of the westerly extension of Pre-Cambrian metamorphic basement at depth in Africa.

Inspection of Fig. 60 shows that it is possible to construct a mega-axis of symmetry for Hercynian thrusting directions throughout the southern part of the North Atlantic domain. It should be recalled that, in the northern part of the North Atlantic, this symmetry axis is seen to connect with the South Atlas Fault Zone in Morocco (Fig. 51). Offshore, this axis follows the westerly extension of the South Atlas Fault before swinging South to follow broadly the outline of the West African Craton. South of 35°N, it bends sharply West to join the "Carolina-Mississippi transform fault". Once again, it is clear that the mega-axis of thrust symmetry is distinct from the Hercynian suture zone.

To conclude this review of the concealed Hercynides in the southern part of the North Atlantic region, it is useful to indicate those terrains which could have

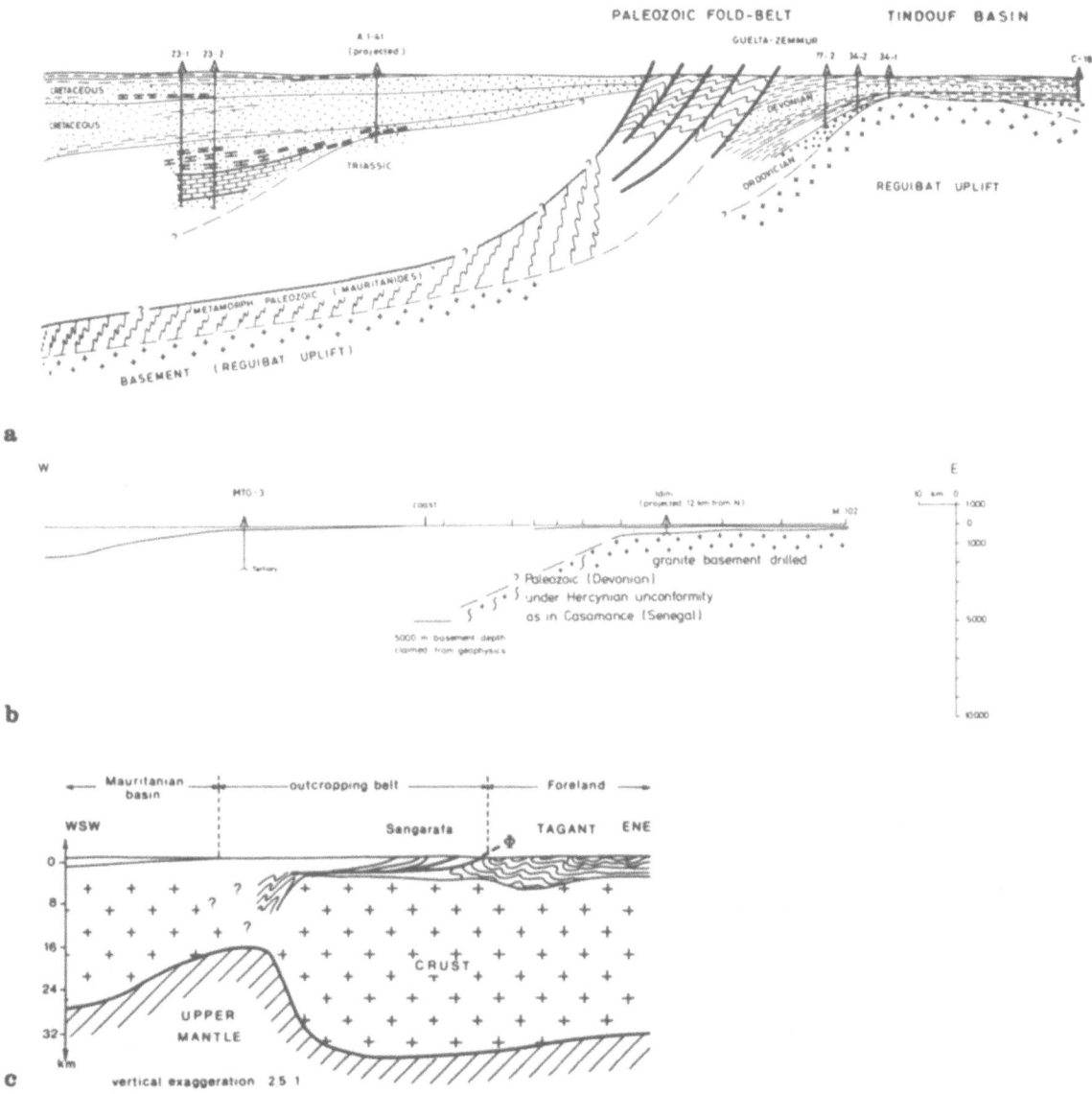

Fig. 63. Comparative cross-sections of the Mauritanian Front at the eastern limit of the W. African basins. **a** Geological sketch section of the Mauritanian Front to the East of the Aaioun-Tarfaya Basin (Lat. 27° N). (After Ranke et al. 1982); **b** Sketch section showing depth to basement to the East of the Senegal Basin (Lat. 18° N). (After Wissmann 1982); **c** Combined geological and gravity section of the Mauritanian suture (Lat. 17° N). (After Lecorché et al. 1983; reproduced by permission of the Geological Society of America)

suffered some degree of metamorphism (see hatched zones on Fig. 60). Due to lack of sufficient information, it is not possible to separate different zones of intensity. According to on-land observations in the U.S.A., it is fairly clear that the area affected by Hercynian deformation is much larger than the high-grade zone (Glover et al., 1983; Dallmeyer et al., 1986) which closely follows the suture to the East of the outcropping Appalachians. Even though there are few data available from beneath the Coastal Plain or beneath the Senegal basin, it is likely that the American metamorphic belt is continuous with the African one. Beneath the Coastal Plain, metamorphic ages ranging from 301–230 Ma are generally concordant with African ages, except that late-stage Hercynian effects in Africa did not continue after 254 Ma. This suggests that the youngest deformations in Africa are older than in America and that Late Permian metamorphism is lacking.

Finally, it remains to discuss the significance of the "East Coast Magnetic Anomaly", a very strong and linear magnetic anomaly which follows the eastern edge of the American continental margin.

This anomaly has given rise to numerous investigations, some of which (e.g. Nelson et al., 1985) have interpreted it as a Hercynian suture. Whatever the origin of this structure, there appears to be a simple argument which rules out the possibility that the East Coast Magnetic Anomaly belongs to the pre-Mesozoic basement. In fact, it is out of question that such a perfectly linear body should remain completely underformed between the Reguibat Spur and the Chesapeake Bay identation if it were really part of the Palaeozoic basement. Without doubt, the linear trace of this anomaly is ample proof that the body corresponds to a Mesozoic or Tertiary intrusion.

7.3 Post-Hercynian Deformation Before the Opening of the North Atlantic

It has been noted several times in this study that there is a dextral shear system which systematically offsets pre-existing structures over an area extending from South of Brittany across the Western Channel Approaches to the Porcupine Bank. The faults in this system follow a N 130° trend which is constant from northern Spain to the Rockall Trough. Furthermore, it is established that this type of shearing is post-Stephanian in age (evidence off W. Brittany) but previous to the intrusion of dolerites dated at between 190 and 205 Ma (Leutwein et al., 1972). Movements or tensional re-activations on these faults are common throughout the Mesozoic and Tertiary. In this part of the North Atlantic domain, this fault system is omni-present and is nowadays considered as the result of Permo-Triassic shearing developed across the Biscay-Labrador segment before the present phase of ocean spreading (Lefort, 1973). The displacements produced by this activity are difficult to estimate. It is true to say that this "shear zone" is characteristic of the West European margin since numerous studies have failed to show an influence on the Canadian side. Several transcurrent zones of the same age are found in Morocco along the South Atlas "Fault" (Tixeront, 1973), but cannot be traced offshore.

Fractures in the offshore basement were frequently re-utilized during the Permo-Triassic in a tensional tectonic regime – this phenomenon has been noted many times in preceding chapters, and is particularly well studied on the American margin (Ballard and Uchupi, 1975; Klitgord et al., 1983; Klitgord and Hutchinson, 1985; Hutchinson et al., in press 1987). However, it would appear that such extensional structures were only clearly developed in a transcurrent regime off Western Europe. The major N 130°-trending faults which traverse Florida (curiously parallel to the European fault system) played a very important role in the opening of the central part of the North Atlantic and can be integrated into a model of Permo-Triassic distension (Klitgord and Popenoe, 1984). As described in Chapter Six, these fractures probably resulted from the rejuvenation of ancient structures having the same trend, and were not produced initially during the onset of North Atlantic Ocean spreading.

7.4 Conclusions Concerning Hercynian-Alleghanian Orogenesis Before the Opening of the North Atlantic

In contradiction to what is normally assumed, Hercynian tectonic effects extend northward far beyond the Variscan Front, even if the intensity is somewhat attenuated (e.g. North Porcupine Bank, NE Newfoundland). To the South of the Front, tectonic effects are variable according to whether one is discussing the northern or western boundary of Gondawana. In fact, the concealed basement is fundamentally different on the African-European flank of Gondwana compared with the African-American flank. In the northern part of the North Atlantic domain, it is possible to distinguish structures that have been produced by the re-activation of pre-existing discontinuities (i.e. Cadomian or pre-Hercynian structures such as the Iberian-Armorican Arc or the Ligerian-Acadian suture) from structures that originated entirely from Carboniferous movements. These latter structures are parallel amongst themselves and follow the northern boundary of the West African Craton – they are generally organized in E-W-trending shear zones. In the southern part of the North Atlantic domain, by contrast, the two continental margins show a different Hercynian history. On the western side, in the Appalachians, horizontal re-activations are so important that it is nearly always difficult to recognize the previous lines of weakness. On the eastern side (e.g. Morocco), the main concealed structures are often more or less superimposed onto the margins of the Pan-African blocks.

The Hercynian tectonics of the Mauritanides show an intermediate style since there are not only tangential displacements (mainly in the North) but also an obvious control from ancient structures.

A further difference between the northern and southern parts of the North Atlantic area is apparent in the suture geometries. To the North of Gondwana, the Acadian and Hercynian sutures are practically overlapping whereas they cut across each other South of Long Island (Figs. 43 and 60). Furthermore, Hercynian tectono-metamorphic events are restricted to the Carboniferous in the North but continue into the Permian in the South. This pattern is reflected in the decrease in metamorphic ages along the Lizard Thrust from SW England to New England (Fig. 51).

The style of deformation around the Gondwanan Continent also appears to vary greatly as a function of latitude; in this way, the Ligerian-Acadian suture seems to have been mainly re-worked as a transcurrent fault during the Carboniferous, whereas the upper part of the suture located to the West of the Reguibat Spur could be partly allochthonous and associated with the onset of thin skin tectonics. Further South again, the suture is only visible as a thrust front, thus indicating an increase in crustal shortening (Fig. 62). This certainly explains the regular southward decrease in the importance of basic/ultrabasic rocks detected in plan view by geophysics in the suture zone.

From another point of view, the N 40/50°-trending dextral shears, which play the same role along the Appalachians as the dextral E-W shears observed to the North of Gondwana, demonstrate an anticlockwise rotation of the entire Gondwanan continent during Hercynian crustal shortening. If, in addition, the ages of tectonic and metamorphic events are taken into consideration, it is apparent that there is a similar rotation in the locus of orogenic activity with time; deformation began in early Carboniferous times in the North and continued through into Permian times in the West.

The presence of a Devonian phase of deformation (Ligerian-Acadian) related to subduction under Armorica and the northeastern Appalachians, as well as the existence of a Carboniferous deformation, also related to subduction further South along the same belt, shows clearly that closure of the Theic Ocean was diachronous. Taken with other considerations, this explains why it is inappropriate to compare the European Hercynides with the Southern Appalachian Piedmont and Coastal Plain — the former was produced by the tightening-up of an intraplate suture zone, whereas the latter is the direct result of continental collision.

The Lizard Thrust (sensu stricto) has a particular significance within the European Hercynides — it represents a kind of diachronous Variscan front which is more important in terms of metamorphic gradient than the classic Variscan Front.

The Iberian-Armorican Arc has long attracted speculation from European geologists, but it really represents only a small fragment of ancient re-activated basement within the Hercynian foldbelt. It is too small to have modified the general organization of the Hercynian foldbelt in this northern part of the North Atlantic domain. Other fragments of Pan-African basement beneath the Coastal Plain of the U.S.A. have probably undergone a similar style of tectonic tightening and even buckling.

The connection between the E-W transcurrent fault systems of Europe and Africa and faults within the Appalachian foldbelt is seen to occur through curved thrust fronts — the formation of these shear zones is an important event in the tectonic development of the Hercynian domain since it can be observed in at least five separate areas between Labrador and Pennsylvania.

Two ancient transform faults can be recognized within the Hercynian domain, but their rejuvenation did not take place in the same tectonic setting. The Porto—Badajoz—Cordoba Fault simply accommodated the intracontinental stresses caused by the northward convergence of Gondwana during the Carboniferous; at the same time, this accentuated the curvature on the Iberian-Armorican Arc. The Carolina-Mississippi fault, on the other hand, was a transform plate margin between Gondwana and Laurentia which behaved both as a shearing and as a thrusting contact. At first, it must have taken up the convergence between Africa and America during the Carboniferous. Then it probably behaved as a thrust contact during the Permian when Gondwana moved towards the North (see Chapter 8).

Finally, it is interesting to recall the classic interpretation of the significance of the 40°N Fault in America. At the beginning of modern oceanography, it was proposed that the South Atlas Fault was correlated with a fault on the other side of the Atlantic situated along the 40th parallel. A transform fault mechanism has been occasionally attributed to this structure, by analogy with oceanic fracture zones. However, such a trans-Atlantic correlation need not have existed if the other established comparisons between Africa and America are taken into account.

Chapter 8

Tectonic Mechanisms Contributing to the Structure of Pangaea

Since stratigraphic information concerning the concealed basement on either side of the North Atlantic is very unequally distributed, it is out of the question to propose a series of palaeogeographic reconstructions for the various domains under consideration in this book. On the other hand, the different methods employed to detect the concealed basement units discussed in this study can be used for a tectonic analysis of the terrains defined here. This is the reason why it appears preferable to conclude with an overview of the main tectonic mechanisms which have led to the structural development of Pangaea. At the same time, this enables a better understanding of certain processes which have not been fully discussed in the preceding chapters. In addition, this provides some insight into the problems of observing tectonic phenomena on the regional scale. Several geodynamic hypotheses are also proposed for the construction of the orogenic belts bordering the North-Atlantic Ocean.

8.1 Ancient Plate Drift

Laurentia can be considered as a fixed plate (palaeomagnetic evidence) throughout the Palaeozoic; the tectonic mechanisms discussed in this chapter have their origin in the continuous and progressive "northward" drift of Gondwana (Scotese, 1984), or in displacement of some of its detached fragments. The Avalon Spur (Armorica plate of Perroud et al., 1984) constitutes, as described above, a major element in this process of continental break up. The presence of certain fragments which would have been previously separated from Gondwana has been envisaged beneath the Coastal Plain and Piedmont of the U.S.A. (Carolinas Block and SE New England). Certain other fragments have failed to drift away from Gondwana (as is possibly the case for the Senegal plate) whereas others are simply the result of plate margin fracturing with no lateral dis-

placement away from the main plate (Wendt, 1985) (this may explain, for example, the cutting up of the Moroccan basement). Such dissimilar histories of development imply that there could be a minimum spreading rate necessary to enable a micro-continent to escape from its parent continent. If the Avalon spur is taken as an example, it would appear that drift rates are greater for larger plates and that small plates have little chance of distancing themselves from the supercontinents. In reality, the situation is more complex, with the existence of large plates (e.g. Senegal) that have hardly moved at all, and other microplates (such as the Carolinas) that have succeeded in migrating a great distance despite their reduced size.

Put in this manner, one is forced to admit that there is no general relationship between drift rate and plate size. In fact, this merely reflects a confusion between ancient and modern plate boundaries. For example, a "Palaeozoic microplate" as described in the literature really only corresponds to the continental part of a lithospheric plate. A large part of this plate may have been made up of oceanic crust that has nowadays disappeared. Thus, we have no real idea of the size of ancient plates.

In any case, the various observations given in this study show that there was never any unity linking the Avalonian terrains; the only point in common between the different outcrops of this age is the presence of a basement of similar age, comparable to the Gondwanan basement. It would appear notoriously difficult to use this criterion to maintain the existence of a single Avalonian plate during the Mid-Upper Palaeozoic (as sometimes proposed).

8.2 Different Types of Suture

The term suture as used here requires a definition, since in 90% of cases they can only be recognized by geophysics. In this study, a suture brings together all the material produced as a result of the opening and closure of an ocean. The material is nearly always basic or ultrabasic in composition and is generally easy to detect at depth by magnetism and gravity. One should be able to find fragments of dyke complex derived from the initial phases of spreading as well as slices of ocean floor of various ages. Blueschist facies rocks may also be found, in addition to fragments of volcanic arc or marginal basin if the crustal shortening has been very great. These materials form elongate geophysical markers which are often several hundred kilometres long and rather narrow. They generally border cordilleran-type orogenic belts and are seen to separate basement domains of contrasting character. Such a definition, which is typical of sutures between plates, would be incomplete without the addition of two other features. Very often, the suture zone corresponds to an alignment of earthquake epicentres and it is always near a major thrust fault that can sometimes be recognized on seismic reflection profiles. These suture zones are generally developed beneath the continental margins or coastal basins surrounding the North Atlantic – such an arrangement is quite remarkable and reveals some aspects of their genesis. The fact that sutures are often concealed beneath shelf areas in this region implies some connection between the present phase of opening of the North Atlantic and previous history of ocean spreading and consumption. Even though this relationship is not discussed here in terms of tectonic processes, it can nevertheless be noticed that there is a link between the dip and strike of ancient subduction zones and the location and width of present-day margins (Lefort, in preparation).

It is certainly seismic reflection which has contributed most to an improved understanding of suture zones, at least as regards their cross-sectional structure. By contrast, no method has yet replaced magnetism or gravimetry for horizontal mapping of these bodies (see Fig. 64), since deep seismic reflection profiles are still uncommon. The seismic profiles illustrated in previous chapters have suggested a system of classification for the sutures described in this book. At this stage, no distinction is made between intraplate sutures developed within the same structural domain and interplate sutures which reflect pre-existing oceans and typically separate basements of contrasting type.

Type 1: the simplest type of suture is represented by the Lizard thrust (CL in Fig. 23) where a thrust fault traverses the entire crust and there are no directly associated décollement structures at the front. The basic/ultrabasic material which marks the outcrop of the suture is on top of the thrust front (Fig. 22) – however, fragments of ocean floor do not persist at depth, which probably implies the existence of a cryptic suture. This type of suture "cuts" the Moho without provoking any displacement, probably because the Moho was formed after the suture. The Outer Isles and Flannan thrusts, which appear to have affected the Moho due to late-stage re-activation, can nevertheless be classified as type 1 sutures.

Type 2: this type of suture is also associated with structures that traverse the entire crust, but there are antithetic faults as well. Such is the case, for example, with the Cadomian suture in the Western English Channel; here, it is known (Lefort, in press 1988b) that the antithetic faults are contemporaneous with suture formation (Fig. 23). The Ossa Morena suture (Fig. 40), before its Hercynian reactivation, the Bay of Maine suture and the Block Island Fault (Fig. 41) (Hutchinson et al., 1985; Phinney, 1986) are all probably type 2 sutures. The point in common is that the antithetic divergence associated with the ancient subduction plane originates in the lower or middle crust. There is one case where the associated antithetic fault is probably representative of a pre-existing suture. This occurs North of Florida (Fig. 62.1), where the antithetic fault appears to correspond to a Pan-African suture that has been cut off by later tectonic activity at depth. Such an hypothesis does not exclude the possibility of Hercynian reactivation of the southward-directed thrust, thus leading to the creation of a "pop-up structure" on the crustal scale (Coward and Smallwood, 1984; Nelson et al., 1987). The divergence which exists between Caledonian surface thrusts on the limits of the Dunnage zone (Fig. 13, section 1) could repeat a situation which is known for the Cadomian in the Western English Channel, provided that the suture affects the entire crust (Keen et al., 1986). In nearly all the examples cited here, a major thrust fault appears to have brought ultrabasic material to the surface.

Type 3: The third type of suture traverses the entire crust but there is development of thin skin-type tectonics in the external foreland (Fig. 62.2); this is seen in the suture which follows the axis of the Appalachian foldbelt (Hatcher, 1981; Hatcher and Zietz, 1980; Thomas, 1983). The basic body detected in plan view may, in part, be highly allochthonous as observed for the SW Appalachians. The Mauritanide suture may also belong to type 3 (Fig. 63) (Ritz and Robineau, 1986). When there has been important erosion of the thrust zone situated in the

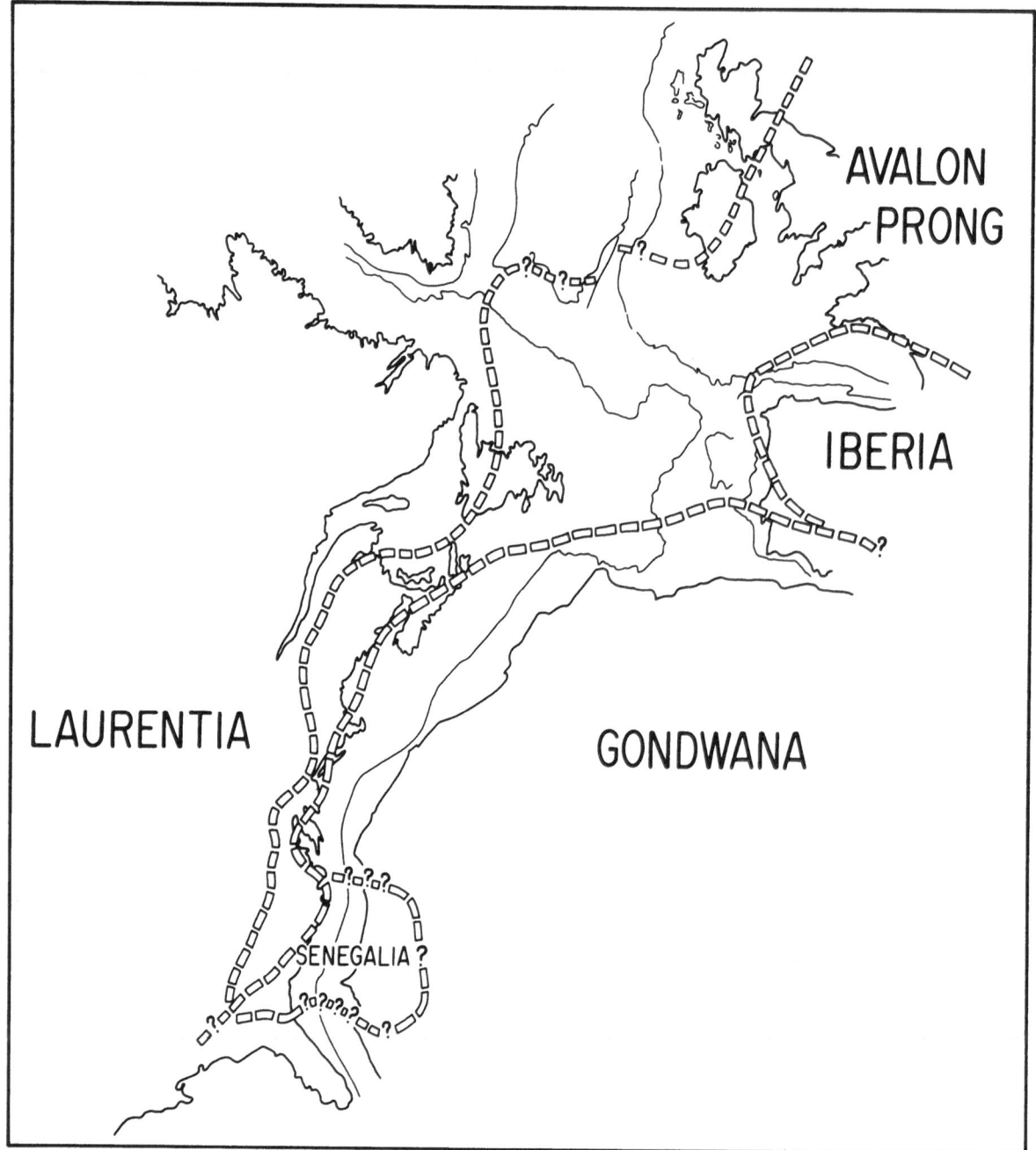

Fig. 64. Distribution of Palaeozoic sutures in the peri-Atlantic domain according to magnetic and gravimetric data. The Acadian suture as represented on Fig. 43 has not been separated here from the Nashoba Zone in New England. (After Lefort, in press 1988b)

foldbelt foreland, type 1 sutures may represent only remnants of a type 3 suture.

Type 4: These sutures are rooted only in the interface zone between lower and upper crust. They include the Hercynian suture which has reworked the Ligerian-Acadian suture in southern Portugal (Ribeiro et al., 1983) (Fig. 39), and the suture described off Wales which extends the Variscan Front (Fig. 52). The same decoupling process has been invoked by Cook (1984) for the SE Appalachians, in opposition to the interpretation due to Hatcher and Zietz (1980).

Type 4 sutures do not always bring up ultrabasic material along the thrust front. They are rarely seen in isolation and are generally part of the tectonic régime typical of type 3 suture forelands (Cook 1986).

Type 5: Finally, the last type corresponds to a

beheading of the suture zone caused by an imbrication of crustal layers. In this case, basic material does not appear at outcrop. This is the interpretation given by Hall et al. (1984) for the Iapetus suture between Ireland and Scotland (Fig. 13). The same interpretation, with only slight modification, is also proposed for the South Armorican suture on the northern margin of the Aquitaine Basin (Lefort, in press 1988c).

This brief summary of the processes of suture formation is hardly exhaustive from the theoretical point of view, but is based solely on data and interpretations (Fig. 65); it demonstrates that certain processes are not mutually incompatible. Furthermore, it shows that seismic interpretations are rarely unequivocal.

It is possible to rationalize the system of classification given here by reducing it to three basic types of suture:

a) Those which traverse the entire thickness of the crust.
b) Those which originate only at the lower-upper crust interface, maybe corresponding to the ductile-fragile transition zone.
c) Those which result from the imbrication of thrusts at different levels, even involving the sub-Moho mantle.

If we extend the arguments put forward by Brewer (1984), it is even possible to consider types a) and b) as identical, but representing different levels of a type 3 suture. Thus, types a) and b) taken together correspond to suture formation by upward ejection, whereas type c) reflects frontal confrontation and imbrication of the crust.

Certain authors have speculated on the original geometry of currently observed sutures; that is, whether they belong to type a), b) or c). In this study, it is considered more prudent to take account of the important isostatic re-adjustments that have taken place in crustal segments throughout the North Atlantic domain — this is corroborated by the systematic horizontality of the present-day Moho beneath ancient foldbelts. For this reason, it appears indispensible to have some idea of the original dip of these structures before proposing any crustal deformation models. The study of a seismic profile across the Himalayas (Hirn et al., 1984) serves as an example of the difficulties which, in certain cases, may arise during any attempt to reconstruct pre-Mesozoic suture geometries. In this example, the Moho shows a stepped topography with successive changes of 20 km at each step. The imbrication phenomena observed in the Himalayas (Matthews and Cheadle, 1986) are taken up in the tectonic

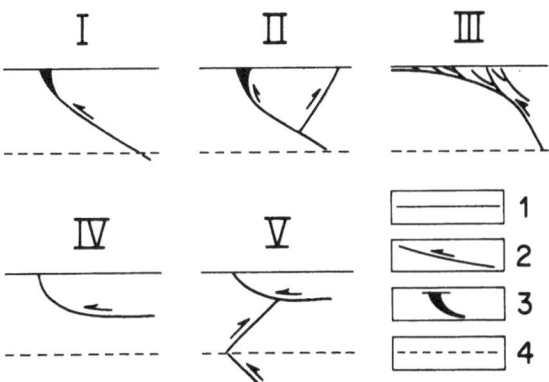

Fig. 65. Simplified classification of sutures bordering the North Atlantic. Suture-types in Roman numerals. 1: sea-level; 2: suture plane; 3: ultrabasic rocks; 4: Moho

interpretation given by Hall et al. (1984) for the Caledonides.

If the deep reflectors dipping away from the Avalon Peninsula in Canada (cf. profile of Keen et al., 1986) are of Caledonian age (Fig. 13), then a similar tectonic interpretation could be adopted for the northern Appalachians. The regional-scale interpretation of Cook (1984) suggests that the Southern Appalachians could also be indirectly linked to a type c) imbricate suture zone.

8.3 The Dip of Pre-Mesozoic Subduction Zones

It is practically impossible to determine the original dip of subduction zones which developed before the sutures shown in Fig. 64 because the geological literature contains numerous and contradictory models for each segment of orogenic belt studied. In some cases, this may be explained by a change in the direction of subduction during the period of plate convergence, but in general this arises from difficulties of interpretation related to the complexity of the different processes involved in collision. Anyway, certain sutures cannot be recognized from surface observations since it has been shown that they are sometimes concealed beneath volcanic arcs (e.g. South Portugal) or even marginal basins (e.g. possibly Nova Scotia). This is the reason for choosing to discuss subduction zone geometry in the light of the direction of tectonic transport recognized by geophysics beneath known or suspected suture zones. Except where other evidence is contradictory, structural criteria should lead to some idea of the dip of ancient subduction zones. The results of such an approach are summarized in Table 2, which gives the probable direction of subduction, the geophysical methods used to recognize the suture and the final date of collision for a number of sutures studied in the literature. As far as imbricate sutures are concerned, it is rather easy to confuse — if there is no other criterion — a seismic reflector corresponding to thrusts with a seismic reflector which really represents the ancient subduction plane. In these particular cases, a preferred model is put forward which does not necessarily exclude other interpretations. Various remarks should be made about Table 2. Firstly, it is not necessarily the most recent or important subduction that has left the best-defined geophysical marker, since tectonic transport may have changed sense and intensity during convergence. However, it is generally apparent that subduction zone dips derived from geophysical studies are consistent with broadly-based geological models — that is, models that have been established from a multi-disciplinary approach. Finally, the term "intra-plate subduction" used in Table 2 does not generally refer to a Himalayan-type subduction, but rather to rejuvenation of a previous inter-plate suture.

— Despite the critical review of Moseley (1977) and the possible imbrication of crustal layers, the Iapetus suture in Europe can be identified with a southward-rising reflector (Beamish and Smythe, 1986) that does not actually reach the surface (Fig. 13c).

— In the Canadian Appalachians, the suture may be imbricated at depth. A direction of subduction is suggested nearer the surface by seismic refraction (Sheridan and Drake, 1968; Haworth et al., 1978), since slices of seismically fast ophiolite can be followed from their surface outcrop (Dunnage Zone) eastwards under the Avalon Peninsula; this dip direction is otherwise confirmed by teleseismic methods (Stewart, 1978). Further West, however, the presence of other reflectors (Keen et al., 1986) suggests a westward dipping subduction zone (Fig. 13).

— Whichever model is adopted for the on-land part of the Southern Appalachians (Hatcher and Zietz, 1980; Cook, 1984), the resulting subduction zone dip is the same. This is further confirmed by gravimetric (Hutchinson et al., 1983; Thomas, 1983) and magneto-telluric studies (Thompson, 1982). In this region, it would seem (Cook, 1986) that Taconic subduction resulted from the underthrusting of

Pre-Cambrian crust beneath a volcanic arc (The Carolinas Arc).

— According to which interpretation is given to the Nashoba Zone in S.E. New England (see Chapter Six) — resulting from the closure of either a back-arc basin or an ocean — the suture can be considered as either a within-plate structure or a true Acadian (or older) interplate suture. Such considerations, however, do not affect the direction of thrusting (Skehan, 1983) which suggests the existence of a major crustal fault plane rising to the East.

— It is not known whether the impingement of the Reguibat Spur on the concealed part of the Appalachians during the Alleghanian orogeny (Chesapeake Bay suture, Fig. 46) provoked the reworking of a possible southward extension of the Acadian suture or whether post-Acadian oceanic crust was caught up during the final collision. In the latter case, the arc of basic rocks detected by magnetism and gravimetry beneath the Coastal Plain could possibly be of Carboniferous age. The radiometric date of 384 Ma, cited by Scott and Cole (1975) from rocks just North of Cape Hatteras (close to the suture zone), implies an Acadian metamorphic event possibly resulting from plate collision at this latitude. But this metamorphism, also recognized elsewhere under the Coastal Plain, could equally well have resulted from a previous microplate collision (e.g. collision of the Carolinas microplate). Since no re-heating effect of Acadian age is known in West Africa, it is fairly certain that, if metamorphism is related to plate collision effects, this collision was restricted to the American side. The intense deformation of the Petersburg granite, which borders the ultrabasic suture-zone material on the West, implies an important Alleghanian tectono-metamorphic event. This granite is curved towards the West (see Fig. 46), shows amphibolite facies metamorphism and yields an age of 262–300 Ma (Glover et al., 1983). The fact that the Reguibat Spur cuts the Ligerian-Acadian suture South of New England (Fig. 43) lends support to the idea that the Alleghanian suture passes beneath the Chesapeake Bay area. One of the criteria which enables a determination of the subduction zone dip is provided by geochemical and geophysical studies (Sinha and Zietz, 1982), which show convincing evidence for a westerly dip at this time. Seismic reflection data do not show any westward-dipping reflector in the upper crust beneath the suture (Pratt et al., 1987), but instead a "transparent" crust, which is often the case when mafic rocks are recorded. The deep reflectors, which are facing East just West of the thickened crust marking the collision zone or the Reguibat

Table 2. Attempt to reconstruct pre-Mesozoic subduction dips around the North Atlantic using suture zone geometries filled circles: interplate subductions; Stars: intraplate "subductions"

Name of subduction	N	E	S	W	Method used	Reference	Age
	(dip direction)						
Grenvillian		•			Gravimetry	Thomas and Tanner (1975)	1000 Ma
Senegal-Rokelides			•		Gravimetry	Villeneuve et al. (1984)	650–550 Ma
Rosselare-Anglesey		•			Tectonics & Geochemistry	Piasecki et al. (1981)	600 Ma
Western Channel		•			Seismic reflection	Lefort (1975); Lefort (1987)	560 Ma
Iapetus (U.K.)	•		?		Seismic reflection & magnetotelluric	Hall et al. (1984); Beamish and Smythe (1986)	420 Ma
Iapetus (Canada)		•/?		•/*?	Seismic refraction & S. reflection & teleseismic	Haworth et al. (1978); Stewart (1978); Keen et al. (1986)	450 Ma
Iapetus (outcropping Southern Appalachians)		•			Gravimetry & Seismic reflection & magnetotelluric	Thompson (1982); Thomas (1983); Cook (1984); Hutchinson et al. (1984)	450 Ma
Nashoba Zone			•/*?		Geology	Skehan (1983)	380 Ma
South Armorican	•				Magnetism	Lefort (1979); de Poulpiquet (1985)	400 Ma
"Collector anomaly" — Southern Ossa Morena	•				Magnetism & Gravimetry	Lefort (1983)	375 Ma
Block Island — Bay of Maine			•		Seismic reflection	Hutchinson et al. (1985; 1987); Phinney (1986); Unger et al. (1987)	375 Ma
Chesapeake Bay (Concealed Appalachians)			•		Geophysics & Geochemistry	Sinha and Zietz (1982) Pratt et al. (1987)	300 Ma
Lizard		*			Seismic reflection	Leveridge et al. (1984); Lefort (1987)	380 Ma
South Armorican	*				Structural geology & Seismic reflection	Lefort et al. (1982) Lefort (1987)	300 Ma
South Portugal	*				Seismic refraction	Mueller et al. (1973)	300 Ma
Morocco				•/?	Geochemistry	Kharbouch et al. (1985)	330 Ma
Variscan Front		*			Seismic reflection	Cheadle et al. (1986)	300 Ma
Mauritanian				*	Gravimetry & electric conductivity	Roussel and Liger (1983); Ritz and Robineau (1986)	360 Ma
"Collector Anomaly"	*				Magnetism	Lefort (in prep.)	300 Ma
Southwestern Appalachians (Concealed)			•		Magnetism & gravimetry & Seismic reflection	Klitgord and Popenoe (1983)	255 Ma
Northern Florida		?•/*			Magnetism & Seismic reflection	Klitgord et al. (1984) Nelson et al. (1985)	235 Ma

Prong (Fig. 62.3), are probably the only remains of this subduction zone. This conclusion is compatible with the southerly extension of subduction directions inherited from the "Acadian".

On the other hand, westerly subduction would imply that all the Alleghanian thrusts in the Appalachians are antithetic with respect to Hercynian subduction. Such a paradox is probably not a major obstacle, since many authors consider that Carboniferous thrusts are partly the result of re-activation along previous discontinuities. Antithetic thrusting is also accepted by Secor et al., (1986). In any case, it is certain that traces of this subduction plane are not easy to recognize on seismic profiles (Behrendt, 1986) due to westward Alleghanian thrusting and shearing (Pratt et al., 1987) which has severely disrupted the terrain up to 260 Ma ago (Glover et al., 1983).

— As far as the late Hercynian discontinuity North of Florida is concerned (which may have acted as a dextral shear), it is possible to imagine that the Theic ocean crust was in contact with Northern Florida at some stage, whilst Gondwana was clearly separated from Laurussia. On the contrary, the Carolina-Mississippi lineament must have progressively juxtaposed Grenvillian and Pan-African continental basements throughout the Carboniferous and part of the Permian. Because the East Coast magnetic anomaly is now known to be related to a Mesozoic or Cenozoic body (see Chapter Seven) and the negative New Brunswick anomaly (Fig. 49) was possibly produced by Triassic grabens, it is difficult to accept that these anomalies really represent a Palaeozoic suture as proposed by Nelson et al. (1985). There is no evidence for the Permo-Carboniferous volcanic arc that should have formed North of Florida if the Brunswick magnetic low had really been a subduction zone. The present author considers that, in this area, we are dealing not only with "Taconic" rejuvenation of the Senegal Pan-African suture (dipping North and following the Brunswick magnetic high) but also the Alleghanian northward thrusting of the Senegal microplate onto Laurentia. This thrusting developed above a fault zone that was primarily transform in character (Horton et al., 1984).

It is possible that the difficulties in locating a Hercynian suture to the SE of the Piedmont derive from the complex composite nature of its basement, which appears to be partly formed of Gondwanan fragments (such as the Carolinas microplate) and segments of volcanic arcs (such as the Carolinas belt).

Lastly, the subduction zone dip proposed for Mo-

rocco is entirely hypothetical due to the fact that geophysical evidence is lacking. If there had ever been subduction in this area, it must have been very restricted in time and space.

8.4 Obduction

The most widely developed obduction features involve the thrusting of oceanic crust onto the western margins of Iapetus (Chapter Three) — examples can be found from the Shetlands (Flinn et al., 1979) through Newfoundland and the Gulf of St Lawrence (Haworth and Keen, 1979) to the Southern Appalachians (Williams and Hatcher, 1983). In most cases, the obducted material is probably Pre-Cambrian and has been transported onto Grenvillian basement (in the case of the Shetlands, the basement is Lewisian). Between the Shetlands and Newfoundland, the allochthonous displacement is generally considered not to exceed a few tens of kilometres. By contrast, the displacement is much larger in the Southern Appalachians, where Hercynian re-activation has had some influence. In Europe, there are few obductions presenting these same geometrical relations, unless the klippes of NW Iberia are taken as comparable (Fig. 46). The ophiolitic klippe of NW Spain contains material dated at 460 Ma. (Bernard-Griffiths et al., 1985) which represents the remnants of oceanic crust belonging to the South Armorican Ocean (Lefort and Ribeiro, 1980) that were tectonically emplaced during the Devonian (Iglesias et al., 1983). Furthermore, it is probable that, as in the Appalachians, this material underwent a second displacement during the Carboniferous. Detailed gravimetric (Bayer and Matte, 1979) (Fig. 66) and seismic refraction (Hirn et al., 1982) studies on these klippes have enabled an estimate of the thickness of remaining oceanic crust. There appears to be a 6.4 km/s layer above terrains characterized by a 5.6 km/s velocity. Attempts to model the data from the northern Spanish margin have met with failure — post-Carboniferous erosion of the basement on the continental shelf has all but eliminated the allochthonous units.

It is also possible that obduction occurred in the Mauritanides, since wedges of ultrabasic material have been found in the external zones; certain ophiolites (Dia, 1984), of probable Upper Proterozoic age (Le Page, 1986) and affected by the Hercynian deformation, were perhaps already obducted

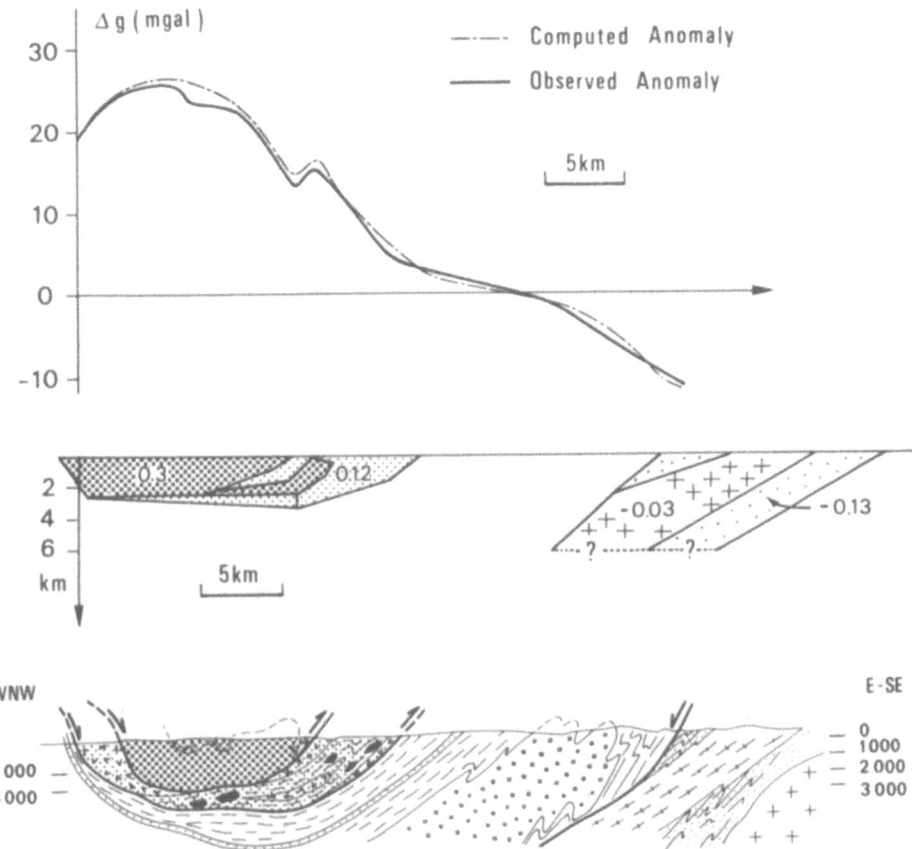

Fig. 66. Gravimetric modelling of an obducted oceanic klippe observed in N.W. Spain. The heavy dotted ornament indicates oceanic crust; numbers on model section refer to density contrasts. For locality, see Fig. 40. (After Bayer and Matte 1979)

onto the West African Craton during the Pan-African orogeny. All the examples cited so far are related to collisions; that is, the obduction of the oceanic material is caused by the closure of an ocean and occurs during convergence. Furthermore, the tectonic transport is often developed along thrusts which are synthetic with respect to the direction of subduction.

For the Ile de Groix Group (Fig. 40), the situation is rather different, as already discussed in Chapter Five. Here, the blueschists and associated greenschist amphibolites display microscopic and outcrop scale evidence (Quinquis, 1980; Cannat, 1983) for an antithetic displacement with respect to the northward-dipping South Armorican subduction (Quinquis and Choukroune, 1981; Lefort et al., 1982). In addition, it is known that this klippe is not thick (Lefort and Segoufin, 1978) and that it probably lies directly on Silurian volcanic arc material generated by subduction of the South

Armorican Ocean (Lefort, 1979). Finally, geochronological arguments suggest that the emplacement of the Ile de Groix klippe could have been before the end of plate convergence, i.e. around 400 Ma ago (Peucat and Cogné, 1977; Peucat, 1983). In this case, therefore, we are dealing with a typical example of syn-subduction obduction similar to the "flake tectonic" process envisaged by Oxburgh (1972). The small onland outcrop of blueschists at Le Bois-de-Céné further South in the Vendée (Fig. 34) shows the same tectonic mechanism.

8.5 Transcurrent Shearing

Three main structural types can be recognized amongst the different examples of shear zone described in this book — this classification takes account of the relations which exist with surrounding terrains.

I – The rarest type appears to be represented by the South Armorican Shear Zone (Fig. 40). On-land studies of the plutonic-metamorphic Ligerian domain (Cogné et al., 1983; Audren, 1986) and associated sedimentary formations (Diot and Blaise, 1978) have shown that the shear zone acted in a sinistral sense during the Devonian, and probably even before. Because of its parallelism with the South Armorican suture (Fig. 34) and its known displacements, contemporaneous with the end of convergence, it is proposed that this intracontinental shear zone has been induced by oblique convergence (as described by Roeder, 1975). Such a process (see Fig. 67.1) could eventually lead to some temporary distensions followed by compressions and may explain the uplift of eclogite slices along the shear zone. It also implies the existence of convergence with a sinistral component during the Siluro-Devonian. This sense of convergence is incompatible with the general configuration of the Iberian-Armorican Arc as deduced for the Carboniferous (Fig. 40). But it is evident that the curvature of this arc was far less pronounced before the Hercynian orogeny (Fig. 35) and that the northerly drift of the Iberian plate along the Porto-Badajoz-Cordoba transform fault led almost inevitably to an oblique convergence (Cogné and Lefort, 1985). Similar structures may have existed on the edge of the Iapetus suture, but reactivation there is so intense that no evidence is preserved at outcrop. By contrast, it is possible that the dextral shears known along the Southern Appalachians are partly due to the oblique closure of the Theic Ocean during the Carboniferous (Fig. 72).

II – The second type of transcurrent structure is represented by the long, post-tectonic faults which cut the Caledonian foldbelt (Fig. 12) and the Appalachians (Fig. 60). The comparison of these two map figures is highly instructive, especially since they are on the same scale and use a projection which does not deform the orientation of tectonic structures. Under these conditions, it can be seen that there is a transcurrent system extending as a straight line from the Gulf of Mexico to the Shetlands; this system cuts all the sinuosities in the foldbelt (Fig. 67.2), but appears to favour the Grenvillian side. The most remarkable feature of this system is its linearity; the progressive accretion of terrains onto the eastern flank of Laurentia, that extends from the Ordovician (in the North) to the Permian (in the South), could not have produced stresses with such constant orientation over such a long time-span. The straightness of this mega-lineament could thus have resulted from a single shear phenomenon. Such a

phenomenon would have affected the entire Caledonian-Appalachian belt before the western part of Gondwana had completely joined the southern part of Laurussia. This is because the mega-shear is cut by the Carolina-Mississippi lineament (Horton et al., 1984).

Van der Voo and Scotese (1981) have also recognized the fundamental unity of this phenomenon, but some authors (e.g. Esang and Piper, 1984; Irving and Strong, 1984) have taken account of other palaeomagnetic data and field observations to contradict the Carboniferous age and the 2000 km displacement originally proposed.

The palaeomagnetic solution given by Kent et al. (1984), based on new data from the West African Craton, confirms the sinistral shear proposed by Van der Voo and Scotese (op. cit.) – however, this applies to the Mid-Upper Devonian. If such an age is more in agreement with geological reality, the scale of this sinistral shear is nonetheless far too great in comparison with the 160 km proven for the Great Glen Fault in Scotland (Winchester, 1973) or the 180 km measured from the schematic map in Fig. 12. In any case, these measurements fall far beneath the power of resolution of palaeomagnetic methods.

As far as the first appearance of this shear belt is concerned, the following dilemma is posed: *either* the sinistral late-Acadian faults which may have existed in the Appalachians – thus extending the Scottish shear belt – have been masked by dextral Hercynian movements, *or* sinistral Devonian-Lower Carboniferous reactivation in the Caledonides and dextral Carboniferous reactivation in the Appalachians has affected a mega-shear zone which inherited its linear trace from a pre-Upper Devonian event.

III – This last type of shear corresponds to the E-W-trending Carboniferous structures which occur as five belts across Pangaea (Figs. 51 and 60) – they are seen at the Variscan Front, in the Armorican Shear Zones, along the Cobequid-Chedabucto Fault and are parallel to the Morocco-Maine and N40° Faults. All these shear zones operated in a dextral sense around Upper Carboniferous times and are seen to cut many early Hercynian structures. Although these belts show a fairly regular spacing of about 250 km, their most remarkable feature is that they all terminate westwards against major bends in the Appalachian foldbelt – that is, the re-entrants situated near Labrador, Quebec and Pennsylvania (Williams, 1978). Knowing that such irregularities in the foldbelt are probably of primary origin and that they reflect the shape of the eastern edge of the Grenvillian plate during the Upper Proterozoic, it is rather obvious that this geometry has controlled the location of Hercynian shear zones after the Gondwanan collision.

The fact that all these fractures show a dextral displacement demonstrates that the drift of Gondwana before its collision was not strictly northward, but rather towards the NW. Finally, the location of these shear zones even within northern Gondwana shows that it is not the boundary of this continent that acted as a buffer, but really the West African Craton which lies within the same plate.

Drake and Woodward (1963) already noticed the close relationship between the N 40° Fault and a major bend in the Appalachians at the same latitude. However, they (op. cit.) were unable to propose a tectonic mechanism due to lack of information. If the late-stage E-W Hercynian shears in Europe were really controlled by the initial outline of the Grenvillian plate, this would provide an unexpected conclusion in contradiction to the model of Arthaud and Matte (1975). This is a salutary example, perhaps the best available, of the importance of studying terrains concealed beneath recent basins and also offshore. A correlation between Grenvillian re-entrants and late Hercynian shear belts would explain why the dextral transform relationship envisaged between the Urals and the Appalachians (Arthaud and Matte, 1977) is incorrect (see Fig. 67.3). The transcurrent fault zones described here are not arranged as Reidel-type faults trending N 130° as suggested by the above authors, but instead define parallel E-W zones. Figure 67 summarizes the three main types of shear zone recognized within the concealed basement.

8.6 Transform Fault Zones

– The problem of the possible transform fault extending from the Carolinas to the Mississippi (Horton et al., 1984) has already been mentioned. This fault shows a relatively complex history since it is associated with a thrust (Nelson et al., 1985). Since this thrust probably only represents a simple late-stage feature, there is no point in trying to locate an ancient volcanic arc North of Florida. Seismic profiles of this isolated deep thrust do not suggest remnants of a subduction zone, but rather an intraplate structure (Fig. 62).

– The Porto-Badajoz-Cordoba transform, recognized in the Iberian Peninsula, is of a different type (Fig. 35) in that it first acted in a dextral sense before its sinistral re-activation. In addition, it does not separate structural domains which are fundamentally different. Furthermore, it is suspected that segments of oceanic crust could have locally bordered this transform (Lefort and Ribeiro, 1980).

– The transform fault identified in Canada, North of Prince Edward Island (Fig. 12), appears to have another origin since it has accommodated an irregularity in the Grenvillian margin during the closure of Iapetus. Even though recognized only from magnetism and gravimetry, this fault may well correspond to a real transform. Physical modelling with sand (Haworth, 1975) points to an offset in the suture around the southerly part of the Labrador coast. It is not out of the question that certain concealed bends in the Caledonian-Appalachian foldbelt could have generated identical transform faults.

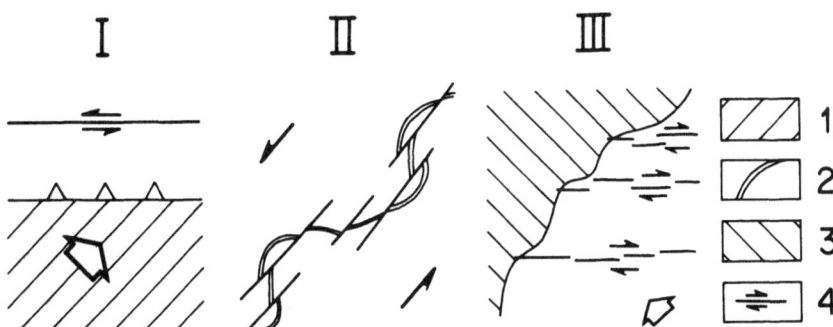

Fig. 67. The three main types (I-III) of shear zone recognized in the peri-Atlantic basement.
1: oceanic crust; 2: Caledonian belt; 3: Grenvillian crust; 4: shearing

8.7 Buckling

For some time, it has been recognized that the arcuate structure on the Newfoundland Grand Banks and the Iberian-Armorican Arc are both homothetic (Fig. 19) (Haworth, 1977; Lefort and Haworth, 1979). It has already been explained why the Grand Banks arcs have broadly conserved their initial Pre-Cambrian geometry; it is cut in the West by the straight Dover-Hermitage Fault (Hanmer, 1981) (Fig. 20), which is itself interrupted by a Devonian granite. This argument, amongst others, reinforces the idea that the Iberian Arc is not an entirely Hercynian structure (Matte and Ribeiro, 1975). The present author considers that the Iberian-Armorican Arc has resulted from simple buckling of a pre-existing Pre-Cambrian structure; this buckling is due to compression which developed between the Avalon Spur and Gondwana during Devonian through to Carboniferous times. A similar interpretation has been put forward (Perroud, 1980) from palaeomagnetic data; this clearly shows that buckling of the Iberian Massif occurred secondarily. A reconstruction of the initial arc curvature (Perroud, 1980) demonstrates that the pre-Hercynian arc in Spain was strictly concentric with the Newfoundland arcuate structures. Figure 68 shows a reconstruction of this phenomenon from a stage when the Iberian arc was already beginning to tighten up. The onset of buckling of this arc was probably more or less contemporaneous with the termination of sinistral shear along the South Armorican Shear Zone.

8.8 Distribution of Small Syn-Tectonic Basins

It is beyond the scope of this study to provide lengthy discussions on the development of the major concealed Palaeozoic basins, as found in Aquitaine (France), Suwannee (Florida) and Bové (Africa). Apart from some stratigraphic information on the nature of formations immediately beneath the Mesozoic cover, there are few available data from the basins themselves. By contrast, it is useful to stress the mode of formation of certain small fault-bounded basins since well-defined examples can be found in the area of study.

The mode of formation of many oblong basins with faulted margins is now recognized as resulting from a pull-apart mechanism (Mann et al., 1983; Aydin and Page, 1984) – many examples of this kind of basin are known in the peri-Atlantic domain. However, it is preferable to discuss only a few well-selected examples in order to stress the synchroneity of certain tectonic processes on both sides of the present North Atlantic (Fig. 69). In fact, it can be shown that the Siddi Bettache and Rehamnas basins in Morocco (Piqué, 1981) were opened by a "pull-apart" mechanism during the Upper Devonian and Dinantian. The associated fractures follow an Appalachian trend if Africa is placed in its Carboniferous location with respect to North America. In such a configuration, the African basins are parallel to the Deer Lake, White Bay, St Georges, Mabou, Minas and Moncton basins in Canada, which show the same age and are controlled by transcurrent dextral shear (Bradley, 1983). This sedimentation was accompanied on both sides of the North Atlantic by contemporaneous igneous activity. Since these basins lie on each side of the Ligerian-Acadian suture, this is evidence for the anticlockwise rotation of Gondwana and the probable close approach of these two zones during the Devono-Carboniferous collision.

There are also small, asymmetric and elongate basins – often fault-bounded on one side only – which run along ancient thrust zones and are particularly common beneath the continental margins. These basins generally result from normal fault re-activation of initially compressive fault zones. Many examples are known from the English Channel and Celtic Sea (Cheadle et al., 1986), in the Eastern U.S.A. (Klitgord and Hutchinson, 1985) and to the West and North of the British Isles (Brewer and Smythe, 1984). In this last area (Fig. 13B), there are even well-preserved sediments belonging to the western margin of Iapetus, buried beneath Mesozoic and Tertiary cover.

A very rare type of offshore basin with parallel boundary faults is represented on Fig. 55. Two possible explanations can be put forward for the formation of this basin. One might imagine that, at first, this type of basin was controlled by a normal fault which lay over the trace of a pre-existing reverse fault at that site, as discussed above. The normal fault was then thrust upward during a compressive phase, resulting in the creation of a "pop-up structure" at the surface and imbricate crustal slices at depth. Alternatively, it is possible that such a basin was initially formed in the classic manner (subsidence above an uplifted Moho) without faulting control. At a later stage, compressive movements would have produced both a "pop-up structure" near the surface and a concentration of deep crustal slices in a zone of weakness

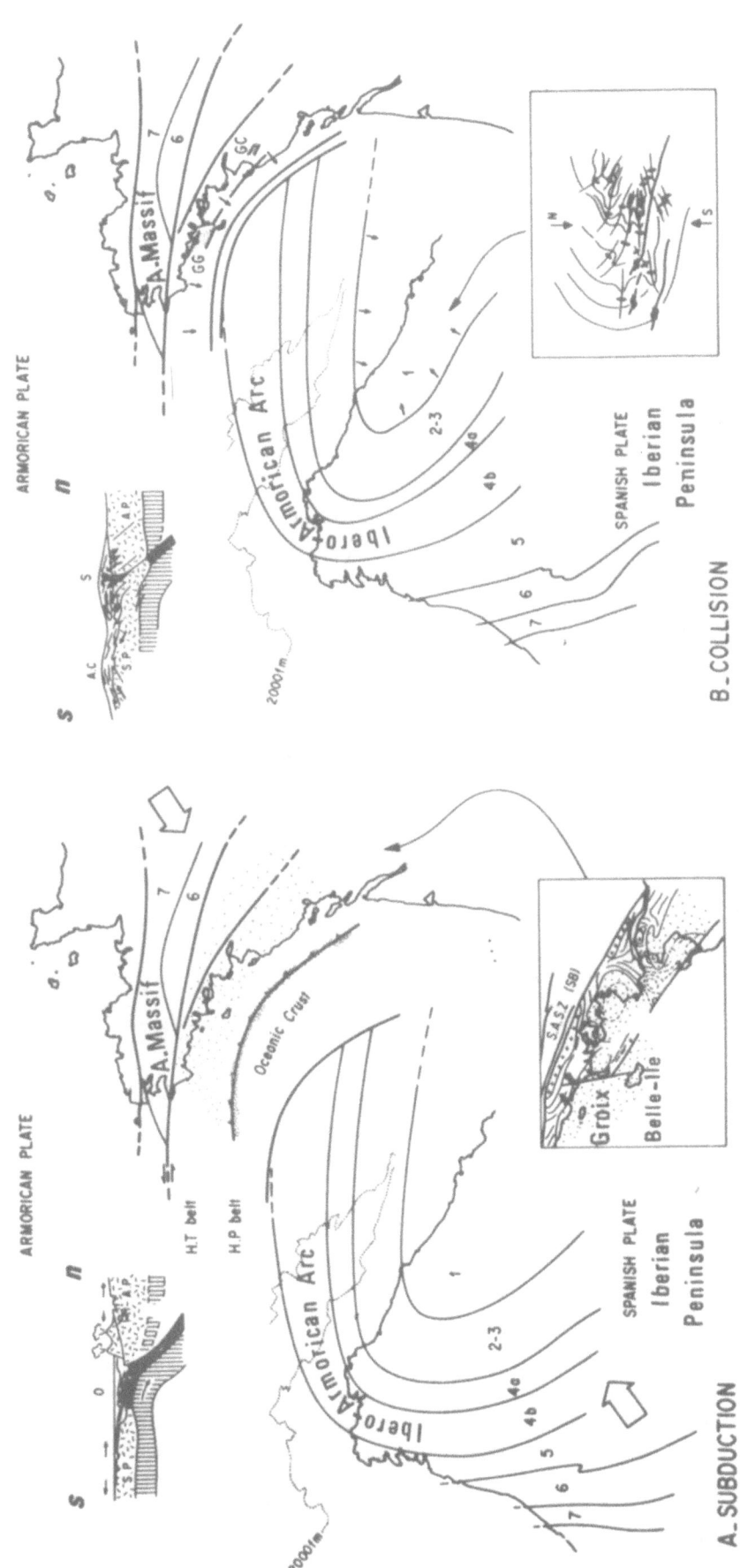

Fig. 68. Buckling of the Iberian microplate between Armorica and Gondwana. The numbers refer to sedimentary zones or structural domains. Arrows give tectonic transport direction during collision. GG: Ile de Groix blueschists; GC: Bois de Céné blueschists. Sketch-sections – AC: tightening of the Galician Arc; AP: Armorican Plate; O: oceanic crust; S: suture; SP: Spanish Plate; divergent arrows: distension zones. Left inset map shows locality of South Armorican Shear Zone. Right inset shows a plan view of buckling in the Galician Arc. (After Lefort 1979)

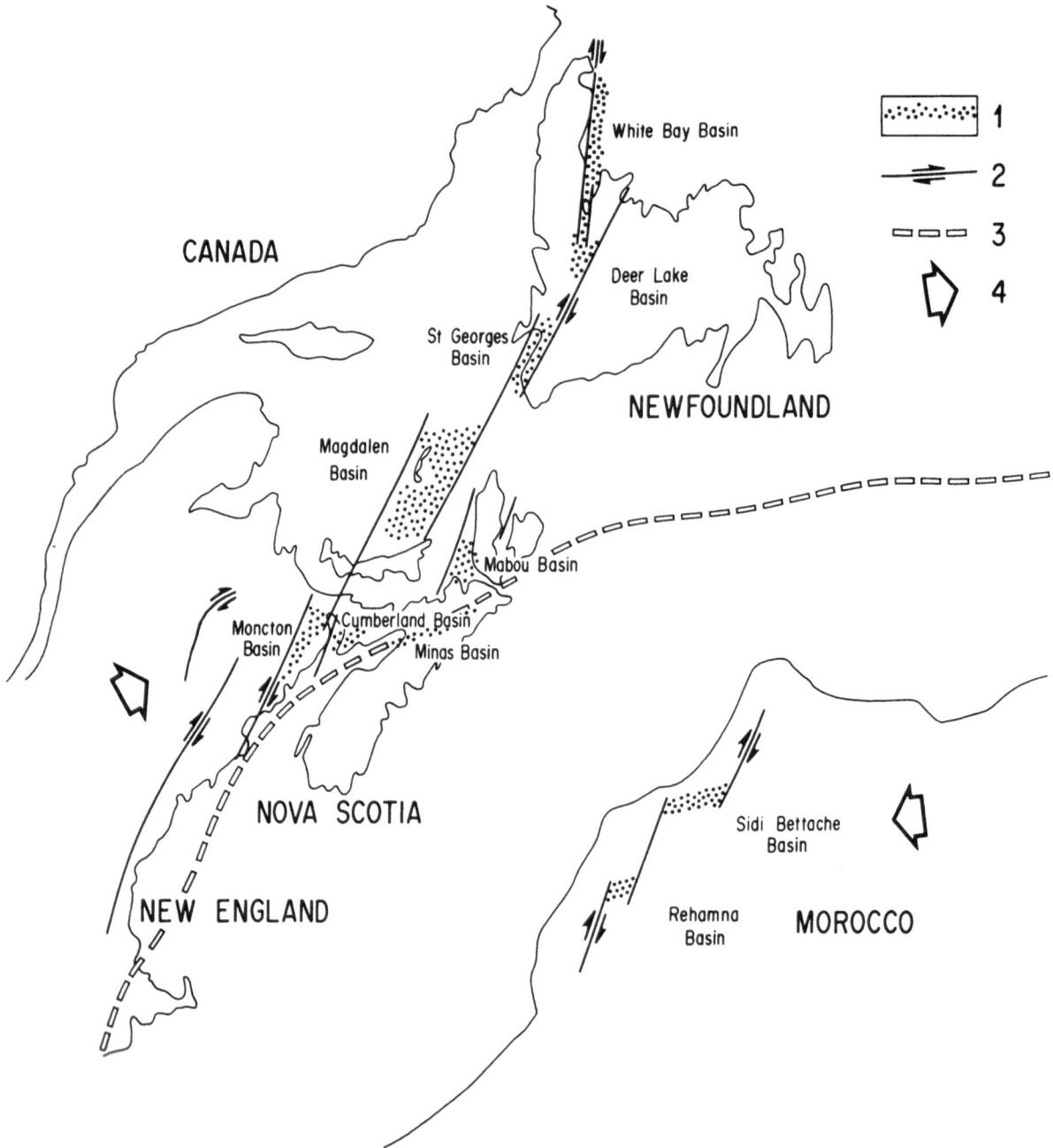

Fig. 69. Interpretative model showing simultaneous opening of "pull-apart" basins in Morocco, Newfoundland, Nova Scotia and New England during the Devono-Dinantian. 1: pull-apart basins; 2: shears; 3: Theic Ocean suture; 4: approximate shortening direction. (After Lefort 1983)

represented by the uplifted Moho. The quality of the available stratigraphic and seismic information on this problem does not enable a choice between the two alternative interpretations.

On the whole, the tectonic development of Pangaea has resulted more from compression than distension; this is probably why most of the sedimentation in small syntectonic basins has taken place during the operation of a pull-apart mechanism. The large Palaeozoic basins concealed beneath cover rocks were either formed before the final assembly of the super continent (e.g. Aquitaine Basin) or are located in stable cratonic areas (Bové and Suwanee basins).

8.9 Rotations at Different Scales

These important phenomena are recognized on all scales from the microscopic to the lithospheric plate, and are often neglected. Only rotations in the horizontal plane will be discussed here, other types of rotation have been more or less covered, albeit indirectly, in the preceding chapters.

As far as large-scale rotations are concerned (e.g. Gondwana) it is possible to distinguish three different kinds of rotational phenomenon. Firstly, palaeomagnetism (Scotese, 1984) has revealed the rotation of the Gondwanan plate as it drifted from the South Pole to the Equator during the Palaeozoic. In this case, the different positions of the plate in longitude are highly speculative and the construction of a finite pole of rotation can only yield random results (Le Pichon et al., 1973).

The list of rotations in Table 1 has a different meaning since it includes only that data necessary to re-assemble Pangaea from the present-day positions of the plates. In contrast to the usual approach of workers interested in continental fits, Lefort and Van der Voo (1981) have proposed rotations that are based on the continuity of trans-Atlantic markers rather than the fitting of continental margins or oceanic magnetic anomalies.

During its "northward" drift, the Gondwanan continent must also have turned on its own axis. The addition of these two rotational effects has brought about an oblique subduction and oblique collision during the phase of convergence with the northern plates (Laurentia being more or less fixed during the whole Palaeozoic). North of Gondwana, however, it is necessary to clearly distinguish, from the structural point of view, the Lower to Mid-Devonian drift, which led to collision, from the Middle to late Carboniferous displacements, which only caused intra-plate deformation.

The anti-clockwise rotation of Gondwana during the Devonian is demonstrated by the westward bends in the Pre-Cambrian ridges mapped out on the southern edge of the Grand Banks of Newfoundland, North of the Theic ocean suture (Fig. 70) (Lefort, 1983); this is corroborated by the eastward bends recognized in the Meguma Zone, which was located South of the Theic ocean suture. This rotation was contemporaneous with a movement of Gondwana towards the "North", which produced sinistral transcurrent shears in S.E. Newfoundland and Scotland (Fig. 72D).

The Late Carboniferous dextral shears on both sides of the Atlantic and along the Appalachian belt are seen to cut not only those structures related to the Devono-Carboniferous collision but also the Ligerian-Acadian suture (Figs. 51, 57 and 60). These shears result also from the anti-clockwise rotation of Gondwana, but at this stage the deformational front

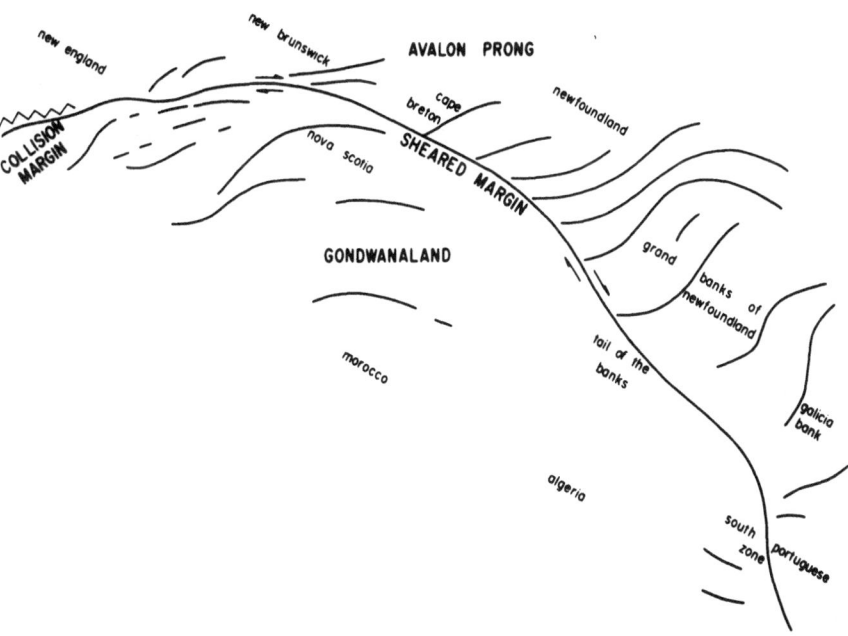

Fig. 70. Schematic map of deformations induced on each side of the Ligerian-Acadian suture during the Devonian. Lineaments are established from magnetic and gravimetric markers (see Figs. 36 and 37). (After Lefort 1983)

was situated at the contact with the West African Craton and no longer at the edge of the Gondwanan plate (Fig. 72.C). It is quite likely that there was continuity between Acadian and Hercynian deformation, but a blockage on the Gondwanan margin caused an enlargement of the zone affected by shearing deformation. Such a blockage could have been brought about by the intrusion of the Reguibat Prong which breaks the continuity of the Ligerian-Acadian suture. In any case, it is certain that Europe came into contact with Gondwana before the collision between Gondwana and North America — this is shown by the ages of tectono-metamorphic events. The opening mechanism of the pull-apart basins described above is a direct result of such a rotation.

The second type of rotation is on a smaller scale (Lefort and Lapointe, work in progress), producing zones in areas where there has been convergence between late-Hercynian E-W shears and Appalachian fractures trending N 50° or N 60° (Figs. 51 and 60). It has been established above that such convergences trace out arcuate shear zones which are characterized by dextral displacements. This leads us to suspect the existence of anti-clockwise rotations within such domains. Recently completed palaeomagnetic studies, combined with new field results, have shown that five of these arcuate shear zones have, in fact, given rise to the development of local anti-clockwise rotations. The importance of these rotations is generally proportional to the cumulative amplitude of the offset measured along the E-W transcurrent shear zones. These rotated domains are situated in N.W. and S.E. Newfoundland, southern Nova Scotia, S.E. New England and to the North of Chesapeake Bay.

The third type of rotation is measured on the microscopic scale — this leads us to the fundamental question as to whether mineral rotations can be linked in some way to megascopic plate rotations. Mineral scale rotations have been studied in detail in S.E. New England (Wintsch and Lefort, 1984), on the edge of the Gondwanan suture. In this area, one can observe a rotation of mineral lineations in relation to temperature history. Figure 71 illustrates the variation of lineation bearing as a function of time and temperature. The main interest of this approach is that the rotation of the mineral lineations (developed during retrogressive metamorphism) are in agreement with the changing tectonic transport directions associated with successive phases of thrusting. Thus, there appears to be a direct relationship between mineral orientation and thrusting direction, both showing a dextral rotation.

Because the thrust directions are here controlled by movements near the border of the neighbouring Gondwanan plate, it is possible that the relationship shown in Fig. 71 is a record of stresses induced by the Gondwanan collision. During the Permo-Triassic, N-S mineral lineations would appear to reflect the continuing passage of Gondwana towards the North defined by palaeomagnetic data.

If all the rotations discussed above are considered together, it is possible to propose a model for the Gondwanan collision; this model shows a strikingly coherent pattern of rotation, thrusting, subduction and shearing from Devonian through to Permian times (Fig. 72). The correlation between micro- and macrotectonics discussed here is probably a special case; the usual deviations of stress trajectories which are observed as one moves away from a plate boundary would certainly prevent any correlation being made a little further West.

The rotation of the West African Craton during Carboniferous times could be indirectly indicated by following the axis of divergence of Hercynian thrusts throughout the North Atlantic domain (Figs. 51 and 60). In fact, it can be observed that this axis, extending from the English Channel to Florida, is clearly deviated near the craton suggesting an anti-clockwise rotation.

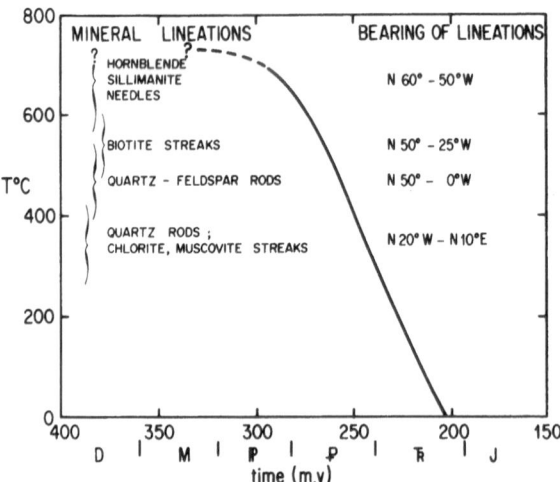

Fig. 71. Rotation of mineral lineations as a function of time and temperature in S.E. New England. (After Wintsch and Lefort 1984; reproduced by permission of the Geological Society of London)

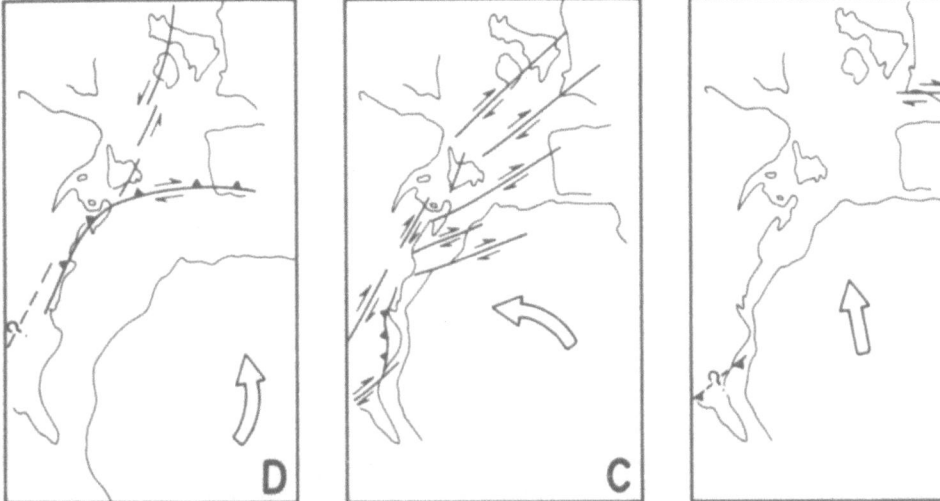

Fig. 72. Reconstruction of plate rotations and associated phenomena around the North Atlantic from Devonian through to Permian times. D: Devonian, C: Carboniferous, P: Permian

8.10 Impingement of Plate Margins

The first-order indenter detected beneath the Mesozoic and Tertiary cover South of the Armorican Massif has been already discussed in a previous chapter. It is considered as posthumous since it results from the tightening-up of an ancient interplate suture zone and is not the direct consequence of an active collision. Figure 73 illustrates (Lefort et al., 1982) the arcuate form of the thrust, the zone injected with Hercynian granites and the position of the offshore impinging block. This block is also responsible for the formation of S-folds observed in

the Groix graben. The development of grabens opposite zones of compression has been explained by Sengor et al. (1978). Off southern Brittany, the offset between indenter and graben is due to late dextral shears since these two structures should initially have been juxtaposed. Strictly speaking, the term "indenter" is not applicable in this case because such effects were originally defined only in terms of brittle deformation (Molnar and Tapponnier, 1977). The South Armorican "indenter", furthermore, is very small compared with normal impingement zones — such an impinging block cannot have a width much less that the thickness of the lithosphere (Chapter Seven); this is the reason

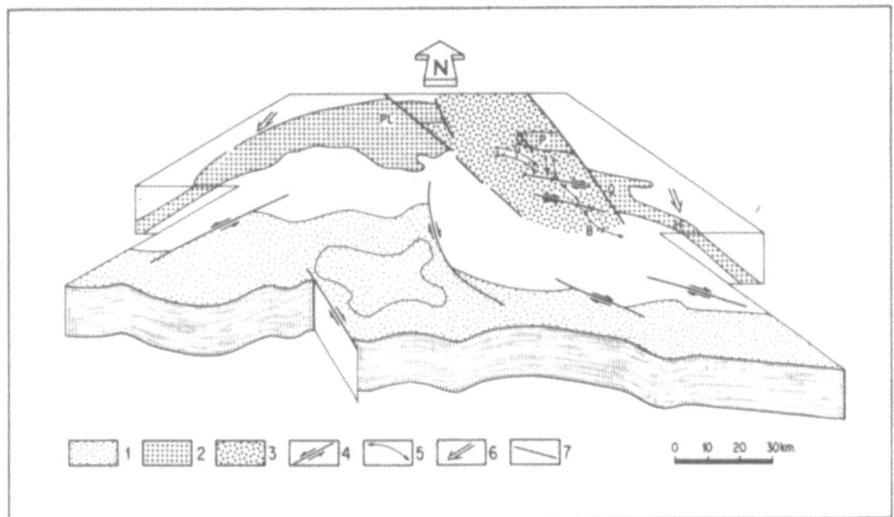

Fig. 73. The posthumous imprint structure off South Brittany (First-order indenter). (After Lefort et al. 1982). 1: basic body; 2: granite ($\approx$ 300 M.a.); 3. graben; 4: shears; 5: fold axes; 6: stress direction; 7: fault

why it is better to consider such a structure as a first-order indenter.

The impact of the Archaean Reguibat Prong onto the Appalachians constitutes a second-order impingement zone extending over about 250 km. A certain number of geophysical characteristics have already been discussed in previous sections; Fig. 74 shows a possible reconstruction of some of the structural features. As in the previous case, the use of the term "indenter" is an extension of the initial definition. The West of the Reguibat Prong is more reminiscent of the plastic stamping model developed by Mattauer (1981); particularly apparent is the divergence of thrusts and the axial planes of folds on each side of a vertical shear belt, as well as the existence of wedges that have been thrown out laterally. These basement wedges have been identified by Tapponnier et al. (1982) in physical modelling studies, but are not recognized with any certitude in the Appalachians — however, such structures possibly exist South of New England (Barosh, 1974) and beneath the Coastal Plain (Bobyarchick, 1981), where dextral and sinistral shearing is contemporaneous. Other E-W-trending shears are situated on either side of the impingement

zone (Figs. 60, 61 and 74) and were active during the westward displacement of the Reguibat Uplift. This has been classically described as a double bend in the foldbelt ("Double virgation forcée" in the sense of Argand).

The most complex impingement zone known at present has resulted from the collision between Gondwana and Laurussia; this kind of impingement occurs over several thousand kilometres and can be classified as a third-order structure. Several overlapping phenomena can be distinguished:

– First of all, there is the typical imprint formed by the rigid West African craton that has led to a simple fracture pattern. This imprint is also responsible for a crescent-shaped orogenic strip running almost continuously from the Mauritanides through the Anti-Atlas to the Ougarta belt. In the East, the zone of brittle deformation directly associated with impingement displays a set of dextral shears — this includes the N 120°-trending "Anti Atlas Main Fault" (Jeannette and Piqué, 1981). In the West, the Zemmours Fault (Rod, 1962) and the N40°/N50°-trending faults affecting the Western part of the Anti-Atlas (Figs. 60 and 75) show a sinistral displacement at this time (Jeannette and Piqué, 1981).

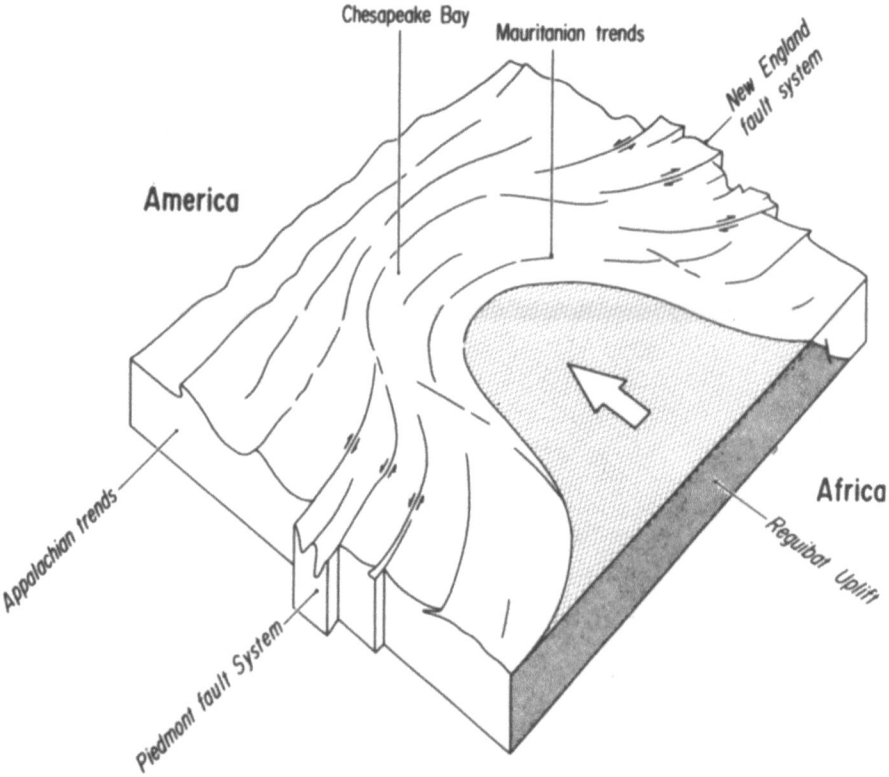

Fig. 74. Block diagram showing imprint of Reguibat Uplift onto the Appalachians during the Carboniferous (Second-order indenter). (After Lefort 1984)

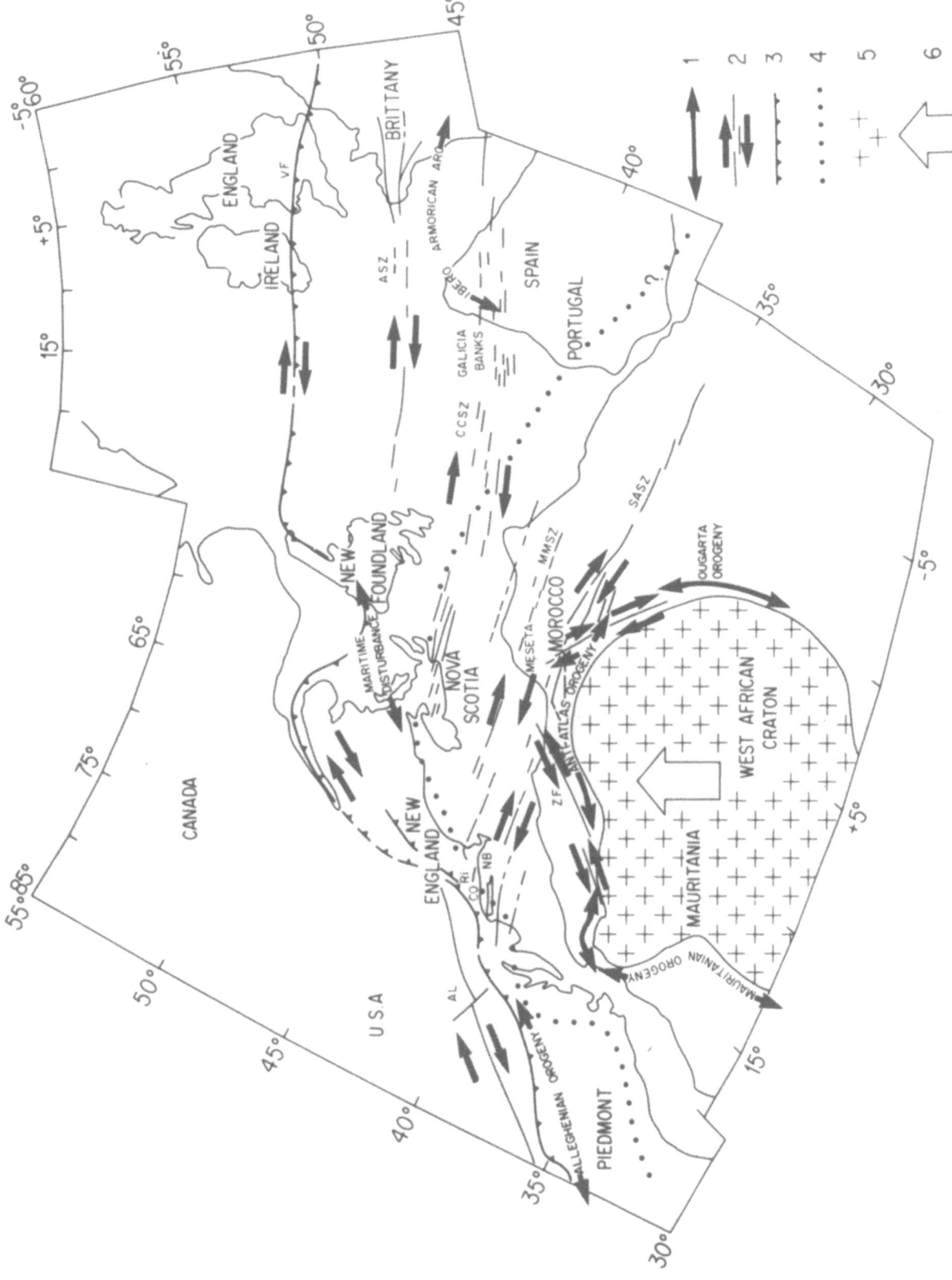

Fig. 75. Map showing influence of impingement of West African craton onto Laurussia during the Upper Carboniferous, 1: axial zones of orogenic belts; 2: shears; 3: thrusts; 4: Acadian-Alleghanian suture; 5: West African craton; 6: northward drift of West African craton

— This convergent set of opposite shears forms part of a larger system which has suffered an anticlockwise rotation as described above. In fact, there appears to have been a superposition of two types of tectonic phenomena in the northern part of the West African Craton; faults caused by impingement are cut by the dextral shear zone of the southern Atlas and by the Bani faults (Fig. 60). On Fig. 75, the arrow does not indicate the resultant of these two mechanisms but merely the probable drift displacement of the West African Craton towards the North.

— At the end of the Carboniferous, the impingement of the West African Craton onto Laurussia was rather unusual in character; this is because, in opposition to the model proposed for the collision between India and Asia (Molnar and Tapponnier, 1977), the true limits of the indenter were not at the boundary of the Gondwanan continent. Instead, the first rigid terrains within this continent appear to

have acted as the indenting block. In addition, there is some evidence that an indenting block turning on its own axis can still have some influence on adjacent terrains that previously belonged to another plate.

The list of tectonic mechanisms described in this Chapter does, in fact, constitute a conclusion since it enables a better understanding of the correlations put forward in this study. However, such mechanisms would not have been understood looking solely at on-land geology. The tectonic processes discussed here operated during the successive periods of continental accretion which characterize the peri-Atlantic domain (Fig. 76). They are certainly not the only mechanisms which could have contributed to the formation of Pangaea, but they are the only ones that can be recognized in the concealed basement by present geological and geophysical methods. Even if this list of possible mechanisms is not exhaustive, the correlations proposed in this book throw new light

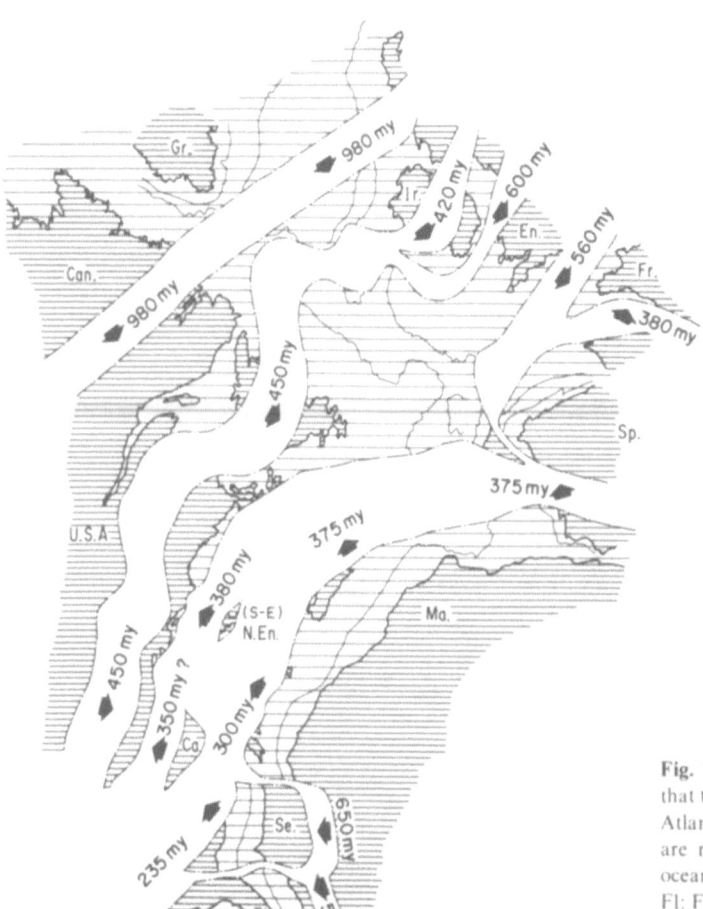

Fig. 76. Possible chronology of continental accretions that took place before opening of the present-day North Atlantic (separations between different continental blocks are not related to the possible widths of pre-existing oceans). Ca: Carolinas; Can: Canada; En: England; Fl: Florida; Fr: France; Gr: Greenland; Ir: Ireland; Ma: Morocco; N.En: New England; Sp: Spain; Se: Senegal; U.S.A.: United States of America

on the concealed basement and may correct a certain number of pre-conceived notions on the continuity of orogenic belts across the North Atlantic. Figure 77 illustrates clearly that the strict separation

that is sometimes maintained between the Mauritanide, Caledonide, Appalachian and Hercynian foldbelts is based simply on a lack of knowledge of the concealed basement.

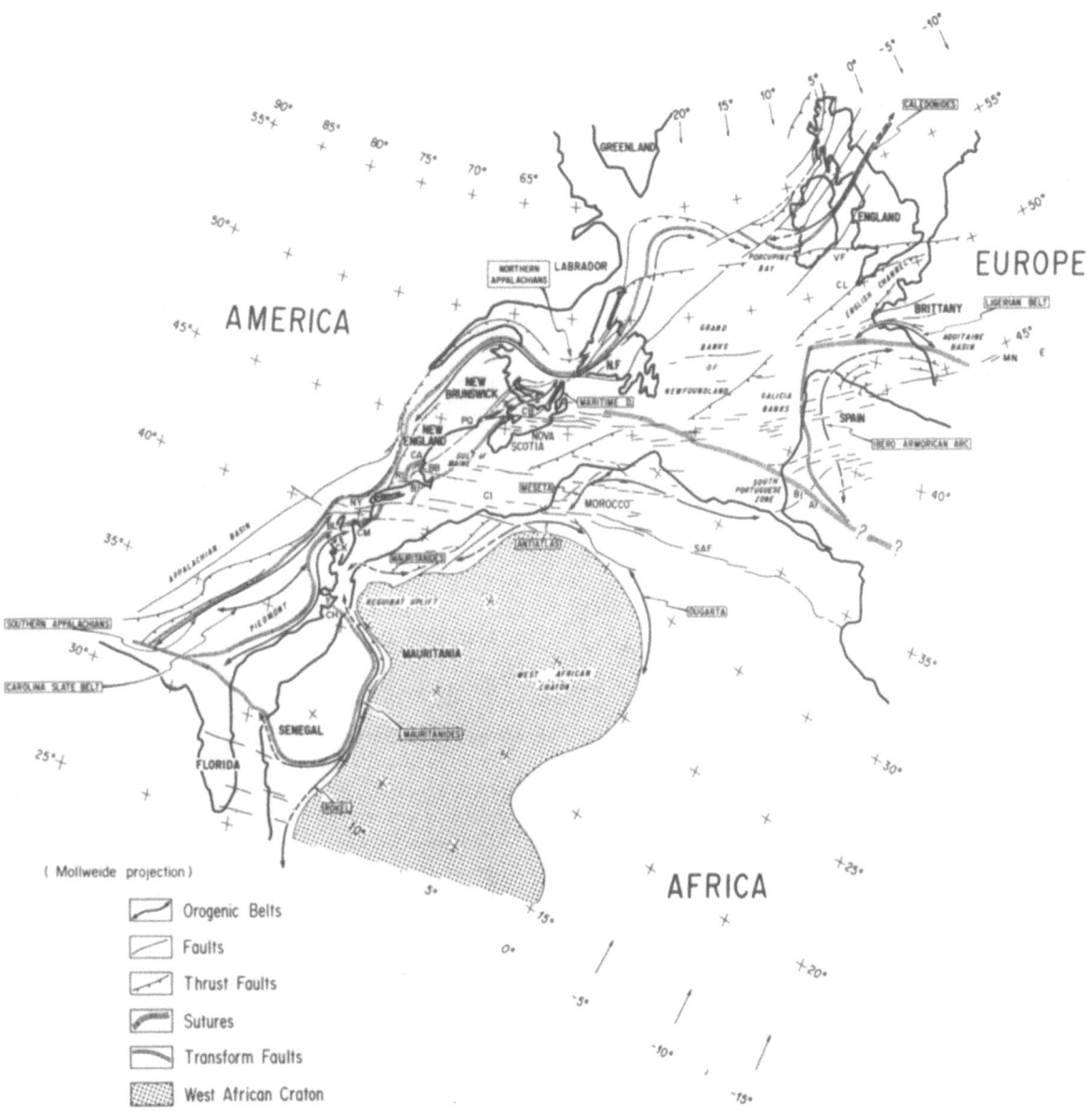

Fig. 77. Palaeozoic structures recognized on each side of the North Atlantic. Ar: Aracena; BB: Boston Basin; Bj: Beja; BL: Baltimore; CAP: Cape Ann; CB: Cobequid Mountains; CI: Canary Islands; CK: Chesapeake Bay; CL: Cape Lizard; CM: Cape May; E: Esterel; MN: Montagne Noire; NF: Newfoundland; NY: New York; PQ: Passamaquoddy Bay; RI: Rhode Island; SAF: South Atlas Fault; VF: Variscan Front. (Modified from Lefort and Van der Voo 1981)

References

Al-Shaikh ZD (1970) The geological structure of part of the central Irish Sea. R Astron Soc Geophys J 20:233-237

Anderson JGC, Owen TR (1968) The structure of the British isles. Pergamon, Oxford

Ando CJ, Cook FA, Oliver JE, Brown LD, Kaufman S (1983) Crustal geometry of the Appalachian orogen from seismic reflection studies. In: Hatcher RD, William H, Zietz I (eds) Contributions to the tectonics and geophysics of Mountain Chains. Geol Soc Am Mem 158:83-101

Andrade AAS (1979) Aspectos Geoquimicos do Ofiolitoides de Beja. Comm Serv Geol Portugal LXIV: 39-48

Andreieff P, Lefort JP, Marec A, Monciardini C (1973) Les terrains antécambriens et paléozoïques au large du Léon et leur relation avec la couverture secondaire et tertiaire de la Manche. Soc Geol Miner Bretagne Bull V: 13-20

Andrews JR (1983) The Iberian pyrite belt, an oblique-slip mobile zone. In: Direction de la Géologie. Proceeding: Le Maroc et l'orogénie paléozoïque, IGCP 27 Symposium, 25-28 août: 32, Ralat

Applin PL (1951) Preliminary report on buried pre-Mesozoic rocks in Florida and adjacent states. US Geol Surv Circ 91:1-28

Arthaud F, Matte Ph (1975) Les décrochements tardihercyniens du Sud-Ouest de l'Europe. Géométrie et essai de reconstitution des conditions de la déformation. Tectonophysics 25:139-171

Arthaud F, Matte Ph (1977) Late Paleozoic strike-slip faulting in southern Europe and Northern Africa: result of a right-lateral shear zone between the Appalachians and Urals. Geol Soc Am Bull 88:1305-1320

Audrain J, Lefort JP (1986) Le lever magmatique de Groix (Morbihan): une aide pour l'interprétation des structures profondes de l'île. Hercynica II: 65-70

Audren CL (1974) Les schistes cristallins de la Vilaine (Bretagne méridionale). Soc Geol Miner Bretagne Bull VI: 1-41

Audren CL (1986) Evolution structurale de la Bretagne méridionale au Paléozoïque. Soc Geol Miner Bretagne Mem 31:1-365

Audren CL, Lefort JP (1977) Géologie du plateau continental sud-armoricain entre les îles de Glenan et de Noirmoutier. Implications géodynamiques. Soc Geol France Bull XIX: 395-404

Audren CL, Le Métour J (1976) Mobilisation anatectique et déformation: les migmatites du golfe du Morbihan (Bretagne méridionale). Soc Geol France Bull XVIII: 1041-1049

Auffret GA, Pastouret L, Guennoc P (1979) Dredged rocks from the Armorican and Celtic margins. Initial Reports of the Deep Sea Drilling Project XLVIII: 995-1013 Washington DC US Government Printing Office

Austin GH, Howie RD (1973) Regional Geology of offshore eastern Canada. Can Geol Surv 71/23:73-107

Autran A (1978) Synthèse provisoire des événements orogéniques calédoniens en France. In: Tozer ET, Schenk PE (eds) Proceedings of the Caledonian Appalachian Orogen of the North Atlantic Regions. Canada Geol Surv Pap 78/13:159-175

Autran A, Cogné J (1980) La zone interne de l'orogène varisque dans l'ouest de la France et sa place dans le développement de la chaîne hercynienne. In: 26th Int Geol Congr Paris Colloq C6 Bur Rech Geol Minières Mem 108:90-111

Auvray B (1979) Genèse et évolution de la croûte continentale dans le Nord du Massif Armoricain. Thesis, University of Rennes

Auvray B, Lefort JP (1971) Etude des terrains antécambriens et paléozoïques immergés au large du petit Trégor (Manche occidentale). Essai cartographique. Soc Geol Miner Bretagne Bull III: 77-82

Auvray B, Lefort JP (1980) Evolution géodynamique du Nord du Massif Armoricain au Protérozoïque supérieur. Geol Soc Lond J 137:213

Auxietre JG, Dunand JP (1978) Géologie de la marge ouest-ibérique (au Nord de 40 = N). Thesis IIIrd cycle, University of Paris VI

Auxini (1969) Correlacion estratigrafica de los sondeos perforados en el Sahara espanol. Bol Geol Minero LXXX: 235-251

Aydin A, Page BM (1984) Diverse pliocen quaternary tectonics in a transform environment, San Francisco Bay region, California. Geol Soc Am Bull 95:1303-1317

Aymé JM (1965) The Senegal salt bassin. In: London Inst Petroleum (ed) Salt Basins around Africa. Elsevier, Amsterdam Lond pp 83-90

Bacon M, Gray F (1970) A gravity survey in the eastern part of the Bay of Biscay. Earth Planet Sci Lett 10:101-105

Bailey RJ (1975) Sub-Cenozoic geology of the British continental margin (Lat 50°N to 57°N) and the reassembly of the North Atlantic late paleozoic supercontinent. Geology 3:591-594

Bailey RJ (1979) The continental margin from 50°N to 57°N. Its geology and development. In: Banner FT, Collins MB, Massie KS (eds) The NW European shelf seas. Elsevier, Amsterdam pp 11-24

Bailey RJ, Grzywacz JM, Buckley JS (1974) Seismic reflection profiles on the continental margin bordering the Rockall Trough. Geol Soc Lond J 130:55-69

Bailey RJ, Buckley JS, Kielmas MM (1975) Geomagnetic reconnaissance on the continental margin of the British isle between 54° and 57°N. Geol Soc Lond J 131:275-282

Baker JW (1971) Intra-lower Paleozoic faults in the Southern Irish Sea. Geol Mag 108:501-509

Balé P, Brun JP (1983) Les chevauchements cadomiens de la Baie de Saint-Brieuc (Massif Armoricain). Paris Acad Sci C R 297:359-362

Ballard RD, Uchupi E (1975) Triassic Rift structure in Gulf of Maine. Am Assoc Petroleum Geol Bull 59:1041-1072

Bamford D (1979) Seismic constraints on the deep geology of the Caledonides and northern Britain. In: Harris AL, Holland CH, Leeake BE (eds) The Caledonides of British isles-reviewed. Geol Soc Lond pp 93-96

Bamford D, Nunn K, Prodehl C, Jacob B (1977) LISPB-III — Upper crustal structure of northern Britain. Geol Soc Lond J 133:481-488

Banks JE (1978) Southern Florida-subsurface features related to oil exploration. Gulf Coast Assoc Geol Soc Trans 28:25-30

Barbu A (1977) Le concept de zone pétrolière potentielle dans l'exploration du bassin des Doukkala (Maroc occidental). Mines Géol 42:49-57

Bard JP (1971) Sur l'alternance des zones métamorphiques et granitiques dans le segment hercynien sub-ibérique; comparaison de la variabilité des caractères géotectoniques de ces zones avec les orogènes orthotectoniques. Bol Geol Minero LXXXII: 324-345

Bard JP (1977) Signification tectonique des métatholéites d'affinité abyssale de la ceinture métamorphique de basse pression d'Aracena (Huelva, Espagne). Soc Geol France Bull XIX: 385-393

Bard JP, Moine B (1977) Variations géochimiques et affinités tholéitiques abyssales des orthoamphibolites d'Acebuches dans la ceinture métamorphique de basse pression d'Aracena (Huelva), Espagne. In: Reun Ann Sci Terre Rennes. Soc Geol France p 41

Bard JP, Capdevila R, Matte Ph (1971) La structure de la chaîne hercynienne, de la meseta ibérique: comparaison avec les segments voisins. In: Technip (ed) Histoire structurale du Golfe de Gascogne. Inst Fran Pet Publ 4:1-68

Bard JP, Burg JP, Matte Ph, Ribeiro A (1980) La chaîne hercynienne d'Europe occidentale en termes de tectonique des plaques. Bur Rech Geol Min Mem 108:233-246

Bardy Ph, Lefort JP (1987) Structure et stratigraphie des formations antémésozoïques du Golfe normano-breton d'après l'interprétation des données sismiques. Paris Acad Sci C R 304:997-1000

Barker AJ, Gayer RA (1985) Caledonides-Appalachian Tectonic analysis and evolution of related oceans. In: Gayer RA (ed) The tectonic evolution of the Caledonide-Appalachian orogen. Vieweg Braunschweig pp 126-165

Barnett RS (1975) Basement structure of Florida and its tectonic implications. Gulf Coast Assoc Geol Soc Trans 24:223-230

Barosh PJ (1974) Lineament studies in New England and their tectonic implications. In: Hodgson RA, Parker Gay S, Benjamins JY (eds) Proceedings of the First International Conference on the New basement Tectonics. Utah Geol Assoc Publ 3:218-235

Barosh PJ, Hermes DO (1981) General structural setting of Rhode Island and Tectonic history of Southeastern New England. In: Guidebook to geologic field studies in Rhode Island and adjacent areas. University of Rhode Island, pp 1-16

Barss MS, Bujak JP, Williams GL (1979) Palynological zonation and correlation of sixty-seven wells, eastern Canada. Can Geol Surv Pap 78:24:7-101

Bass MN (1969) Petrography and age crystalline basement rocks in Florida- some extrapolations. Am Assoc Petroleum Geol Mem 11:283-310

Bassot JP (1969) Aperçu sur les formations précambriennes et paléozoïques du Sénégal oriental. Soc Geol France Bull 7:160-169

Bayer R, Matte Ph (1979) Is the mafic/ultramafic massif of Cabo Ortegal (Northwest Spain) a nappe emplaced during a variscan obduction? a new gravity interpretation. Tectonophy! sics 57:9-18

Bayley RW, Muehlberger WR (compls) (1968) Basement Rockmap of the United States. US Geol Surv 1st edn, Scale 1:2500000

Beamish D, Smythe DK (1986) Geophysical images of the deep crust: the Iapetus suture. Geol Soc Lond J 143:489-497

Behrendt JC (1986) Structural interpretation of multichannel seismic reflection profiles crossing the southeastern United States and the adjacent continental margin. Decollement, faults, Triassic (?) basins and Moho reflections. In: Barazangi M, Brown L (eds) Reflection seismology The continental crust. Geodynamics Ser Am Geophys Union 14:210-213

Behrendt JC, Schlee J, Robb JM, Silverstein MK (1974) Structure of the continental Margin of Liberia, west Africa. Geol Soc Am Bull 85:1143-1158

Bell K, Blenkinsop J, Strong DF (1977) The geochronology of some granitic bodies from eastern Newfoundland and its bearing on Appalachian evolution. Can J Earth Sci 14:456-476

Belt ES (1968) Post-Acadian rifts and related facies, eastern Canada. In: Zen E-An (ed) Studies in Appalachian geology — Northern and Maritime. Interscience, New York, pp 95-113

Bennison GM, Wright AE (eds) (1970) The geological history of the British Isles. Edward Arnold, London

Berden JM (1964) Stratigraphy and faunas of subsurface lower Paleozoic rocks, Florida and adjacent states (abstr). Geol Soc Am Sp Pap 82:10

Bernard-Griffiths J, Peucat JJ, Cornichet J, Iglesias Ponce de Léon M, Gil-Ibarguchi (1985) U-Pb, Nd Isotope and REE geochemistry in eclogites from Cabo Ortegal complex, Galicia, Spain: An example of REE immobility conserving MORB-like patterns during high-grade metamorphism. Chem Geol 52:217-225

Bessoles B (1977) Geologie de L'Afrique. Le craton Ouest Africain. Bur Rech Geol Minièr Mem 88:1-402

Betz G (1965) Réinterprétation sismique n°181 Golfe de Gascogne. Compagnie d'étudés pétrolières (Intern Rept)

Bigg PJ, Crosby A, Davey RJ, Dingwall RG, Evans CDR, Harland R, Hughes MJ, O'B Knox RW, Lott GK, Medd AW, Morton AC, Turner RE, Warrington G, Wilkinson IP, Coleman BE, Ivimey-Cook HC, Campbell FA, Cooles GP, Dungworth G, Goodwin NS, Lister SW, Lowe SP, Taylor S (1981) The Zephyr (1977) wells, South-western Approaches and western English Channel. Lond Inst Geol Sci Rep 81/8:1-44

Biju-Duval B, Dercourt J, Le Pichon X (1977) From the Thetys ocean to the Mediterranean seas: a plate tectonic model of the evolution of the western alpine system. In: Biju-Duval B, Montadert (eds) Structural history of the Mediterranean basins. Editions Technip, Paris pp 143-759

Billing MP (1976) Geology of the Boston Basin. Geol Soc Am Mem 16:5-30

Binns PE, Mc Quillin R, Kenolty N (1974) The geology of the Sea of the Hebrides. Lond Inst Geol Sci Rep 73/4:1-44

Binns PE, Mc Quillin R, Fannin NGT, Kenolty N, Ardus DA (1975) Structure and stratigraphy of sedimentary basin in the Sea of the Hebrides and the Minches. In: Woodland AW (ed) Petroleum and the continental shelf of north west Europe. Applied Science, Barking 6:93-102

Black M, Hill M, Laughton AS, Matthews DH (1964) Three non magnetic seamount off the Iberian coast. Geol Soc Lond Q J 120:477-517

Blundell DJ (1979) The geology and structure of the Celtic Sea.

In: Banner FT, Collins MB, Massie KS (eds) The NW European shelf seas: geology and sedimentology. Elsevier, Amsterdam, pp 43-60

Blundell DJ, Parks R (1969) A study of the crustal structure beneath the Irish sea. R Astron Soc Geophys J 17:45-62

Bobyarchick AR (1981) The Eastern Piedmont fault system and its relationship to Alleghanian tectonics in the Southern Appalachians. J Geol 89:335-347

Boillot G, Capdevila R, Hennequin-Marchand J, Lamboy M, Lepretre JP (1973) La zone nord-pyrénéenne, ses prolongements sur la marge continentale nord-espagnole et sa signification structurale. Paris Acad Sci C R 277:2629-2632

Boillot G, Dupeuble PA, Malod J (1979) Subduction and tectonics on the continental margin off northern Spain. Mar Geol 3:53-70

Bott MHP, Holder AP, Long RE, Lucas AL (1970) Crustal structure beneath the granites of South-West England. In: Newale G, Rast N (eds) Mechanism of igneous intrusion. Geol J Spec issue 7:93-102

Bott MHP, Armour AR, Himsworth EM, Murphy T, Wylie G (1979) An explosion seismology investigation of the continental margin west of the Hebrides, Scotland, at 58 = N. Tectonophysics 59:217-231

Bott MPH, Young DGG (1971) Gravity measurements in the north Irish Sea. Geol Soc Lond Q J 126:413-434

Bouchez JL, Guillet P, Chevalier F (1981) Structures d'écoulement liées à la mise en place du granite de Guerande (Loire Atlantique). Soc Geol France Bull 23:387-399

Bousquet Ph, Burollet PF, Ferrat J (1977) Nouvelles données sur le paléozoique de la mer du Labrador. Reun Ann Sci Terre Rennes, p 105

Bradley DC (1982) Subsidence in Late Paleozoic basins in the northern Appalachians. Tectonics 1:107-123

Bradley DC (1983) Tectonics of the Acadian orogeny in New England and adjacent Canada. J Geol 91:381-400

Brewer JA (1984) Clues to the deep structure of the European Variscides from crustal seismic profiling in North America. In: Hutton DHW, Sanderson DJ Variscan Tectonics of the North Atlantic Region. Geol Soc Lond Spec Pub 14:253-263

Brewer JA, Smythe DK (1984) MOIST and the continuity of crustal reflector geometry along the Caledonian — Appalachian orogen. Geol Soc Lond J 141:105-120

Bridgwater D, Watson J, Windley BF (1973) The Archean craton of the North Atlantic region. R Soc Lond Philos Trans A 273:493-512

Brown L, Barazangi M, Kaufman S, Oliver J (1984) The first decade of COCORP: 1974-1984. In: Barazangi M, Brown L (eds) Reflection seismology: a global perspective. Geodynamics Ser Am Geophys Union 13:107-120

Brun JP, Burg JP (1982) Combined thrusting and wrenching in the ibero-Armorican arc: a corner effect during continental collision. Earth Planet Sci Lett 61:319-332

Buckley JS, Bailey RJ (1975) A free-air gravity anomaly contour map of the Irish continental margin. Mar Geophys Res 2:184-195

Bufler RT, Watkins JS, Shaub FJ, Worzel L (1980) Structure and early geologic history of the deep central Gulf of Mexico Basin. In: Pilger RH (ed) The origin of the Gulf of Mexico and the early opening of the central north Atlantic. Baton Rouge LA, Louisiana State University, pp 3-16

Bunce ET, Crampin S, Hersey JB, Hill MN (1964) Seismic refraction observations on the continental boundary west of Britain. J Geophys Res 69:3853-3863

Bullard EC, Everett JE, Smith AG (1965) The fit of the continents around the Atlantic. In: Blackett PMS, Bullard EC, Runcorn SK (eds) A Symposium on continental drift. R Soc Lond Philos Trans A 258, pp 41-51

Burke K (1976) Development of graben associated with the initial rupture of the Atlantic ocean. Tectonophysics 36:93-112

Caetano H (1983) Structure crustale de la zone sud portugaise et de la zone Osso-Morena d'après les études de sismologie expérimentale. Thesis, University of Paris VI

Cailleux Y, Deloche Ch, Gonord H, Rolin P (1983) Les zones de cisaillement hercynien en basse Meseta Marocaine. Proceedings of the Symposium: Le Maroc et l'orogénie paléozoique. IGCP 27, 25-28 August 1983, Rabat, p 15

Callame B (1965) Nouvelles observations sur le haut-fond sous-marin de Rochebonne. Soc Sci Natl Charentes-Maritimes IV: 6

Cannat M (1983) Cinématique de charriage ophiolitiques (Klamath mountains, Semail, Groix) Thesis IIIrd cycle University of Nantes

Capdevila R, Vidal Ph (1975) Données géochimiques et radiométriques sur les granulites et charnockites de la marge continentale nord-espagnole (Golfe de Gascogne). Réun Ann Sci Terre Montpellier, p 89

Capdevila R, Corretge G, Floor P (1973) Les granitoides varisques de la meseta ibérique. Soc Geol France Bull XV: 209-228

Capdevila R, Lamboy M, Leprêtre JP (1974) Découverte de granulites, de charnockites et de syénites nepheliniques dans la partie occidentale de la marge continentale nord espagnole. Paris Acad Sci C R 278:17-20

Capdevila R, Boillot G, Lepvrier C, Malod JA, Mascle G (1980) Les formations cristallines du Banc Le Danois (Marge nord Ibérique). Paris Acad Sci C R 291:317-320

Capote R (1983) Los tiempos precambricos: Discusion e interpretaciones de Conjunto. in Geologia de Espana. Publ Inst Geol Miner Espana 1:111-116

Carpenter MSN (1976) Petrogenetic study of the glaucophane schists and associated rocks from the Ile de Groix, Britanny, France. D. Phil Thesis University of Oxford

Carpenter MSN, Civetta L (1976) Hercynian high pressure/low temperature metamorphism in the Ile de Groix blueschists. Nature 262:276-277

Carpenter MSN, Peucat JJ, Pivette B (1982) Geochemical and geochronological characteristics of Palaeozoic volcanism in the Saint-Georges-sur-Loire Synclinorium (S. Armorican Massif). Evidence for pre-Hercynian tectonic evolution. Bur Rech Geol Minièr Bull I/1-2:63-79

Carvalho D (1972) The metallogenetic consequences of plate tectonics and the Upper Paleozoic evolution of Southern Portugal. Serv Fom Miner Portugal Estud Not 3/41:297-320

Caston VND, Dearnley R, Harisson RK, Rundle CC, Styles MT (1981) Olivine-dolerite intrusions in the Fastnet Basin. Geol Soc Lond J 138:31-46

Cazes M, Torreilles G, Bois C, Damotte B, Galdeano A, Hirn A, Mascle A, Matte Ph, Pham Van Ngoc, Raoult JF (1985) Structure de la croûte hercynienne du nord de la France: premiers résultats du profil ECORS. ECORS Rep, pp 925-941

Chacon J, Oliveira V, Ribeiro A, Oliveira JT (1983) La Estructura de la zona de Ossa-Morena. In: Comba JA (ed) Geologia de Espana. Publ Inst Geol Miner Espana, pp 490-503

Chapman CA (1962) Bays-of-Maine Igneous Complex. Geol Soc Am Bull 73:883-888

Charlot R (1978) Caractérisation des événements éburnéens et panafricains dans l'Anti-Atlas marocain. Apport de la méthode géochronologique Rb/Sr. Thesis University of Rennes

Chauvel JJ (1958) Contribution à l'étude des minérais de fer de l'Ordovicien inférieur en Bretagne. Thesis University of Rennes

Chauvel JJ, Robardet M (1971) Le minerai de fer de St Sauveur le Vicomte (Manche). Position stratigraphique, étude pétrographique. Signification paléogéographique. Soc Geol Miner Bretagne Bull 2:61-71

Chauvel JJ, Audren Cl, Auvray B (1975) Mise en évidence d'une série volcano-sédimentaire dans la partie orientale de Belle-Ile-en-Mer (Bretagne méridionale). Réun Ann Sci Terre Montpellier, p 98

Cheadle M, Matthews D, Warner M, Blundell D, Day G, Chadwick A, Mascle A, Gariel O, Montadert L, Lefort JP, Legall B, Sibuet JC, Cazes A, Schroeder M (1986) Deep seismic reflection profiling between England, France and Ireland. Geol Soc Lond J 143:45-52

Cherkis NZ, Fleming HS, Massingill JV, Feden RH (1975) Evidence for the emergence of the Hercynian Front on the north American continent. Can J Earth Sci 12:1474-1479

Chesher JA, Deegan CE, Ardus DA, Binns PE, Fannin NGT (1972) I.G.S. marine drilling with m.v. Whitethorn in Scottish waters 1970-1971. Lond Inst Geol Sci Rep 72/10:1-25

Chesher JA, Smythe DK, Bishop P (1983) The geology of the Minches, Inner Sound and Sound of Raasay. Lond Inst Geol Sci Rep 83/6:1-29

Choubert G (1952) Histoire géologique du domaine de l'Anti-Atlas (Maroc). Assoc Afr Geol Surv, pp 105-116

Chowns TM, Williams CT (1983) Pre-Cretaceous rocks beneath the Georgia coastal plain-Regional implications. US Geol Surv Prof Paper 1313:1-42

Cobbold PR, Quinquis H (1980) Development of sheath folds in shear regimes. J Struct Geol 2:119-126

Cochrane NA, Wright JA (1977) Geomagnetic sounding near the northern termination of the Appalachian system. Can J Earth Sci 14:2858-2864

Cogné J (1957) Schistes cristallins et granites en Bretagne méridionale: le domaine de l'Anticlinal de Cornouaille. Mem Carte Geol France. Ministère de l'Industrie Paris (1960), pp 1-382

Cogné J (1960) Sur l'origine sédimentaire des porphyroïdes de Belle-Ille (Morbihan). Paris Acad Sci CR 250:3350-3352

Cogné J (1974) Le Massif Armoricain. In: Debelmas J (ed) Géologie de la France. Douin Paris I: 105-161

Cogné J (1976) La chaîne hercynienne ouest-européenne correspond-elle à un orogéne par collision? Propositions pour une interprétation géodynamique globale. In: Centre National de la Recherche Scientifique (ed) Colloquium Int CNRS Géologie de l'Himalaya 268:111-129

Cogné J, Lefort JP (1985) The Ligerian orogeny: a protovariscan event, related to the Siluro Devonian evolution of the Thetys I Ocean. In: Gee D, Sturt B (eds) The Caledonide orogene, Scandinavia and related areas. Wiley Chichester, pp 1185-1193

Cogné J, Wright AE (1980) L'orogène cadomien. 26th Int Geol Congr Paris Colloq C6 Bur Rech Geol Miniér Mem, pp 29-55

Cogné JP, Choukroune P, Cogné J (1983) Cisaillements varisques superposés dans le Massif de Lanvaux (Bretagne centrale). Paris Acad Sci CR 296:733-776

Cook FA (1984) Towards an understanding of the southern Appalachian Piedmont crustal transition − A multidisciplinary approach. Tectonophysics 109:77-92

Cook FA (1986) Continental evolution by lithospheric shingling. In: Barazangi M, Brown L (eds) Reflection Seismology: the continental crust. Geodynamics Ser Am Geophys Union 14:13-19

Cook FA, Brown LD, Kaufman S, Oliver JE, Petersen TA (1981) COCORP Seismic profiling of the southern Appalachians orogen beneath the coastal plain of Georgia. Geol Soc Am Bull 92:738-748

Cormier RF, Smith TE (1973) Radiometric ages of granitic rocks, southwestern Nova Scotia. Can J Earth Sci 10:1201-1210

Cortesini A, Minner JR (1972) Petroleum developments in central and southern Africa in 1971. Am Assoc Petroleum Geol Bull 56:1749-1792

Coward MP, Siddans AWB (1979) The tectonic evolution of the welsh Caledonides. In: Harris AL, Holland CH, Leake BE (eds) The Caledonides of the British isles. Geol Soc Lond J Scott Acad Press, pp 187-198

Coward MP, Smallwood S (1984) An interpretation of the variscan tectonics of SW Britain. In: Hutton DH, Sanderson DJ (eds) Variscan Tectonics of the North Atlantic region. Geol Soc Lond Spec Publ 14:89-102

Cramer FH (1971) Position of the North Florida lower paleozoic block in Silurian time-phytoplankton evidence. J Geophys Res 76:4754-4757

Cramer FH, Diez de Cramer MD CR (1972) Subsurface section from Portuguese Guinea dated by palynomorphs as Middle Silurian. Am Assoc Petroleum Geol Bull 56:2271-2272

Crenn Y, Rechenmann J (1965) Mesures gravimétriques et magnétiques au Sénégal et en Mauritanie Occidentale. Publ Office Rech Sci Tech Outre Mer 6:1-59

Cutt BJ, Laving JG (1977) Tectonic elements and geologic history of the South Labrador and Newfoundland continental shelf, Eastern Canada. Can Soc Petroleum Geol 5:1037-1058

Dainty AM, Frazier JE (1984) Bouguer gravity in northeastern Georgia: a buried suture, a surface suture and granites. Geol Soc Am Bull 95:1168-1175

Dainty AM, Keen CE, Keen MJ, Blanchard JE (1966) Review of geophysical evidence on crust and upper mantle structure on the eastern seaboard of Canada. In: Steinhart J, Smith TJ (eds) The earth beneath the continents. Am Geophys Union Monogr 10:349-369

Dallmeyer RD (1982) 40Ar/39Ar ages from the Narragansett Basin and southern Rhode Island basement terrane: their bearing on the extent and timing of Alleghenian tectonothermal events in New-England. Geol Soc Am Bull 93:1118-1130

Dallmeyer RD (1984) 40Ar/39Ar from a pre-Mesozoic crystalline basement penetrated at holes 537 and 538 A of the deep sea drilling project Leg 77, Southeastern Gulf of Mexico: Tectonic implications. Initial Reports of the Deep Sea Drilling Project LXXVII: 497-504. US Government Printing Office Washington DC

Dallmeyer RD, Villeneuve M (1986) A polyphase tectonothermal evolution for the Southwesternmost Mauritanide orogen: Senegal. Proceedings of IGCP 233 Symposium 1-6 September, Oviedo, p 87

Dallmeyer RD, Villeneuve M (1987) 40Ar/39Ar mineral age record of polyphase tectonothermal evolution in the southern Mauritanide orogen, southeastern Senegal. Geol Soc Am Bull 98:602-611

Dallmeyer RD, Wright JE, Secor DT, Snoke AW (1986) Character of the Alleghanian orogeny in the southern Appalachians: PART II, Geochronological constraints on the tectonothermal evolution of the eastern Piedmont in South Carolina. Geol Soc Am Bull 97:1329-1344

Dallmeyer RD, Caen-Vachette M, Villeneuve M (1987) Age of post-tectonic granites in southern Guinea (West Africa) and the peninsular Florida subsurface: implications for constrasting origins of southern Appalachian exotic terranes. Geol Soc Am Bull 99:87–93

Daniels DL, Zietz I (1978) Geological interpretation of geomagnetic maps of the Coastal Plain region of South Carolina and parts of North Carolina and Georgia. US Geol Surv Open File Rep 78/261:1–47

Daniels DL, Ziets I, Popenoe P (1983) Distribution of subsurface lower Mesozoic rocks in the Southern United States, as interpreted from regional aeromagnetic and gravity maps. US Geol Surv Prof Paper 1313:1–24

Dean WT, Martin F (1978) Lower Ordovician acritarches and trilobites from Bell island, Eastern Newfoundland. Can Geol Surv Bull 284:1–35

Delahaye A (1976) Etude de la seismicité recente de la région d'Oléron. Thesis IIIrd cycle, University of Paris VI

De Poulpiquet J (1985) Etude géophysique d'un margeur magnétique situé sur la marge continentale sud-armoricaine. Arguments géologiques en faveur d'un modèle de suture de plaque. Thèse IIIrd cycle, University of Rennes

Deunff J, Lefort JP, Paris F (1971) Le microplancton Ludlovien des formations immergées des Minquiers (Manche) et sa place dans la distribution du paléoplancton silurien. Soc Geol Miner Bretagne Bull 1:9–28

De Voogd B, Keen CE (1987) Lithoprobe East: results from reflection profiling of the continental margin. Grand Banks Region. R Astron Soc Geophys J 89:195–200

Dia O (1984) La chaîne panafricaine et hercynienne des Mauritanides face au bassin protérozoïque supérieur à Dévonien de Taoudeni, dans le secteur-clé de Mejeria (Taganet, sud R.I.M.): lithostratigraphie et tectonique. Un exemple de tectoniques tangentielles superposées. Thesis Univesité of Aix Marseille III

Didier J, Guennoc P, Pautot G (1977) Granodiorites, granulites et charnockites de l'éperon de Goban, au contact du domaine océanique. Paris Acad Sci C R 284:713–716

Dillon WP, Mc Ginnis LD (1983) Basement structure indicated by seismic-refraction measurements offshore from south Carolina and adjacent area. US Geol Surv Prof Paper 1313:1–17

Dillon WP, Sougy, JMA (1974) Geology of West Africa and Canary and Cape Verde Islands. In Nairn AEM, Stehli FG (eds) The ocean basins and margins, the North Atlantic. Plenum, Lond, pp 315–390

Diment WH, Urban TC, Revetta FA (1972) Some geophysical anomalies in the eastern United States. In: Robertson EC (ed) Nature of the solid earth. Mc Graw-Hill, New York, pp 544–572

Dineley DL (1975) North Atlantic old red sandstone. Some implications for Devonian paleogeography In: Yorath CJ, Parker ER, Glass DJ (eds) Canada's continental margin Canadian Soc Petroleum Geolog Mem 4:773–790

Diot H, Blaise J (1978) Etude structurale dans le Précambrien et le Paléozoïque de la partie méridionale du domaine ligérien (SE du Massif Armoricain). Mauges, synclinal d'Ancenis et sillon houiller de la Basse Loire. Soc Geol Miner Bretagne Bull 1:31–50

Dobinson A (1978) Geophysical studies south and west of Kintyre in the solid geology of the Clyde Sheet. Lond Inst Geol Sci Rep 78/9:79–91

Dobson MR, Evans WE, Whittington R (1973) The geology of the south Irish Sea. Lond Inst Geol Sci Rep 73/11:1–36

Douglas RJW (ed) (1970) Geology and economic minerals of Canada. Can Geol Surv Rep 1: Maps and charts

Drake CL, Nafe JE (1968) Geophysics of the North Atlantic region. In: UNESCO Symposium on continental drift emphasizing the history of the Atlantic area. Montevideo-Uruguay October 67, p 55

Drake CL, Woodward H (1963) Appalachian curvature, wrench faulting and offshore structures. N Y Acad Sci Trans 26:48–63

Drake CL, Heirtzler J, Hirshman J (1963) Magnetic anomalies off eastern North America. J Geophys Res 18:5229–5275

Dunham KC (1970) Smoothed aeromagnetic map of Great Britain and Northern Ireland. Lond Inst Geol Sci 1st edn, scale 1/158400

Dunning FW, Max MD (1975) Explanatory notes to the geological map of the exposed and concealed Precambrian basement of the British Isles. Geol Soc Lond Sp Rep 6:11–14

Eden RA, Deegan CE, Rhys GH, Wright JE, Dobson MR (1973) Geological investigations with a manned submersible in the Irish Sea and off western Scotland 1971. Lond Inst Geol Sci Rep 73/2:1–27

Edmonds EA, Mc Keown MC, William M (1975) South-West England. In: British regional geology, 4th edn. HMS Office, London, pp 1–136

Edwards JWF (1984) Interpretation of seismic and gravity surveys over the eastern part of the Cornubian platform. Geol Soc Lond Sp Publ 14:119–124

Eisbacher GH (1969) Displacement and stress field along part of the Cobequid fault, Nova Scotia. Can J Earth Sci 6:1095–1104

Elias PN, Strong DF (1982) Timing of arrival of the Avalon zone in the northeastern Appalachians: a new look at the Straddling granite. Can J Earth Sci 19:1088–1094

Esang CB, Piper JDA (1984) Palaeomagnetism of caledonian intrusive suites in the Northern Highlands of Scotland: constraints to tectonic movements within the Caledonian orogenic belt. Tectonophysics 104:1–34

Evans D, Kenolty N, Dobson MR, Whittington RJ (1980) The geology of the Malin Sea. Lond Inst Geol Sci Rep 79/15:1–44

Evans D, Chesher JA, Deegan CE, Fannin NGT (1982) The offshore geology of Scotland in relation to the IGS shallow drilling programme 1970-1978. Lond Inst Geol Sci Rep 81/12:1–36

Ewing M, Worzel JL, Steenland NC, Press F (1950) Geophysical investigations in the emerged and submerged Atlantic coastal plain. Geol Soc Am Bull 61:877–892

Felix Cl (1972) Interprétation d'une paragenèse à glaucophane-épidote/lawsonite-grenat dans les glaucophane-schistes plurifacials de l'île de Groix (Morbihan, France). Paris Acad Sci C R 275:317–320

Fitch FJ, Forster SC, Miller JA (1984) The 40Ar/39Ar age spectrum of a rock from Gerrans Bay, Cornwall. Geol Soc Lond J 141:21–25

Flinn D, Frank PL, Brook PL, Pringle M (1979) Basement-cover relations in Shetland. In: Harris A, Holland L, Leake CH (eds) The Caledonides of the British Isles-Reviewed. Geol Soc Lond Sp Publ 8:109–116

Franke W, Engel W (1986) Synorogenic sedimentation in the Variscan belt of Europe; Soc Geol France Bull 8:25–33

Freire de Andrade C (1937) Os vales submarinos portugueses e o diastrofismo das Berlengas e da Estremadura. Serv Geol Portugal Mem

Freshney E, Taylor R (1980) The Variscides of southwest Britain. In: Bordas, 26th International Geological Congress (eds) Geology of the European countries U K. Dunod, Paris, pp 379–387

Fritsch J, Hinz K, Von Rad U, Roeser H, Weigel W, Wissman G (1978) Ergebnisse geowissenschaftlicher Untersuchungen der Valdivia Westafrika. Fahrt VA 10/1975 BMFT-Forschungsbericht M 78-03:1–60

Gaibar-Puertas C (1976) Variaciones del espesor crustal y grado de equilibro isostatico associables a las anomalias de Bouguer en la Espana Peninsular. Bol Geol Minero 87/84:371-401

Gapais D, Le Corre C (1980) Is the Hercynian belt of Brittany a major shear zone? Nature 288:574-575

Gardiner PRR, Mac Carthy AJ (1981) The late paleozoic evolution of southern Ireland in the context of tectonic basins and their transatlantic significance. in: Kerr J Wn, Ferguson AJ (eds) Geology of the North Atlantic borderlands. Can Soc Petroleum Geol Mem 7:683-725

Gardiner PRR, Sheridan DJR (1981) Tectonic framework of the Celtic Sea and adjacent areas with special references to the location of the Variscan front. J Struct Geol 3:317-331

Gates O (1969) Lower Silurian-Lower Devonian volcanic rocks of New England coast and Southern New Brunswick. in: Kay M (ed) North Atlantic geology and continental drift. Am Assoc Petroleum Geol Bull Mem 12:484-503

Gates AE, Simpson C, Glover III L (1986) Appalachian carboniferous dextral strike-slip faults: an example from Brookneal, Virginia. Tectonics 1:119-133

Gaudette HE (1980) Zircon isotopic age from the Union ultramafic complex, Maine. Can J Earth Sci 18:405-409

Gebauer D, Bernard-Griffiths J, Krebbs, O, Grunenfelder M (1978) U-Pb systematics of zircons and monazite from a mafic complex and its country rocks (Sauviat, French Central Massif). US Geol Surv Open File Rep 78/701:131-132

Gérard A (1975) La tectonique du socle sous la Manche occidentale d'après les données du magnétisme aéroporté. R Soc Lond Philos Trans A 279:55-68

Gérard JP (1979) Etude géologique et géophysique de la mer Malin. Thesis IIIrd cycle, University of Rennes

Gérard JP, Boillot G (1977) Geology of the north Irish continental shelf: Mar Geol 23:171-179

Gibbs AK, Barron CN (1983) The Guinea Shield reviewed. Episode 2:7-14

Gil-Ibarguchi I (1983) Revision de los datos radiometricos de rocas hercynicas y prehercynicas de la parte N del macizo iberico. Publ Inst Geol Miner Espana, pp 601-607

Gil-Ibarguchi I, Julivert M, Martinez FJ (1983) La evolucion de la cordillera Hercynica en el tiempo. Publ Inst Geol Miner Espana, pp 607-612

Given MM (1977) Mesozoic and early Cenozoic geology offshore Nova Scotia. Can Soc Petroleum Geol Bull 1:63-91

Glover III L, Speer A, Russell GS, Farrar SS (1983) Age of regional metamorphism and ductile deformation in the central and southern Appalachians. Lithos 16:223-245

Gohn GS (ed) (1983) Studies related to the Charleston, South Carolina, earthquake of 1886, Tectonic and seismicity. US Geol Surv Prof Pap 1313

Grant AC (1972) The continental margin off Labrador and Eastern Newfoundland — morphology and geology. Can J Earth Sci 9:1399-1430

Grau G, Montadert L, Delteil R, Winnock KE (1973) Structure of the European continental margin between Portugal and Ireland from seismic data. Tectonophysics 20:319-339

Graviou P, Auvray B (1985) Caractérisation pétrographique et géochimique des granitoïdes cadomiens du domaine nord-armoricain: implications géodynamiques. Paris Acad Sci C R II: 315-318

Grimaud S (1981) La marge ibérique au nord et à l'ouest du banc de Galice (Espagne). Thesis IIIrd cycle University of Paris VI

Groupe Galice (1976) Les anomalies magnétiques du champ terrestre dans la région des bancs de Galice. Reun Ann Sci Terre Paris, p 81

Groupe Galice (1979) The continental margin off Galicia and Portugal: acoustical stratigraphy, dredge stratigraphy and structural evolution. Initial Reports of the Deep Sea Drilling Project VXLII: 633-662. US Government Printing Office Washington DC

Grow JA, Bowin CO, Hutchinson DR (1979) The gravity field of the US continental margin. Tectonophysics 54:27-52

Guennoc P (1978) Contribution à l'étude des marges passives: structure et évolution géologique de la pente continentale d'un secteur de l'Antlantique nord-est: de la terrasse de Mariadzek à l'eperon de Goban. Thesis IIIrd cycle, University of Brest

Guetat Z (1981) Etude gravimétrique de la bordure occidentale du craton ouest-africain. Thesis IIIrd cycle, University of Montpellier

Guezou JC, Michard A (1976) Note sur la structure du môle cotier mesetien dans l'ouest des Rehamna (Maroc hercynien). Sci Geol Bull 29:171-182

Guillou JJ (1976) Role possible du mécanisme transgressif dans la genèse des gites marins de fer et de manganèse. Un exemple ordovicien (Sierra de Caurel, Lugo, Espagne). Paris Acad Sci C R 282:2021-2024

Hall J, Brewer JA, Matthews DH, Warner MR (1984) Crustal structure across the Caledonides from the "WINCH" seismic reflection profiles: influences on the evolution of the Midland Valley of Scotland. R Soc Edinburg Earth Sci 75:97-109

Hanmer S (1981) Tectonic significance of the northeastern Gander zone, Newfoundland: an Acadian ductile shear zone. Can J Earth Sci 18:120-135

Haszeldine RS (1984) Carboniferous North Atlantic paleogeography: stratigraphic evidence for rifting, not megashear or subduction. Geol Mag 121:443-463

Hatcher RD (1972) Developmental model for the Southern Appalachians. Geol Soc Am Bull 83:2735-2760

Hatcher RD (1981) Thrusts and nappes in the North American Appalachian Orogen. In: Mc Clay KR, Price NJ (eds) Thrust and Nappe Tectonics. Geol Soc Lond Spec Publ 8:491-500

Hatcher RD, Zietz I (1980) Tectonic implications of regional aeromagnetic and gravity data from the southern Appalachians. In: Wones D (ed) The Caledonides in the USA. Virginia Polytechnic Institute and State University Blacksburg VA Mem 2:317-325

Haworth RT (1975) The development of Atlantic Canada as a result of continental collision. Evidence from offshore gravity and magnetic data. In: Yorath CJ, Parker ER, Glass J (eds) Canada's continental margins and offshore petroleum exploration. Can Soc Petroleum Geol Mem 4:59-77

Haworth RT (1975) Paleozoic continental collision in the northern Appalachians in light of gravity and magnetic data in the Gulf of St Lawrence. In: Van der Linden WJM, Wade JA (eds) Offshore geology of Eastern Canada. Can Geol Surv Pap 74/30:1-10

Haworth RT (1977) The continental crust northeast of Newfoundland and its ancestral relationship to the Charlie Fracture zone. Nature 266:246-249

Haworth RT (1980) Appalachian structural trend northeast of Newfoundland and their trans-Atlantic correlation. Tectonophysics 64:111-130

Haworth RT (1981) Geophysical expression of Appalachian-Caledonide structures on the continental margins of the North Atlantic. In: Kerr JW, Ferguson AJ, Mahan LC (eds) Geology of the North Atlantic borderlands. Can Soc Petroleum Geol Mem 7:429-446

Haworth RT (1983) Geophysics and geological correlation within the Appalachian-Caledonian-Hercynian-Mauritanide orogens — An introduction. In: Schenk PE (ed) Regional trends in the geology of the Appalachian-Caledonian-Hercynian-Mauritanide orogen. NATO ASI Series: 1-10 Reidel, Dordrecht

Haworth RT, Jacobi RD (1983) Geophysical correlation between the geological zonation of Newfoundland and the British Isles. In: Hatcher RD, Williams H, Zietz I (eds) Contributions to the tectonic and geophysics of mountain chains. Geol Soc Am Mem 18:25-32

Haworth RT, Keen CE (1979) The Canadian Atlantic margin: a passive continental margin encompassing an active past. Tectonophysics 59:83-126

Haworth RT, Lefort JP (1979) Geophysical evidence for the extent of the Avalon zone in Atlantic Canada. Can J Earth Sci 16:552-567

Haworth RT, McIntyre JB (1975) The gravity and magnetic field of Atlantic offshore Canada. Can Geol Surv Pap 75/9:1-22

Haworth RT, McIntyre JB (1977) The gravity and magnetic fields of the Gulf of St Lawrence, Canada. Can Geol Surv Pap 75/42:1-11

Haworth RT, Miller HG (1982) The structure of Paleozoic oceanic rocks beneath Notre Dame Bay, Newfoundland. In: St Julien P, Beland J (eds) Major structural zones and faults of the northern Appalachians. Geol Assoc Can Spec Pap 24:130-149

Haworth RT, Sanford BV (1976) Paleozoic of Northeast Gulf of St Lawrence. Can Geol Surv Pap 76/1A:1-6

Haworth RT, Grant AC, Folinsbee RA (1976) Geology of the continental shelf off southeastern Labrador. Can Geol Surv Pap 76/1C:61-70

Haworth RT, Poole WH, Grant AC, Sanford BV (1976) Marine Geoscience Survey northeast off Newfoundland. Can Geol Surv 76/1A:7-15

Haworth RT, Lefort JP, Miller HG (1978) Geophysical evidence for an east-dipping Appalachian subduction zone beneath Newfoundland. Geology 6:522-526

Hermes DO, Zarman RE (1985) Late proterozoic and devonian plutonic terrane within the Avalon zone of Rhode Island. Geol Soc Am Bull 96:272-282

Hermes DO, Ballard RD, Banks PO (1978) Upper Ordovician peralkalic granites from the Gulf of Maine. Geol Soc Am Bull 89:1761-1774

Higgins MW, Zietz I (1983) Geologic interpretation of geophysical maps of the pre-Cretaceous "basement" beneath the coastal Plain of the Southeastern United States. Geol Soc Am Mem 158:125-130

Hinz K, Schulter HV, Grant AC, Srivastava SP, Umpleby D, Woodside J (1979) Geophysical transects of the Labrador sea: Labrador to southwest Greenland. Tectonophysics 59:151-183

Hinz K, Dostmann H, Fritsch J (1982) The continental margin of Morocco: seismic sequences, structural elements and geological development. In: Von Rad U, Hinz K, Sarnthein M, Seibold E (eds) Geology of the Northwest African continental margin 3:34-60. Springer Berlin Heidelberg New York

Hirn A, Senos L, Sapin M, Mendes Victor L (1982) High to low velocity succession in the upper crust related to tectonic emplacement: Tras os Montes Galicia (Iberia), Brittany and Limousin (France). R Astron Soc Geophys J 70:1-10

Hirn A, Lepine JC, Jopert G, Sapin M, Wittlinger G, Xin XZ, Yuan GE, Jing WX, Wen TJ, Bai XS, Pandey MR, Tater JM (1984) Crustal structure and variability of the Himalayan border of Tibet. Nature 307:23-25

Hobson GD, Overton A (1973) Sedimentary refraction seismic surveys, Gulf of St Lawrence. In: Hood PJ (ed) Earth Science Symposium offshore Canada. Can Geol Surv Pap 71/23:325-336

Holder AP, Bott MHP (1971) Crustal structure in the vicinity of south-west England. R Astron Soc Geophys J 23:465-489

Hollard H (1967) Le Dévonien du Maroc et Sahara Nord-occidental. In: Oswald DH (ed) International Symposium on the Devonian System Calgary. Alberta Soc Petroleum Geol, pp 203-244

Hollard H, Michard A, Piqué A (1976) L'orogénie acadienne dans les hercynides marocaines. Reun Ann Sci Terre Paris, p 225

Horn R, Münck F, Muraour P (1974) Quelques remarques sur la tectonique du socle sous la plateforme continentale atlantique d'après le magnétisme. Intl Colloq Ocean Expl Bordeaux, pp 1-6

Horton JW, Zietz I, Nearthy TL (1984) Truncation of the Appalachian piedmont beneath the coastal Plain of Alabama: Evidence from new magnetic data. Geology 12:51-55

Howie RD (1984) Carboniferous evaporites in Atlantic Canada. In: Belt ES Macqueen RW (eds) 9eme Congr International sur la stratigraphie et la géologie du Carbonifère Comptes-Rendus. Washington and Urbana – Champaigne. 3:131-142

Howie RD, Barss MS (1975) Upper Paleozoic rocks of the Atlantic provinces – Gulf of St Lawrence and adjacent continental shelf. In: Van der Linden WJM, Wade JA (eds) Offshore geology of eastern Canada. Can Geol Surv Pap 74/30:37-50

Hughes CJ, Brückner WD (1971) Late Precambrian Rocks of Eastern Avalon Peninsula, Newfoundland, A volcanic island complex. Can J Earth Sci 8:899-914

Hutchinson DR, Grow JA, Klitgord KD, Swift BA (1983) Deep structure and evolution of the Carolina Trough. In: Watkins JS, Drake CL (eds) Studies in continental margin geology. Am Assoc Petroleum Geol Mem 34:129-152

Hutchinson DR, Grow JA, Klitgord KD (1984) Crustal structure beneath the southern Appalachians: non uniqueness of gravity modeling. Geology 11:611-615

Hutchinson DR, Klitgord KD, Detrick RS (1985) Block island fault: a Paleozoic crustal boundary on the Long island platform. Geology 13:875-879

Hutchinson DR, Klitgord KD, Trehu AM (1987) Structure of the lower crust beneath the Gulf of Maine. R Astron Soc Geophys J 89:189-1984

Hutchinson DR, Klitgord KD, Detrick RS. Rift basins of the Long island platform. Geol Soc Am Bull (in press)

Hyde RS (1984) Geologic History of the carboniferous Deer Lake Basin, West Central Newfoundland Canada. In: 9ème Congrès International sur la stratigraphie et la géologie du Carbonifère Comptes-Rendus. Washington and Urbana – Champaigne 3:85-104

Hydrographic Department UK Defence (1973) Gravity-free air anomalies, quarter million plotting maps 016-14 GC, 016-15 GC and 016-20 G5

Iglesias M, Ribeiro ML, Ribeiro A (1983) La interpretacion aloctonista de la estructura del Noroeste peninsular. In: Comba JA (ed) Geologia de Espana. El N.O de la zona centroiberica con metamorfismo de alto grado. Publ Inst Geol Minero Espana, pp 459-467

Instituto Geografico e Cadastral (1960) Carta gravimetrica de Portugal (scale 1/1000 000)

Instituto Geographico y Cadastral (1972) Avance del mapa gravimetrico de la Peninsula Iberica, scale 1/2 000 000:1-30

Irving E, Strong F (1984) Evidence against large-scale carboniferous strike-slip faulting in the Appalachian-Caledonian orogen. Nature 310:762-764

Irving E, Emslie RF, Ueno H (1974) Upper Proterozoic Pole from Laurentia and the history of the Grenville structural province. J Geophys Res 79:5491-5502

Jacobi RD, Kristoffersen Y (1981) Transatlantic correlations of geophysical anomalies on Newfoundland, British isles, France and adjacent continental shelves. In: Kerr JW, Ferguson AJ, Mahan LC (eds) Geology of the North Atlantic borderlands. Can Soc Petroleum Geol Mem 7:197-229

Jansa L, Wade JA (1975) Geology of the continental margin off Nova Scotia and Newfoundland. In: Van der Linden WJM, Wade JA (eds). Can Geol Surv Paper 74/30:51-105

Jansa L, Wiedmann J (1982) Mesozoic-Cenozoic development of the Eastern North American and Northwest African Continental Margins: A comparison. In: Von Rad U, Hinz K, Sarnthein M, Seibold E (eds) Geology of the northwest African continental margin. Springer, Berlin Heidelberg New York 11:215-269

Jansa LF, Mamet B (1984) Offshore Visean of Eastern Canada: paleogeographic and plate tectonic implications. In: Belt ES, Macqueen RW (eds) 9ème Congrès International sur la stratigraphie et la géologie du Carbonifère Comptes Rendus. Washington and Urbana − Champaigne 3:205-216

Jeannette D, Piqué A (1981) Le Maroc hercynien: plate-forme disloquée du craton ouest-africain. Paris Acad Sci C R 293:79-82

Jegouzo P, Peucat JJ, Audren C (1986) Caractérisation et signification géodynamique des orthogneiss calco-alcalins d'âge ordovicien de Bretagne méridionale. Soc Geol France Bull 5:839-848

Johnson GL, McMillan NJ, Egloff J (1975) East Greenland continental margin. In: Yorath CJ, Parker ER, Glass DJ (eds) Canada's continental margins. Can Soc Petroleum Geol Mem 4:205-224

Johnson GL, McMillan NJ, Rasmussen M, Campsie J, Dittmer F (1975) Southwest Greenland continental margin. In: Yorath CJ, Parker ER, Glass DJ (eds) Can Soc Petroleum Geol Mem 4:391-409

Jones EJW (1981) Seismic refraction shooting on the continental margin west of the Outer Hebrides, northwest Scotland. J Geophys Res 86:553-574

Jones EJW, Mgbatogu CCS (1982) The structure and evolution of the west African continental margin off Guiné Bissau, Guinée, and Sierra Leone. In: Scrutton RA, Talwani M (eds) The ocean Floor. John Wiley, New York, pp 165-201

Julivert M (1971) L'évolution structurale de l'arc asturien. In: Technip (ed) Histoire structurale du Golfe de Gascogne. Inst Fran Pet Publ 1:2-28

Julivert M, Truyols J, Verges J (1983) El Devonico en el Macizo Iberico. In: Comba JA (ed) Geologia de Espana. Publ Inst Geol Minero Espana, pp 265-311

Kane MF, Yellin MJ, Bell KG, Zietz I (1972) Gravity and magnetic evidence of lithology and structures in the Gulf of Maine region. US Geol Surv Prof Paper 726B:1-22

Keen CE, Haworth RT (1986) North American continent-ocean Transects program Transect D3. Rifted continental margin off Nova-Scotia (offshore Eastern Canada). In: The Decade of North American Geology. Geol Soc Am DNAG:D3

Keen CE, Hyndeman RD (1979) Geophysical review of the continental margin of eastern and western Canada. Can J Earth Sci 16:712-747

Keen CE, Keen MJ, Nichols B, Reid I, Stockmal GS, Colman-Sadd SP, O'Brien JO, Miller H, Quinlan G, Williams H,

Wright J (1986) Deep seismic reflection profile across the Northern Appalachians. Geology 2:141-145

Kennedy MJ (1976) Southeastern margin of the northeastern Appalachians: Late Precambrian orogeny on a continental margin. Geol Soc Am Bull 87:1317-1325

Kennedy MJ (1979) The continuation of the Canadian Appalachians into the Caledonides of Britain and Ireland. In: Harris AL, Holland CH, Leake BE (eds) The Caledonides of the British isles-reviewed. Geol Soc Lond, pp 33-64

Kennedy MJ, Blackwood RF, Colman-Sadd SP, O'Driscoll CF, Dickson WL (1982) The Dover-Hermitage Bay fault: Boundary between Gander and Avalon zones, Eastern Newfoundland. In: St Julien P, Beland J (eds) Major structural zones and faults of the northern Appalachians. Geol Assoc Can Spec Pap 24:231-248

Kent DV, Dia O, Sougy JMA (1984) Paleomagnetism of lower-middle Devonian and Upper proterozoic-cambrian (?) rocks from Mejeria (Mauritania, West Africa). In: Van der Voo R, Scotese CR, Bonhommet N (eds) Plate reconstruction from Paleozoic paleomagnetism. Geodynamics series Am Geophys Un 12:99-115

Keppie JD (1977) Tectonics of southern Nova Scotia: Nova Scotia. Dept Mines Energy Nova Scotia Pap 77/1:1-34

Keppie JD (1979) Geological map of the Province of Nova Scotia. Dept Mines Energy Nova Scotia Scale 1/500 000

Keppie JD, Dostal J (1979) Paleozoic volcanic rocks of Nova Scotia. In: Wones DR (eds) Proceedings of the Caledonides in the USA IGCP 27 Symposium 5-6 September Blacksburg − Virginia Polytechnic Institute and State University Mem 2:249-259

Keppie JD, Ruitenberg AA, Fyffe L, Mc Cutcheon S, P St Julien, Skidmore B, Beland J, Hubert C, Williams H, Bursnall J (1982) Structural map of the Appalachian orogen in Canada. Memorial University Newfoundland Map Series: 4 1st edn Scale 1/2000 000

Keppie JD, St Julien P, Hubert C, Beland J, Skidmore B, Fyffe LR, Ruitenberg, Mc Cutcheon SR, Ruitenberg AA, Williams H, Bursnall J (1983) Times of deformation in the Canadian Appalachians. In: Schenk P (ed) Regional trends in the geology of the Appalachian − Caledonian − Hercynian − Mauritanide orogen. NATO ASI Series. Reidel Dordrecht 116:307-314

Kharbouch F, Juteau T, Treuil M, Joron JL, Piqué A, Hoepffner CH (1985) Le volcanisme dinantien de la Meseta Marocaine nord-occidentale et orientale. Caractères pétrographiques et géochimiques et implications géodynamiques. Sci Geol 2:155-164

King AF (1977) Subdivision and paleogeography of Late Precambrian and Early Paleozoic rocks of the Avalon Peninsula − Newfoundland. Geol Soc Am Abstr Prog 9:284

King EL (1982) Depositional environment of the lower paleozoic sediments on the Grand Banks of Newfoundland. BSD University of Dalhousie, Halifax, pp 1-81

King AF, Brueckner WD, Anderson MM, Fletcher J (1974) Late Precambrian and Cambrian sedimentary sequences of eastern Newfoundland. Fieldtrip Manual B6 (GAC/MAC)

King LH, Poole WH, Wanless RK (1985) Geological setting and age of the Flemish Cap granodiorite. East of the Grand Banks of Newfoundland. Can J Earth Sci 22:1286-1298

Kirby GA, (1984) The petrology and geochemistry of dykes of the Lizard ophiolites Complex, Cornwall. Geol Soc Lond J 141:53-59

Klitgord KD, Behrendt JC (1977) Aeromagnetic anomaly map of the United States Atlantic continental margin. US Geol Surv Map Ser MF 913 scale 1/1 000000

Klitgord KD, Behrendt JC (1979) Basin structure of the US Atlantic margin. In: Watkins JS, Montadert L, Dickerson PW (eds) Geological and geophysical investigations of the continental margins. Am Assoc Petroleum Geol Mem 29:85-112

Klitgord KD, Hutchinson R (1985) Distribution and geophysical signatures of early Mesozoic rift basins beneath the US Atlantic continental margin. In: Robinson GR, Froelich AJ (eds) Proceeding of the Second US Geological Survey workshop on the Early Mesozoic Basins of the Eastern United States. US Geol Surv Circ 946:45-61

Klitgord KD, Popenoe P (1983) Geophysical tectonic studies of the United States Atlantic coastal plain and continental margin. US Geol Surv Open file Rep 83/843:185-199

Klitgord KD, Schouten H (1982) Early mesozoic atlantic reconstruction from sea floor-spreading data. EOS (Am Geophys Union Trans) 63/18:307

Klitgord KD, Schouten H (1984) Atlantic maximum closure reconstruction. US Geol Surv (unpublished)

Klitgord KD, Schouten H (1986) Plate kinematics of central Atlantic. In: Vogt PR, Tucholke BE (eds) The geology of North America volume M. The western North Atlantic Region. Geol Soc Am 22:351-378

Klitgord KD, Dillon WP, Popenoe P (1983) Mesozoic tectonics of the southeastern United States coastal Plain and continental margin. US Geol Surv Prof Pap 1313:1-15

Klitgord KD, Popenoe P, Schouten H (1984) Florida: A jurassic transform plate boundary. J Geophys Res 89:7753-7772

Klitgord KD, Hutchinson DR, Grow JA, Schouten H (1985) Baltimore canyon Trough: Tectonic features. US Geol Surv continental Margin map series MF scale 1/1 000 000

Kreuzer H, Müller P, Wissman G (1984) Petrography and K. Ar dating of the Mazagan granodiorite, deep sea drilling project Leg 79, Hole 544 A and 547 B. Initial Reports of the Deep Sea Drilling Project LXXIX: 543-549. US Government Printing Office Washington DC

Kristoffersen Y (1977) Sea floor spreading and the early opening of the North Atlantic. Earth Planet Sci Lett 38:273-290

Laboratoire de détection géophysique (1977) Seismes proches, localisés par le réseau LDG France. Distance épicentrale 1500 km. Laboratoire de détection géophysique du CEA Bull January, Montrouge.

Lagarde JL (1985) Cisaillements ductiles et plutons granitiques contemporains de la déformation hercynienne post-viséenne de la Meseta Marocaine. Hercynica 1:29-37

Lamboy M (1976) Géologie marine et sous-marine du plateau continental au nord-ouest de l'Espagne, genèse des glauconies et des phosphorites. Thesis University of Rouen

Laughton AS (1975) Tectonic evolution of the Northeast Atlantic Ocean. A review. Norges Geol Undersokelse 316:169-193

Lecorché JP (1983) Structure of the Mauritanides. In: Schenk PE (ed). Regional trends in the geology of the Appalachian — Caledonian — Hercynian — Mauritanide orogen. NATO ASI Series 116:347-353 Reidel Dordrecht

Lecorché JP, Sougy J (1978) Les Mauritanides, Afrique occidentale. Essai de synthèse. In: Tozer ET, Schenk PE (eds) Proceedings of the Caledonian — Appalachian Orogen in the North Atlantic Region IGCP 27 Symposium. Can Geol Surv Pap 78/13:231-239

Lecorché JP, Roussel J, Sougy J, Guetat Z (1983) An interpretation of the geology of the Mauritanides orogenic belt (West Africa) in the light of geo0hysical data. In: Hatcher RD, Williams H, Zietz I (eds) Contributions to the tectonics and geophysics of mountain chains. Geol Soc Am Mem 158:131-147

Lefort JP (1970) Etude géologique de la Manche au Nord du Trégor — III, Géologie du substrat rocheux et morphologie. Soc Geol Miner Bretagne Bull 11:89-103

Lefort JP (1973) La "zonale" Biscaye-Labrador: mise en évidence de cisaillements dextres antérieurs à l'ouverture de l'Atlantique nord. Mar Geol 14:33-38

Lefort JP (1975) Le socle périarmoricain: étude géologique et géophysique du socle submergé à l'Ouest de la France. Thesis University of Rennes

Lefort JP (1975) Etude géologique du socle anté-mésozoïque au nord du Massif Armoricain: Limite et structure de la Domnonée. R Soc Lond Philos Trans A 279:123-135

Lefort JP (1979) Iberian-Armorican arc and Hercynian orogeny in Western Europe. Geology 7:384-388

Lefort JP (1983) La rotation du Gondwanaland et ses effets dans les maritimes et au Maroc au Dévono-Carbonifère. In: Direction de la Géologie, Proceeding of Le Maroc et l'orogénie paléozoïque, IGCP 27 Symposium, 25-28 août Rabat: 27

Lefort JP (1983) A new geophysical criterion to correlate the Acadian and Hercynian orogenies of western Europe and eastern America. In: Hatcher RD, Williams H, Zietz I (eds) Contribution to the tectonics and geophysics of mountain chains Geol Soc Am Mem 158:3-18

Lefort JP (1984) Mise en évidence d'une virgation carbonifère induite par la dorsale Reguibat (Mauritanie) dans les Appalaches du Sud (USA). Arguments géophysiques. Soc Geol France Bull XXVI: 1293-1303

Lefort JP (1984) The main basement features recognized in the northern part of the north atlantic area. Initial Reports of the Deep sea Drilling Project LXXX: 1103-1114. US Government Printing Office Washington DC

Lefort JP (1988a) Evidence for the imprint of the Reguibat uplift (Mauritania) onto the Central and southern Appalachians of the USA. Possible influence on the geometry of the Mauritanian, Theic and Iapetus suture. Afr Earth Sci J (in press)

Lefort JP (1988b) L'organisation structurale du socle profond de la Manche occidentale d'après l'interprétation des profils sismiques SWAT: un exemple d'intersection entre une suture cadomienne et des structures hercyniennes divergentes. In: Technip (ed) SWAT Rep. Inst Fran Pet Publ (in press)

Lefort JP (1988c) Corrélations entre les profils de sismique réflexion profonde "ECORS-GASCOGNE" et "ECORS-MASSIF ARMORICAIN". In: Technip (ed) ECORS-GASCOGNE Rep. Inst Fran Pet Publ (in press)

Lefort JP, Bardy Ph (1987) Mise en évidence de chevauchements cadomiens sur le profil sismique SWAT n°10 levé en Manche. Soc Geol France Bull (in press)

Lefort JP, Deunff J (1970) Découverte de paléozoïque à microplancton au sud de la Manche occidentale. Paris Acad Sci C R 270:271-274

Lefort JP, Deunff (1971) Esquisse géologique de la partie méridionale du golfe normano-breton (Manche). Paris Acad Sci C R 272:16-19

Lefort JP, Deunff J (1974) Etude du socle antémosozoïque de la partie septentrionale du golfe normano-breton. Bur Rech Geol Minièr V:73-83

Lefort JP, Haworth RT (1978) Geophysical study of basement fractures of the western European and eastern Canadian shelves: transatlantic correlations and late Hercynian movements. Can J Earth Sci 15/3:397-404

Lefort JP, Haworth RT (1979) The age and origin of the deepest correlative structures recognized off Canada and Europe. Tectonophysics 59:139-150

Lefort JP, Haworth RT (1981) Geophysical correlation between basement features in North Africa and Eastern New England: their control over North Atlantic structural evolution. Soc Geol Miner Bretagne Bull XIII: 103–116

Lefort JP, Haworth RT (1984) Geophysical evidence for the extension of the variscan front onto the Canadian continental margin: Geodynamic and paleogeographic consequences. In: Hutton DHW, Sanderson DJ (eds) Variscan tectonics of the North Atlantic region. Geol Soc Lond Spec Publ 14:219–231

Lefort JP, Max MD (1984) Development of the Porcupine Seabight: use of magnetic data to show the direct relationship between early oceanic and continental structures. Geol Soc Lond J 141:663–674

Lefort JP, Peucat JJ (1974) Le socle antémésozoïque submergé à l'Ouest de la baie d'Audierne. Paris Acad Sci C R 279:635–637

Lefort JP, Ribeiro A (1980) La faille Porto-Badajoz-Cordoue-a-t-elle contrôlé l'évolution de l'océan paléozoïque sud-armoricain? Soc Geol France Bull XXII: 455–462

Lefort JP, Segoufin J (1978) Etude géologique de quelques structures magnétiques reconnues dans le socle péri-armoricain submergé. Implications géodynamiques concernant la fracturation proto-atlantique et l'orogenèse hercynienne dans le domaine armoricain. Soc Geol France Bull XX: 185–192

Lefort JP, Segoufin J (1978) Etude comparée des structures profondes et des anomalies magnétiques allongées reconnues en Manche occidentale et en Baie d'Audierne: existence possible d'une suture cryptique au nord-ouest du Massif armoricain (France). Tectonophysics 46:65–75

Lefort JP, Van der Voo R (1981) A kinematic model for the collision and complete suturing between Gondwanaland and Laurussia in the Carboniferous. J Geol 85:537–550

Lefort JP, Alveirinho Dias J, Monteiro JH, Ribeiro A (1981) L'organisation des structures profondes du socle à l'Ouest de la faille Porto-Tomar-Badajoz: apport des données géophysiques. Comm Serv Geol Portugal 67:57–63

Lefort JP, Audren Cl, Max MD (1982) The southern part of the Armorican orogeny a result of crustal shortening related to reactivation of a pre-Hercynian mafic belt during Carboniferous time. Tectonophysics 89:359–377

Lefort JP, Audren Cl, Jegouzo P, Max MD, Grant P, Rattey P (1982) Disposition of structures in the South Brittany (France). Belt of high pressure metamorphism. In: Blanchard J, Mair J, Morrison I (eds) Proceedings of the 6th Symposium on the Confederation Mondiale des Activités Subaquatiques. 14–8 September Heriot-Watt University Edinburg, pp 285–291

Lefort JP, Peucat JJ, Deunff J, Le Hérissé A (1984) The Goban spur Paleozoic basement. Initial Reports of the Deep Sea Drilling Project LXXX: 677–679. US Government Printing office Washington DC

Lefort JP, Max MD, Roussel J (in press) The north-west boundary of Gondwanaland and its relationship with two older satellite sutures: geophysical evidences. In: Fettes DJ, Harris AL (eds) Synthesis of the Caledonian Rocks of Britain. NATO ASI Series, Reidel Dordrecht

Lefort JP, Vidal Ph, Cabioch L (1978) First dating of a submerged granite in the western Channel: appearance of an intrusion of the lower Paleozoic age. Estuarine Coastal Mar Sci 7:373–379

Le Gall B (1983) La chaine hercynienne en Cornouaille anglaise: étude géologique du front du Lizard et des formations à blocs des zones du Meneage et de Roseland. Implications géodynamiques. Thesis IIIrd cycle University of Brest

Le Gall B (1984) Les formations chaotiques au front du Lizard

(Cornouaille anglaise): nouvelles données sur leur lithologie, leur structure et leur signification géodynamique. Soc Geol France Bull XXVI: 1357–1364

Legault JA (1982) First report of Ordovician (Caradoc-Ashgill) polynomorphs from Orphan Knoll, Labrador Sea. Can J Earth Sci 19:1851–1856

Lehner P, de Ruiter PAC (1977) Structural history of Atlantic margin of Africa. Am Assoc Petroleum Geolog Bull 61:961–981

Le Mouël JL, Le Borgne E (1971) La cartographie magnétique du Golfe de Gascogne. In: Technip (ed) Histoire structurale du Golfe de Gascogne. Inst Fran Pet Publ 2:1–12

Le Page A (1986) La lithostratigraphie des grandes zones structurales des Mauritanides entre le 14e et le 16e parallèle nord (Sénégal oriental et Rep. Isl. de Mauritanie) Essai d'interprétation géodynamique. Afr J Earth Sci 5:119–139

Le Page A, Campredon R (1981) Contribution des données des images LANDSAT à l'étude structurale d'un secteur des Mauritanides en Mauritanie du Sud et au Sénégal oriental. Rev Inst Français Petrole 36/1:17–33

Le Pichon X, Fox PJ (1971) Marginal offsets, fracture zones and the early opening of the north Atlantic. J Geophys Res 76:6294–6308

Le Pichon X, Francheteau J, Bonnin J (1973) Plate tectonics. Elsevier Scientific Amsterdam

Le Pichon X, Sibuet JC, Francheteau J (1977) The fit of the continents around the North Atlantic ocean. Tectonophysics 38:169–209

Le Pochat G (1984) Bassins paléozoïques cachés sous l'Aquitaine. In: Colloque national sur la géologie profonde de la France. Bur Rech Geol Minièr 81/7:1–39

Leprêtre JP (1974) Traitement et utilisation de données magnétiques. Application à la marge continentale Nord-espagnole située entre 3° et 6° de longitude Ouest. Thesis IIIrd cycle, University of Rennes.

Lesquer A, Beltrao JF, De Abreu FAM (1984) Proterozoic links between northeastern Brazil and west Africa: a plate tectonic model based on gravity data. Tectonophysics 110:9–26

Leutwein F, Sonet J, Zimmermann JL (1972) Dyke basique du Massif armoricain septentrional. Paris Acad Sci C R 275:1327–1330

Leveridge BE, Holder NT, Day GA (1984) Thrust nappe tectonics in the Devonian of south Cornwall and the western English Channel. In: Hutton DHW, Sanderson DJ (eds) Variscan tectonics of the North Atlantic region. Geol Soc Lond Spec Publ 14:103–118

Liger JL (1979) Structure profonde du bassin côtier senegalo-mauritanien. Interprétation de données gravimétriques et magnétiques. Thesis IIIrd cycle, University of St Jérome Marseille

Lilly HD (1966) Late precambrian and appalachian tectonics in the light of submarine exploration of the Great Bank of Newfoundland and in the Gulf of St Lawrence: preliminary views. Am J Sci 264:569–574

Long LT, Bridges SR, Dorman LM (1972) Simple Bouguer gravity map of Georgia. Inst of Tech Atlanta Scale 1/500 000

Loring DH (1974) Superficial geology of the Gulf of St Lawrence. In: Van der Linden WJM, Wade JA (eds) Offshore Geology of Eastern Canada. Can Geol Surv Paper 74/30:11–34

Maher JC (1971) Geologic framework and petroleum potential of the atlantic coastal plain and continental shelf. US Geol Surv Prof Paper 659:1–98

Maillet P (1977) Etude géochimique de quelques séries spilitiques armoricaines. Implications géotectoniques. Thesis IIIrd cycle, University of Rennes

Malod JA, Temine D, Boillot G (1984) La marge déformée du Nord-Ouest de l'Espagne. Cent Natl Exploit Oceans Spec Publ 29:1-135

Mann P, Hempton MR, Bradley DC, Burke K (1983) Development of pull-apart basins. J Geol 91:529-554

Mattauer M (1981) La formation des chaînes de Montagne. In: La dérive des continents (la tectonique des Plaques) Pour la Science: 40-55. Diffusion Belin Paris

Mattauer M, Seguret M (1971) Les relations entre la chaîne des Pyrénées et le Golfe de Gascogne. In: Technip (ed) Histoire structurale du Golfe de Gascogne. Inst Fran Pet Publ I: 4-24

Mattauer M, Proust F, Tapponnier P (1972) Major strike-slip fault of late hercynian age in Morocco. Nature 237:160-162

Mattauer M, Proust F, Tapponnier P (1980) Tectonic mechanism of obduction in relation with high pressure metamorphism. In: Centre National de la Recherche Scientifique (ed) Orogenic mafic and ultramafic associations Int Colloq CNRS n°272. Grenoble 1977, pp 197-201

Matte PH (1986) La chaîne varisque parmi les chaînes paléozoïques péri-atlantiques, modèle d'évolution et position des grands blocs continentaux au Permo-Carbonifère. Soc Geol France Bull II: 9-24

Matte Ph, Ribeiro A (1975) Forme et orientation de l'ellipsoïde de déformation finie dans la virgation hercynienne de Galice. Relations avec le plissement et hypothèses sur la genèse de l'arc ibéro-armoricain. Paris Acad Sci CR 280:2825-2828

Matthews DH, Cheadle MJ (1986) Deep reflections from the Caledonides and Variscides west of Britain and comparison with the Himalayas. In: Baraganzy M, Brown L (eds) Reflection seismology: A global perspective. Geodynamics series. Am Geophys Union 13:5-19

Maugis P (1955) Etude de pré-reconnaissance pétrolière dans le bassin du Sénégal. Direct Feder Mines Geol AOF Bull 19:99-128

Max MD (1975) The Precambrian rocks of southeast Ireland. In: Harris et al. (eds). A correlation of the Precambrian rocks in the British Isles. Soc Geol Lond Spec Rep 6:96-101

Max MD (1976) The pre-Paleozoic basement in southeastern Scotland and the Southern Uplands fault. Nature 264:485-486

Max MD (1978) Tectonic control of offshore sedimentary basins to the north and west of Ireland. J Petroleum Geol 1:103-110

Max MD, Barber PL (1978) The westward continuation of the Leannan Fault of Donegal and its bearing on the Great Glen Fault system. Geol Mag 115:215-218

Max MD, Inamdar DD (1983) Detailed compilation magnetic map of Ireland and a summary of its deep geology. Irish Geol Surv Rep 83/1:1-10

Max MD, Lefort JP (1984) Does the variscan front in Ireland follow a dextral shear zone? In: Hutton DHW, Sanderson DJ (eds) Variscan tectonics and the north Atlantic region. Geol Soc Lond Spec Pap 14:177-183

Max MD, Riddihough RP (1975) Continuation of the Highland boundary fault in Ireland. Geology 3:206-210

Max MD, Sonet J (1979) Rb-Sr evidence for Grenville event in the pre-Caledonian basement of northwest Ireland. Geol Soc Lond J 136:379-382

Max MD, Inamdar DD, Mc Intyre T (1982) Compilation magnetic map: the Irish continental shelf and adjacent areas. Irish Geol Surv Rep 82/2

Mc Grath Ph, Hood PJ, Cameron GW (1973) Magnetic Surveys of the Gulf of St Lawrence and the Scotian Shelf. In: Hood PJ (ed) Earth Science Symposium on offshore eastern Canada. Can Geol Surv Pap 71/23:339-358

Mc Lean AC, Deegan CE (1978) The solid geology of the Clyde Sheet. Lond Inst Geol Sci Rep 78/9:1-14

Mc Master RL, De Boer J, Collins BP (1980) Tectonic development of southern Narragansett bay and offshore Rhode Island. Geology 8:496-500

McPherson J (1886) Comparaison des terrains cristallins d'Espagne et du Finistère. Soc Geol France Bull 14:828-830

Mc Quillin R, Binns PE (1973) Geological structures in the sea of the Hebrides. Nature Phys Sci 241:2-4

Mc Quillin R, Wright JE, Owens B, Lister TR (1969) Recent geological investigations in the Irish sea. Nature 22:365-366

Michard A (1976) Elements de géologie marocaine. Serv Geol Maroc Mem 252

Michard A (1978) Brève description du segment caledono-hercynien du Maroc. In: Caledonian-Appalachian Orogen of the North Atlantic Region, IGCP 27 Symposium. Can Geol Surv Pap 78/13:213-230

Michard A, Piqué A (1980) The variscan belt in Morocco: structure and developmental model. In: Wones DR (ed) Proceedings of the Caledonides in the USA IGCP 27 Symposium Virginia Polytechnic Institute and State University Blacksburg 5-9 September Virginia Mem 2:317-325

Miller AH, Garland GD (1953) Analysis of local anomalies in southeastern New-Brunswick. In: Garland GD (ed) Gravity Measurement in the Maritime provinces. Dom Obs Publ 16:220-237

Miller JA, Roberts DG, Matthews DH (1973) Rocks of Grenville age from Rockall Bank. Nature Phys Sci 246:61

Milton C, Grasty R (1969) "Basement" rocks of Florida and Georgia. Am Assoc Petroleum Geol Bull 53:2483-2493

Molnar P, Tapponnier P (1977) Relation of tectonics of eastern China to the India-Eurasia collision: application of slip-line field theory to large scale continental tectonics. Geology 5:212-216

Montadert L, Roberts DG, De Charpal O, Guennoc P (1979) Rifting and subsidence of the northern continental margin of the Bay of Biscay. In: Montadert L, Roberts DG (eds) Initial Reports of the Deep Sea Drilling Project 48:1025-1060. US Government Printing Office Washington DC

Moseley F (1977) Caledonian plate tectonics and the place of the English Lake district. Geol Soc Am Bull 8:764-768

Mougenot D, Capdevila R, Polain Ch, Dupeuble PA, Mauffret A (1985) Nouvelles données sur les sédiments anté-rifts et le socle de la marge continentale de Galice. Paris Acad Sci CR 301:323-328

Mueller S, Prodehl C, Mendes AS and Sousa-Moreira V (1973) Crustal structure in the southwestern part of the Iberian Peninsula. Tectonophysics 20:307-318

Munha J (1979) Blue amphiboles, metamorphic regime and plate tectonic modelling in the Iberian Pyrite Belt. Contrib Mineral Petrol 69:279-289

Murphy T (1981) Geophysical evidence. In: Holland CH (ed) Geology of Ireland. Scottish Academic Press Edinberg, pp 225-229

Musellec P (1974) Géologie du plateau continental portugais au nord du Cap Carvoeiro. Thesis IIIrd cycle University of Rennes

Naylor D, Shannon PM (1982) The geology of offshore Ireland and west Britain. Graham Trotman London

Nelson KD, Arnow JA, Mc Bride JH, Willemin JH, Huang J, Zheng L, Oliver JE, Brown LD, Kaufman S (1985) New

COCORP profiling in the southeastern United States. Part I: Late Paleozoic suture and Mesozoic rift basin. Geology 13:714–718

Nelson KD, Mc Bride JH, Arnow JA, Oliver JE, Brown LD, Kaufman S (1985) New COCORP profiling in the southeastern United States. Brunswick and East coast magnetic anomalies, opening of the north-central Atlantic ocean. Geology 13:718–721

Nelson KD, Mc Bride JH, Arnow JA, Wille DM, Brown LD, Oliver JE, Kaufman S (1987) Result of recent COCORP profiling in the Southeastern United States. R Astron Soc Geophys J 89:141–146

Odom AL, Brown JF (1976) Was Florida a part of North America in the Lower Paleozoic? Geol Soc Am Abstr Progr 8:237–238

O'Hara K, Gromet LP (1985) Two distinct late precambrian avalonian terranes in southern New England and their late Paleozoic juxtaposition. Am J Sci 285:673–709

Olivet JL, Bonnin J, Beuzart P, Auzende JM (1984) Cinematique de L'Atlantique nord et central. Cent Natl Exploit Oceans Publ 54:1–8

Olszewski WJ, Gaudette HE (1982) Age of the Brookville gneiss and associated rocks, southeastern New-Brunswick. Can J Earth Sci 19:2158–2166

Olszewski WJ, Gaudette HE, Keppie JD, Donohoe HV (1981) Rb-Sr whole rock age of the Kellys Mountain basement complex Cape-Breton Island. Geol Soc Am Abstr Progr 13:169

Osberg PH (1978) Synthesis of the geology of the northeastern Appalachians, U.S.A. In: Tozer ET, Schenk PE (eds) Proceedings of Caledonian – Appalachian – Orogen of the North Atlantic Region IGCP 27 Symposium. Can Geol Surv Pap 78/13:137–148

Osberg PH (1983) Timing of orogenic events in the U.S. Appalachians. in: Schenk PE (ed) Regional trends in the geology of the Appalachian – Caledonian – Hercynian – Mauritanide orogen. NATO ASI Series 116:315–337. Reidel Dordrecht

Oxburgh ER (1972) Flake tectonics and continental collision. Nature 239:202–204

Panchon Ruiz A (1978) Anomalias geomagneticas de Espana: Interpretacion Geologica y II. Rev Geofis 1:15–101

Paris F, Robardet M (1977) Paléogéographie et relations ibero-armoricaines au Paléozoïque antécarbonifère. Soc Geol France Bull XIX: 1121–1126

Paris F, Robardet M (1985) Evaluation des affinités entre le paléozoïque cache sous l'Aquitaine et les formations armoricaines contemporaines. Bur Rech Geol Miniér Doc Géologie profonde de la France 2, 95/7:11–36

Paris F, Peucat JJ, Chalet M (1985) U-Pb zircons dating of volcanic rocks nearby the Silurian – Devonian boundary in southern Armorican Massif (France). Terra Cognita 2/3:237

Parson LM (1979) The state of strain adjacent to the Great Glen fault. In: Harris AL, Holland CH, Leake BE (eds) The Caledonides of the British Isle-Reviewed. Geol Soc Lond Scott Acad Press, pp 287–289

Parson LM, Masson DG, Rothwell RG (1983) Remnants of a submerged pre-Jurassic (Devonian ?) landscape on Orphan Knoll offshore Eastern Canada. Can J Earth Sci 21:61–66

Pastouret L, Maury RC (1982) Présence de pillow-lavas tholéïtiques dans le canyon Shamrock (marge continentale armoricaine); leur place dans l'histoire du Golfe de Gascogne. Paris Acad Sci C R 294:669–674

Pelletier BR (1971) A granodioritic drill core from the Flemish Cap, eastern canadian continental margin. Can J Earth Sci 8:1499–1503

Perconig E (1960-62) Sur la constitution géologique de l'Andalousie occidentale en particulier du bassin du Guadalquivir. In: Livre à la mémoire du Professeur P. Fallot. Soc Geol France Mem 1:229–256

Perroud H (1980) Contribution à l'étude paléomagnétique de l'arc ibéro-Armoricain. Thesis IIIrd cycle University of Rennes

Perroud H, Van der Voo R, Bonhommet N (1984) Paleozoic evolution of the Armorican plate on the basis of paleomagnetic data. Geology 12:579–582

Peucat JJ (1983) Géochronologie des roches métamorphiques (Rb-Sr et U-Pb). Exemples choisis au Groënland, en Laponie, dans le Massif armoricain et en Grande Kabylie. Soc Geol Miner Bretagne Mem 28

Peucat JJ, Cogné J (1977) Geochronology of some blueschists from Ile de Groix (France). Nature 268:131–132

Peucat JJ, Charlot R, Mifdal A, Chantraine J, Autran A (1979) Définition géochronologique de la phase bretonne en Bretagne centrale. Etude Rb/Sr de granites du domaine centre armoricain. Bur Rech Geol Miniér Bull 4:349–356

Peucat JJ, Vidal Ph, Godard G, Postaire B (1982) Precambrian U-Pb zircon ages in eclogites and garnet pyroxenites from South Britanny (France): an old oceanic crust in the west european hercynian belt ?. Earth Planet Sci Lett 60:70–78

Phillips WEA (1978) The Caledonide orogen in Ireland. In: Proceedings of Caledonian – Appalachian orogen of the North Atlantic region IGCP 27 Symposium. Can Geol Surv Pap 78/13:97–103

Phillips WEA, Stillman CJ, Murphy T (1976) A Caledonian plate tectonic model. Geol Soc Lond J 132:579–609

Phinney A (1986) A seismic cross section of New England Appalachians: the orogen exposed. In: Barazangi M, Brown L (eds) Reflection seismology: the continental crust. Geodynamics Series. Am Geophys Union 14:157–172

Piasecki MAJ, Van Breemen O, Wright AE (1981) Late precambrian geology of Scotland, England and Wales. In: Kerr JW, Ferguson AJ, Mahan LC (eds) Geology of the North Atlantic Borderlands. Can Soc Petroleum Geol Mem 7:57–94

Piqué A (1979) Evolution structurale d'un segment de la chaîne hercynienne: la Meseta marocaine nord-occidentale. Thesis University of Strasbourg

Piqué A (1983) Structural domains of the Hercynian belt in Meseta marocaine nord-occidentale. Influence des fractures du socle précambrien sur la sédimentation et la déformation de la couverture paléozoïque. Soc Geol France Bull XXIII: 3–10

Piqué A (1983) Structural domains of the hercynian belt in Morocco. In: Schenk PE (ed) Regional trends in the geology of the Appalachian – Caledonian – Hercynian – Mauritanide. NATO ASI Series 116. Reidel Dordrecht, pp 339–345

Piqué A, Chalouan A, Fadli D (1985) Des rides paléogéographiques aux anticlinaux. Permanence des directions structurales dans la Meseta marocaine occidentale au cours du Paléozoïque. Sci Geol Bull 38:147–153

Pitcher WS (1969) Northeast – trending faults of Scotland and Ireland, and chronology of displacements. In: Kay M (ed) North Atlantic – Geology and continental drift. Am Assoc Petroleum Geol Mem 12:724–733

Pitman WC III, Talwani M (1972) Sea-floor spreading in the North Atlantic. Geol Soc Am Bull 83:619–632

Poag CW (1982) Stratigraphic reference section for Georges

Bank Basin – Depositional model for New England passive margin. Am Assoc Petroleum Geol Bull 66:1021–1041

Pojeta J, Kriz J, Berden JM (1976) Silurian – Devonian pelecypods and Paleozoic stratigraphy of subsurface rocks in Florida and Georgia and related Silurian pelecypods from Bolivia and Turkey. US Geol Surv Prof Pap 879:1–32

Ponsard JF (1984) La marge du craton ouest-africain du Senegal à la Sierra-Leone: interprétaton géophysique de la chaîne panafricaine et des bassins du Protérozoïque à l'actuel. Thesis IIIrd cycle University of Marseille

Poole WH (1976) Plate tectonic evolution of the Canadian Appalachian Region. Can Geol Surv Pap 76/1 B: 113–126

Popenoe P, Zietz I (1977) The nature of the geophysical basement beneath the coastal plain of South Carolina and northeastern Georgia. US Geol Surv Prof Pap 1028:119–137

Portugal Ferreira M (1972) Rochas metamorphicas. University of Coimbra, Coimbra

Powell D (1983) Time of deformation in the British Caledonides. In: Schenk PE (ed) Regional Trends in the geology of the Appalachian – Caledonian – Hercynian – Mauritanide orogen. NATO ASI Series 116 Reidel Dordrecht, pp 293–299

Pratt TL, Coruh C, Costain JK (1987) Lower crustal reflection in Central, Virginia USA. R Astron Soc Geophys J 89:163–170

Prodehl C, Sousa Moreira V, Mueller S, Mendes AS (1976) Deep-seismic sounding experiments in central and southern Portugal. In: XIVth General Assembly of European Seismology Community Berlin, pp 261–266

Proust F, Petit JP, Tapponnier P (1977) L'accident du Tizi n'Test et le rôle des décrochements dans la tectonique du Haut-Atlas occidental (Maroc). Soc Geol France Bull 19:541–551

Pruvost P (1949) Les mers et les terres de Bretagne aux temps paléozoïques. Hebert and Haug Ann VII: 345–360

Quinquis H (1980) Schistes bleus et déformation progressive. L'exemple de l'île de Groix (Massif armoricain). Thesis IIIrd cycle of Rennes

Quinquis H, Choukroune P (1981) Les schistes bleus de l'île de Groix dans la chaîne hercynienne: implication cinématique. Soc Geol France Bull 23:409–418

Ranke V, Von Rad U, Wissmann G (1982) Stratigraphy, facies and tectonic development of the on- and offshore Aaiun – Tarfaya basin – A review. In: Von Rad U, Hinz K, Sarnthein M, Seibold E (eds) Geology of the Northwest African continental margin. Springer, Berlin Heidelberg New York 61:86–105

Rankin D (1976) Appalachian salients and recesses: late Precambrian continental break-up and the opening of the Iapetus ocean. J Geophys Res 81:5605–5619

Rast N (1984) The Alleghenian orogeny in eastern North America. In: Hutton DHW, Sanderson DJ (eds) Variscan tectonics of the North Atlantic region. Geol Soc Lond Spec Publ 14:197–217

Rast N, Grant RH, Parker JSD, Han Chang Teng (1979) The carboniferous succession in southern New Brunswick and its state of deformation. In: 9ème Congrès International sur la stratigraphie et la géologie du Carbonifère Comptes-Rendus. Washington and Urbana-Champaigne 3:13–22

Rattey PR, Sanderson DJ (1984) The structure of SW Cornwall and its bearing on the emplacement of the Lizard complex. Geol Soc Lond J 141:87–95

Reynolds PH, Zentilli M, Mueke GK (1980) K-Ar and 40Ar/39Ar geochronology of granitoid rocks from Southern Nova Scotia: its bearing on the geological evolution of the

Meguma zone of the Appalachians. Can J Earth Sci 18:386–394

Ribeiro A, Oliveira JT, Brandao Silva J (1983) La estructura de la zona sur-Portuguesa. In: Comba JA (ed) Geologia de Espana. Publ Inst Geol Minero Espana, pp 504–512

Riddihough, RP (1975) A magnetic map of the area 51–55 N, 10–16 W. Dublin Inst Adv Stud Geophys Bull 33:1–34

Riddihough RP, Max MD (1976) A geological framework for the continental margin to the west of Ireland. Geol J 11:109–120

Ritz M, Robineau B (1986) Crustal and upper mantle electrical conductivity structures in West Africa: Geodynamic implications. Tectonophysics 124:115–132

Roberts DG (1970) The Rif-Betic orogen in the Gulf of Cadiz. Mar Geol 9:1131–1137

Roberts DG (1970) Recent geophysical studies on the Rockall Plateau and adjacent areas. Geol Soc Lond Proc 1662:87–92

Roberts DG (1975) Tectonic and stratigraphic evolution of the Rockall Plateau and Trough. In: Woodland AW (ed) Petroleum and the Continental shelf of North West Europe. Applied Science, Barking, 5:77–91

Roberts DG (1975) Marine geology of Rockall Plateau and Trough. R Soc Lond Philos Trans A 278:447–509

Roberts DG, Jones MT (1978) A bathymetric and gravity survey of the Rockall Bank, HMS Hecla. Admirality Marine Sci Publ 19:1–45

Roberts DG, Ardus DA, Dearnley R (1973) Pre-Cambrian rocks drilled on the Rockall Bank. Nature 244:21–23

Roberts DG, Montadert L, Searle RC (1979) The western Rockall Plateau: stratigraphy and structural evolution. Initial Reports of the Deep Sea Drilling Project XLIII: 1061–1087. US Government Printing Office Washington DC

Robinson P, Hall LM (1979) Tectonic synthesis of southern New-England. In: Wones DR (ed) Proceedings of the Caledonides in the USA IGCP 27 Symposium, 5–9 September Blacksburg VA. Virginia Polytechnic Institute and State University 2:73–90

Rod E (1962) Fault pattern, northwest corner of the Sahara Shield. Am Assoc Petroleum Geol Bull 46:529–552

Rodgers J (1970) The tectonics of the Appalachians. John Wiley, New York

Roeder DH (1975) Tectonic effects of dip changes in subduction zones. Am J Sci 275:252–264

Root SI, Hoskins DM (1977) Latitude 40 N° fault zone, Pennsylvania: a new interpretation. Geology 5:719–729

Rousseau A (1971) Géologie du plateau continental Nord-Espagnol entre 2°20 et 3°35. Thesis IIIrd cycle, University of Rennes

Rousseau A (1980) Apport de la gravimétrie à la connaissance de la lithosphère du Bassin d'Aquitaine. Thesis University of Bordeaux I

Roussel J, Liger JL (1983) A review of deep structures and ocean-continent transition in the Senegal Bassin (West Africa). Tectonophysics 91:183–211

Ruellan E (1985) Evolution de la marge Atlantique du Maroc (Mazagan): étude par submersible, seabeam et sismique reflexion. Thesis IIIrd cycle University of Brest

Ruellan E, Auzende JM (1985) Structure et évolution du plateau sous-marin de El Jadida (Mazagan, Ouest Maroc). Soc Géol France Bull I: 103–114

Ruffman A, Van Hinte JE (1973) Orphan Knoll- a "chip off the North American plate". In: Hood PJ (ed) Earth Science Symposium of offshore Eastern Canada. Can Geol Surv Paper 71/23:407–449

Ruitenberg AA, Mc Cutcheon SR (1982) Acadian and Hercynian structural evolution of southern New-Brunswick. In: St Julien P, Beland J (eds) Major structural zones and faults of the Northern Appalachians. Geol Assoc Can Spec Pap 24:131–148

Ruitenberg AA, Fyffe LR, Mc Cutcheon SR, St Peter CJ, Irrinki RR, Venugopal DV (1977) Evolution of Pre-Carboniferous tectonostratigraphic zones in the New-Brunswick Appalachians. Geosci Can 4:171–181

Russel MJ (1968) Structural controls of base metal mineralization in Ireland in relation to continental drift. Inst Mining Metallurgy Trans 77:117–128

Saadi M (1975) Oil possibilities in the Rif area. Mines Géol 37:41–56

Sabine PA (1965) The petrology of specimens from Haig Fras. Colston Pap 17:299–300

Sabine PA, Snelling NJ (1969) The Seven Stones granites, between Land's End and the Scilly Isles. Geol Soc Lond Proc 1654:47–50

Sabine PA, Watson J (1965) Introduction to isotopic age-determinations of rocks from the British isles. Geol Soc Lond Q J 121:477–533

Sanderson DJ (1984) Structural variation across the northern margin of the Variscides in NW Europe. In: Hutton DHW, Sanderson DJ (eds) Variscan Tectonics of the North Atlantic region. Geol Soc Lond Spec Publ 14:149–165

Sapin M, Prodehl C (1973) Long range profiles in western Europe I. Crustal structures between the Bretagne and the Central Massif of France. Ann Geophys 29:127–145

Sarkar PK (1978) Petrology and geochemistry of White Rock metavolcanic suite, Yarmouth, Nova Scotia. Thesis University of Dalhousie, Halifax

Saupé F (1973) La géologie du gisement de mercure d'Almaden. Science Terre Mem 29

Schenk PE (1970) Regional variation of the flysch-like Meguma group (Lower Paleozoic) of Nova-Scotia compared to recent sedimentation off the Scotian Shelf. Geol Assoc Can Spec Pap 7:127–153

Schenk PE (1978) Synthesis of the Canadian Appalachians. In: Caledonian — Appalachian orogen of the North Atlantic region IGCP 27 Symposium. Can Geol Surv Pap 78/13:111–136

Schlee JS (1980) A comparison of two Atlantic-type continental margins. US Geol Surv Prof Pap 1167:1–21

Scholle PA (1979) Data summary and petroleum potential. US Geol Surv Circ 800:18–23

Schultz LK, Grover RL (1974) Geology of Georges Bank Basin. Am Assoc Petroleum Geol Bull 58:1159–1168

Sclater JG, Hellinger S, Tapscott C (1977) The paleobathymetry of the Atlantic ocean from the Jurassic to the present. J Geol 85:509–662

Scotese ChR (1984) An introduction to this volume: Paleozoic paleomagnetism and the assembly of Pangea. In: Van der Voo R, Scotese CR, Bonhommet N (eds) Plate reconstruction from Paleozoic paleomagnetism Geodynamic Series. Am Geophys Union 12:1–10

Scotese ChR, Bambach RK, Barton C, Van der Voo R, Ziegler AM (1979) Paleozoic base maps. J Geol 87:217–277

Scott KR, Cole JM (1975) Eastern US continental margin. In: Yorath CJ, Parker ER, Glass DJ (eds) Canada's continental margins and offshore petroleum exploration. Can Soc Petroleum Geol Mem 4:33–43

Scrutton RA (1970) Results of a seismic refracton experiment on Rockall Bank. Nature 227:826–827

Scrutton RA (1972) The structure of Rockall plateau microcontinent. R Astron Soc Geophys J 27:259–275

Secor DT, Samson SL, Snoke AW, Palmer AR (1983) Confirmation of the Carolina Slate belt as an exotic terrane. Science 221:649–650

Secor DT, Snoke AW, Bramlett KW, Costello OP, Kimbrell OP (1986) Character of the Alleghanian orogeny in the southern Appalachians: Part I. Alleghanian deformation in the eastern Piedmont. Geol Soc Am Bull 97:1319–1328

Secor DT, Snoke AW, Dallmeyer RD (1986) Character of the Alleghanian orogeny in the southern Appalachians: Part III. Regional Tectonic relations. Geol Soc Am Bull 97:1347–1353

Segoufin J (1975) Structure du plateau continental armoricain. R Soc Lond Philos Trans A 279:109–121

Sengor AMC, Burke K, Dewey JF (1978) Rifts at high angles to orogenic Belts: tests for their origin and the Upper Rhine Graben as an example. Am J Sci 278:24–40

Servicos Geologicos de Portugal (1979) Intensidade Total do campo magnetico (copia provisoria)

Sevastopulo GD (1981) Hercynian structures (II). In: Holland CH (ed) Geology of Ireland. Scott Acad Press, Edinburg, pp 189–199

Sheridan RE, Drake CL (1968) Seaward extension of the Canadian Appalachians. Can J Earth Sci 3:337–373

Sheridan RE, Houtz RE, Drake CL, Ewing M (1969) Structure of continental margin off Sierra Leone, West Africa. J Geophys Res 74:2512–2530

Sheridan RE, Grow JA, Behrendt JC, Bayer KE (1979) Seismic refraction study of the continental edge off the Eastern United States. Tectonophysics 59:1–26

Sheridan RE, Grosby JT, Kent KM, Dillon WP, Paul CK (1981) The geology of the Blake Plateau and Bahamas region. In: Kerr J.Wm, Ferguson AJ, Mahan LC (eds) Geology of the North Atlantic Borderlands. Can Soc Petroleum Geol Mem 7:487–502

Shride AF (1976) Stratigraphy and structural setting of the Newsburry volcanic complex, northeastern Massachusetts. In: Geology of Southeastern New England. New England intercoll Geol Conf Guidebook, pp 291–300

Sibuet JC (1972) Histoire structurale du Golfe de Gascogne. Thesis University of Strasbourg

Sibuet JC (1972) Contribution de la gravimétrie à l'étude de la Bretagne et du plateau adjacent. Soc Geol France C R Sommaires 3:124–129

Simon B, Popoff M, Villeneuve M, Muller J (1981) A.T.P. Teledection n°80–231. Rapport mi-parcours. Lab Sci Terre St Jérôme Marseille X: 1–27

Simpson RW, Bothner WA, Shride AF (1979) Offshore extension of the Clinton-Newburry and Bloody Bluff fault system of Northeastern Massachusetts. In: Wones DR (ed) Proceedings of the Caledonides in the USA. IGCP 27 Symposium 5–9 Septembre Blacksburg Virginia, Mem 2:229–234

Sinha AK, Zietz I (1982) Geophysical and geochemical evidence for a Hercynian magmatic arc, Maryland to Georgia. Geology 10:593–596

Skehan JW (1973) Subduction zone between the paleo-american and Paleo-African plates in New England. Rev Union Geophys Mexicana 13 (4):291–308

Skehan JW (1983) Geological profiles through the Avalonian terrain of Southeastern Massachusetts Rhode Island and Eastern Connecticut, USA. In: Profiles of orogenic belts. Geodynamics Series. Am Geophys Union 10:275–300

Skehan JW, Murray DP (1980) A model for the evolution of eastern margin of the northern Appalachians. In: Wones DR (ed) Proceedings of the Caledonides in the USA. IGCP 27 Symposium 5-9 September Blacksburg VA. Virginia Polytechnic Institute and State University Mem 2:229-234

Smith DL (1982) Review of the tectonic history of the Florida basement. Tectonophysics 88:1-22

Smith PJ, Bott MHP (1975) Structure of the crust beneath the Caledonian foreland and Caledonian belt of the north scottish shelf region. R Astron Soc Geophys Soc 40:187-205

Smith DL, Gregory RG, Emhof JW (1981) Geothermal measurements in the southern Appalachian mountain and southeastern coastal plains. Am J Sci 281:282-298

Smith AG, Hurley AM, Briden JC (1981) Phanerozoic paleocontinental world maps. Harland WB, Cook AH, Hughes NF, Sclater JG, Richardson SW (eds) Cambridge Earth Sc Series, pp 1-102. Cambridge University Press, Cambridge

Snelling NJ (1968) Potassium-argon dating of the Wolf Rock phonolite. Lond Inst Geol Sci Rep (1967): 152

Sorensen H (1974) The alkaline rocks. John Wiley, New York

Sougy J (1964) Les formations paléozoïques du Zemmour Noir (Mauritanie Occidentale). Etude pétrographique, stratigraphique et paléontologique. Thesis University of Nancy

Sousa Moreira V, Mueller ST, Mendes HS, Prodehl CL (1977) Crustal structure of Southern Portugal. Inst Geophys Pol Acad Sci Publ 115:413-426

Spariosu DJ, Kent DV (1983) Paleomagnetism of the Lower Devonian Traveler felsite and the Acadian orogeny in New England Appalachians. Geol Soc Am Bull 94:1319-1328

Spariosu DJ, Kent DV, Keppie JD (1984) Late paleozoic motions of the Meguma Terrane, Nova Scotia: New paleomagnetic evidence. In: Van der Voo R, Scòtese CR and Bonhommet N (eds) Plate reconstruction from paleozoic paleomagnetism. Geodynamic Series, Am Geophys Union 12:82-98

Srivastava SP (1978) Evolution of the Labrador sea: its bearing on the early evolution of the north Atlantic. R Astron Soc Geophys J 52:313-357

Srivastava SP, Falconner RKH, Mac Lean B (1981) Labrador sea, Davis Strait, Baffin Bay: geology and geophysics — a review. In: Kerr JW, Ferguson AJ, Manhan LC (eds) Geology of the north Atlantic borderlands. Can Soc Petroleum Geol Mem 7:333-398

Stephens LE, Goodacre AK, Cooper RV (1971) Results of underwater gravity surveys over the Nova Scotia continental shelf. Can Geol Surv Earth Phys Branch, Gravity Map Series 123:1-9

Stewart ICF (1978) Teleseismic reflections and the Newfoundland lithosphere. Can J Earth Sci 15:175-180

Stillman CJ (1981) Caledonian igneous activity. In: Holland CH (ed) Geology of Ireland. Scottish Acad Press, Edinburg, pp 83-106

Strong DF, O'Brien SJ, Taylor SW, Strong PG, Wilton DH (1978) Aborted proterozoic rifting in eastern Newfoundland. Can J Earth Sci 15:117-131

Styles MT, Rundle CC (1984) The Rb-Sr isochron age of the Kennack gneiss and its bearing on the age of the Lizard complex, Cornwall. Geol Soc Lond J 141:15-20

Swanson MT (1982) Preliminary model for an early transform history in central Atlantic rifting. Geology 10:317-320

Tamain G (1978) L'évolution calédono-varisque des Hespérides. Proceeding of Caledonian — Appalachian orogen of the North Atlantic region IGCP 27 Symposium. Can Geol Surv Pap 78/13:183-241

Tapponnier P, Peltzer G, Le Dain Y, Armijo R, Cobbold P (1982) Propagating extrusion tectonics in Asia: New insight from simple experiment with plasticine. Geology 10:611-616

Taylor PT, Zietz I, Dennis LS (1968) Geologic implications of aeromagnetic data for the eastern continental margin of the United States. Geophysics XXXIII: 755-780

Ters M (1979) Les synclinoriums paleózoïques et le Précambrien sur la façade occidentale du Massif Vendéen — Stratigraphie et structure. Bur Rech Geol Minièr Bull 1/4:293-301

Ters M, Chantraine J (1980) The Eodevonian metamorphism in the south-eastern part of the armorican massif. 26th Int Geol Congr Paris Field Conf book 1303:1-15

Thomas WA (1977) Evolution of Appalachian — Ouachita salients and recesses from reentrants and promontories in the continental margin. Am J Sci 277:1233-1278

Thomas M (1983) Tectonic significance of paired gravity anomalies in the southern Appalachians. In: Hatcher R, Williams H, Zietz I (eds) Contributions to the Tectonics and Geophysics of Mountain Chains. Geol Soc Am Mem 158:113-124

Thomas MD, Tanner JG (1975) Cryptic suture in the eastern Grenville Province. Nature 256:392-394

Thomas WA, Chowns TM, Daniel DL, Nearthery TL, Glover L III, Gleason RJ (1988) Pre-Mesozoic paleogeologic map of Appalachians-Quachita orogen beneath Atlantic and Gulf coastal Plain. In: Hatcher RD Jr, Thomas WA, Viele GW (eds) The Appalachians-Quachita orogen in the United States. Geol Soc Am Mem V F-2 Plate 6

Thompson B (1982) Crustal electrical conductivity measurements in the Georgia Piedmont. Thesis University of Cornell Ithaca, New York

Tixeront M (1973) Lithostratigraphie et minéralisations cuprifères et uranifères stratiformes syngénétiques et familiières des formations détritiques permo-triasiques du couloir d'Argana, Haut-Atlas occidental (Maroc). Serv Geol Maroc Notes Mém 249:147-177

Triboulet CL (1974) Les glaucophanites et roches associées de l'Ile de Groix (Morbihan, France): étude minéralogique et pétrogénétique. Contrib Mineral Petrol 45:65-90

Uchupi E, Austin JA (1979) The geologic history of the passive margin off New England and the Canadian Maritime Provinces. Tectonophysics 59:53-69

Uchupi E, Emery KO, Bowin CO, Phillips JD (1976) Continental margin off Western Africa: Senegal to Portugal. Am Assoc Petroleum Geol Bull 60:809-878

Umpleby DC (1979) Geology of the Labrador shelf. Can Geol Surv Pap 79/13:1-39

Unger JD, Stewart DB, Phillips JD (1987) Interpretation of migrated seismic reflection profiles across the Northern Appalachians in Maine. R Astron Soc Geophys J 89:171-176

Vaillant FX (1972) Contribution à l'étude géologique de la plateforme continentale Sud-Armoricaine et de la zone Ouest du Golfe de Gascogne par les méthodes sismique et magnétique. Thesis IIIrd Cycle University of Bordeaux I

Valliant HD, Mac Nab RF, Stephen LE, Grant ST, Cooper RV (1975) Results of underwater and surface regional gravity surveys off the coast of Labrador 1975. Can Geol Surv Earth Physics branch Gravity Map Series, pp 1-18

Van Breemen O, Johnson MRW, Bowes Dr (1978) Crustal additions in late Precambrian times. In: Bowes DR and Leake BE (eds) Crustal evolution in north western Britain and adjacent region. Geol J Spec Issue 10:81-106

Van den Bosh JWH (1981) Mémoire explicatif de la carte gravimétrique du Maroc (provinces du Nord) au 1/500 000. Serv Geol Maroc Notes Mém 234b:1-219

Van der Linden WJM, Srivastava SP (1975) The crustal structure of the continental margin off central Labrador. Can Geol Surv Pap 74/30:233–241

Van der Voo R, Scotese CH (1981) Paleomagnetic evidence for a large (2000 km) sinistral offset along the Great Glen fault, during Carboniferous time. Geology 9:583–589

Vegas R (1978) Sedimentation and tectonism in the Iberian massif prior to the Hercynian deformation (Late Precambrian to Silurian times). In: Del Castro (ed) Geologia de la parte norte del macizo Iberico. Cuadernos del seminario de estudios ceramicos de Sarguadelos 27:270–284

Vegas R, Munoz M (1977) El contacto entre las zonas Surportuguesa y Ossa – Morena en el SW de Espana: Una nueva interpretacion. Comm Serv Geol Portugal 60:31–51

Veinante A, Santoire JL (1980) Seismicité récente de l'arc sud-armoricain et du Nord-Ouest du Massif Central: mécanismes au foyer et tectonique. Soc Geol France Bull 22:93–102

Vidal Ph (1973) Premières données géochronologiques sur les granites hercyniens du Sud du Massif armoricain. Soc Geol France Bull XV: 239–245

Vidal Ph (1976) L'évolution polyorogénique du Massif armoricain: apport de la géochronologie et de la géochimie du strontium. Thesis University of Rennes

Vigneresse JL (1978) Gravimétrie et granites armoricains: structure et mise en place des granites hercyniens. Thesis University of Rennes

Villeneuve M (1980) Schéma géologique du Nord de la Guinée. Soc Geol France CR 2:54–57

Villeneuve M, Lesquer A, Ponsard JF, Roussel J (1984) Geology and geophysics of the southern Mauritanides and the northern Rokelides – Evidences of a panafrican suture at the western border of the west African craton. In: Klerk J, Michot J (eds) African Geology. Afr Geol Tervuren, pp 57–65

Vogt PR, Avery OE (1974) Detailed magnetic surveys in the northeast Atlantic and Labrador sea. J Geophys Res 79:363–389

Wade JA, Grant AC, Sanford BV, Barss MS (1975) Basement structures: Eastern Canada and adjacent areas. Can Geol Surv Map 1400 A

Watson JV (1977) The Outer Hebrides: a geological perspective. Geol Assoc Proc 88:1–14

Watson J (1978) The basement of the Caledonides orogen in Britain. In: Tozer ET, Schenck PE (eds) Proceedings of Caledonian-Appalachian orogen of the North Atlantic region. IGCP 27 Symposium. Can Geol Surv Pap 78/13:75–77

Watson J, Dunning FW (1979) Basement-cover relations in the British Caledonides. In: Harris AL, Holland CH, Leake BE (eds) The Caledonides of the British isles-reviewed. Geol Soc Lond, pp 67–91

Weaver DF (1967) A geological interpretation of the Bouguer anomaly field of Newfoundland, Dominion observatory Publ 35:233–251

Webb GW (1969) Paleozoic wrench faults in Canadian Appalachians. In: Kay M (ed) North Atlantic Geology and continental drift. Am Assoc Petroleum Geol Mem 12:754–786

Weigel W, Wissmann G, Goldflam P (1982) Deep seismic structure (Mauritania and central Morocco). In: Von Rad U, Hinz K, Sarnthein M, Seibold E (eds) Geology of the Northwest African continental Margin. Springer, Berlin Heidelberg, New York, pp 132–159

Wendt J (1985) Disintegration of the continental margin of northwestern Gondwana: Late Devonian of the eastern Anti-Atlas (Morocco). Geology 13:815–818

Whitmarsh RB, Langford JJ, Buckley JS, Bailey RJ, Blundell DJ (1974) The crustal structures beneath Porcupine Ridge as determined by explosion seismology. Earth and Planet Sci Lett 22:197–204

Whittard WF (1962) Geology of the Western Approaches of the English Channel, a progress report. R Soc Lond Proc A 265:395–406

Willefert S (1963) Graptolites du Silurien et du Lochkovien de Touchchent (anticlinorium de Kasba – Tadla – Azrou, Maroc Central). Serv Geol Maroc Notes Mém 172:69–94

Williams H (1972) The Appalachian structural province – stratigraphy. In: Price RA, Douglas RJW (eds) Variation in Tectonic styles in Canada. Geol Assoc Can Spec Pap 11:188–202

Williams H (1978) Tectonic lithofacies map of the Appalachian orogen. Memorial University of Newfoundland Map 1 A, Scale 1/2000 000

Williams H (1978) Geological development of the Northern Appalachians: its bearing on the evolution of the British isles. In: Bowes DR, Leake BE (eds) crustal evolution in Northwestern Britain and adjacent regions. Geol J Spec Issue 10:1–22

Williams H, Hatcher RD (1983) Appalachian suspect terranes. In: Hatcher RD, Williams H, Zietz I (eds) Contributions to the tectonics and geophysics of Mountain Chains. Geol Soc Am Mem 158:33–53

Williams H, Max MD (1979) Zonal subdivision and regional correlation in the Appalachian – Caledonian orogen. In: Wones DR (ed) Proceedings of the Caledonides in the USA IGCP 27 Symposium. Virginia Polytechnic Institute and State University 5-7 September Blacksburg VA, Mem 2:57–62

Williams H, St Julien P (1978) The Baie Verte – Brompton Line in Newfoundland and regional correlation in the Canadian Appalachians. Can Geol Surv Pap 78/1 A: 225–229

Wilton DH (1983) The geology and structural history of the Cape Ray fault zone in southwestern Newfoundland. Can J Earth Sci 20:1119–1133

Winchester JA (1973) Pattern of regional metamorphism suggests a sinistral displacement of 160 km along the Great Glen fault. Nature 246:81–84

Windley BF (1977) The evolving continents. John Wiley, New York

Winnock E (1971) Géologie succinte du bassin d'Aquitaine. In: Technip (ed) Histoire structurale du Golfe de Gascogne. Inst Fran Pet Publ IV: 1–30

Wintsch RP, Lefort JP (1984) A clockwise rotation of variscan strain orientation in SE New-England. In: Hutton DHW, Sanderson DJ (eds) Variscan Tectonics of the North Atlantic Region. Geol Soc Lond Spec Publ 14:245–252

Wissman G (1982) Stratigraphy and structural features of the continental margin basin of Senegal and Mauritania. In: Von Rad U, Hinz K, Sarnthein M, Seibold E (eds) Geology of the Northwest African continental Margin. Springer, Berlin Heidelberg, New-York 9:160–181

Wissmann G, Roeser HA (1982) A magnetic and halokinetic structural Pangea Fit of Northwest Africa and north America. Geol Jahrb E 23:43–61

Wissmann G, Müller P, Kreuzer H, Harre W, Reinecke Th (1982) Mazagan Granite (off Morocco) the source of Meguma Graywacke (NS)? Vortrags – Kurzfassung für Programm der

Frühjahrstagung 1982 der Deutschen Geophysikalischen Gesellschaft in Hannover

Wones DR (1979) A comparison between granitic plutons of New-England USA and the Sierra Nevada batholith California. In: Wones DR (ed) Proceedings of the Caledonides in the USA, IGCP 27 Symposium Blacksburg 5–9 September. Virginia Polytechnic Institute and State University Mem 2:123–130

Wright AE (1969) Precambrian rocks of England, Wales and SE Ireland. In: Kay M (ed) North Atlantic geology and continental drift. Am Assoc Petroleum Geol Mem 12:93–109

Wright JE, Hull JH, Mc Quillin R, Arnold SE (1971) Irish Sea investigations 1969–1970. Lond Inst Geol Sci Rep 71/19:1–55

Zwart HJ, Dornsiepen VF (1978) The tectonic framework of central and western Europe. Géol Mijnboow 57:627–654

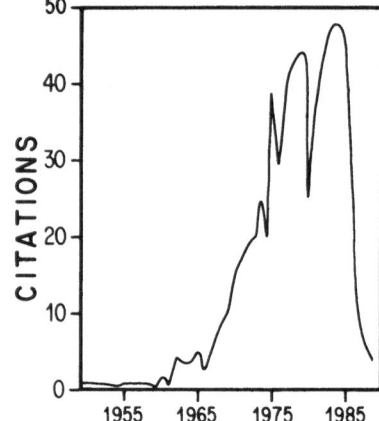

Index